Interior Design

Using Autodesk® Revit® 2025

Daniel John Stine AIA, IES, CSI, CDT, Well AP

SDC PUBLICATIONS

SDC Publications
P.O. Box 1334
Mission, KS 66222
913-262-2664
www.SDCpublications.com
Publisher: Stephen Schroff

Examination Copies
Books received as examination copies are for review purposes only and may not be made available for student use. Resale of examination copies is prohibited.

Electronic Files
Any electronic files associated with this book are licensed to the original user only. These files may not be transferred to any other party.

Trademarks
Autodesk Revit is a registered trademark of Autodesk, Inc. Windows 10 is a trademark of Microsoft. Mac and Boot Camp are trademarks of Apple. Parallels Desktop is a trademark of Parallels. Photoshop is a trademark of Adobe. All other trademarks are trademarks of their respective holders.

The author and publisher of this book have used their best efforts in preparing this book. These efforts include the development, research and testing of the material presented. The author and publisher shall not be liable in any event for incidental or consequential damages with, or arising out of, the furnishing, performance, or use of the material.

ISBN-13: 978-1-63057-674-5
ISBN-10: 1-63057-674-3

Printed and bound in the United States of America.

Preface

Interior Design using Revit 2025 by Daniel Stine is an outstanding and thoroughly organized book to help design students, faculty and practitioners develop Autodesk Revit skills and techniques required in today's dynamic Architecture Engineering and Construction (AEC) industry. The book provides the user a well-rounded knowledge of Revit and skills needed for both academia and industry. The book also includes access to nearly 100 dynamic video tutorials.

As an interior design educator seeking to develop a computer applications course in Revit, I found the book not only instrumental for developing the pedagogical content for my course but also extremely useful for me as an educator to learn and gain expertise in Revit. I taught myself Revit using Daniel Stine's books and in turn developed two computer applications courses in beginning to advanced Revit skills for my students. The book is fantastically structured for various levels of users. The chronological order which ties accurately to the design process fosters great connection and understanding of the construction and detailing process for students as one uses the book in the classroom environment.

The book focuses on helping the user learn Revit while modeling a two-story law office building. The architectural model of the two-story law office building comes with the columns, beams, exterior walls, and structural elements already developed so the learner can focus on developing their knowledge of non-structural elements and details relevant for interior design. This allows more emphasis primarily on interior elements rather than the architectural envelope.

The book follows a chronological order that mimics the design process. The first chapter provides an overview of Revit, Building Information Modeling (BIM) and fundamentals about the software used in the AEC industry. The second chapter covers how to model walls, doors, windows, and other elements in Revit. For interior design students this offers a brief overview to help with understanding the basics of modeling structural elements in Revit. The rest of the book contains detailed instruction on developing the interiors for a two-story law office building given the programmatic requirements. Some of the exercises covered include managing and creating photorealistic materials; developing walls, doors, and windows; modeling floor finishes; ceilings with soffits; casework; custom reception desk; restrooms; furniture and light fixtures. Furthermore, topics such as design options, photorealistic rendering, construction documents, schedules and worksharing processes are covered.

Overall, this exceptional book offers a step-by-step, well-organized opportunity to introduce Revit and Building Information modeling to interior design curriculum with exercises well laid out for courses focused on computer applications to construction to lighting from beginning to advanced interior design studio levels. The exercises are very engaging and easy to follow, and the accompanying videos are a great resource. Upon completion of the exercises in this book the user will have developed a great understanding of Revit in Interior Design and is ready to thrive in the Revit/BIM world.

Abimbola O. Asojo, PhD, AIA, LEED AP, NCIDQ
Professor of Interior Design
Associate Dean for Research, Creative Scholarship and Engagement
College of Design, University of Minnesota

Foreword

The intent of this book is to provide the interior design student a well-rounded knowledge of Autodesk Revit tools and techniques. These skills can then be applied to enhance professional development in both academia and industry.

The overall premise of the book is to learn Revit (or Revit Architecture) while developing the interior of a two-story law office. The reader is provided an architectural model with established columns, beams, exterior walls, minimal interior walls and roofs in which to work. This allows more emphasis to be placed on interior design rather than primary architectural elements. The chapter's chronology generally follows the typical design process. Students will find this book helps them more accurately and efficiently develop their design ideas and skills.

The first chapter introduces the reader to Revit, Building Information Modeling (BIM) and the basics of opening, saving and creating a new project. The second provides a quick introduction to modeling basic elements in Revit including walls, doors, windows and more. This chapter is designed to show students how powerful Revit is and hopefully make them more excited about learning it.

The remainder of the book is spent developing the interior space of the law office with an established space program. A student will learn how to view and navigate within the provided 3D architectural model, manage and create materials and develop spaces with walls, doors and windows. Once all the spaces are added to the model, several areas are explored and used as the basis to cover Revit commands and workflows.

At the end of this tutorial, the reader will be able to model floor finishes, ceilings with soffits, casework, custom reception desk, restrooms, furniture and light fixtures. Additional features such as tags, schedules and photo-realistic rendering will be covered.

Errata:
Please check the publisher's website from time to time for any errors or typos found in this book after it went to the printer. Simply browse to www.SDCpublications.com, and then navigate to the page for this book. Click the **View/Submit errata** link in the upper right corner of the page. If you find an error, please submit it so we can correct it in the next edition.

You may contact the publisher with comments or suggestions at service@SDCpublications.com.
Please do not email with Revit questions unless they relate to a problem with this book.

Online Content:

If the reader has never used Revit before, it is recommended they work through the two "Revit Basics" appendices via the Exclusive Bonus Content after completing Chapter 1 and 2.

Another online appendix is devoted to an introduction to computers. Much of the basics are covered, from computer hardware and software to file management procedures, including step-by-step instruction on using a flash drive.

Software required:

To successfully complete the exercises in this book you will need to have access to Autodesk's *Revit 2025*. A 30-day trial may be downloaded from Autodesk's website. Additionally, qualifying students may download the free student version of the software from **students.autodesk.com**. Both are fully functional versions of the software.

Instructors:

An Instructor's resource guide is available with this book. It contains:

- Answers to the questions at the end of each chapter
- Printed examples of completed work for each chapter (use this when grading)
- Outline of tools and topics to be covered in each lesson's lecture
- Suggestions for additional student work (for each lesson)

About the Authors:

Daniel John Stine, AIA, CSI, CDT, IES, WELL AP, is a distinguished Wisconsin-registered architect with over two decades of professional experience, currently serving as the Director of Design Technology at Lake|Flato, a leading architecture firm based in San Antonio, Texas. Dan is an active member of the national AIA Committee on the Environment (COTE) Leadership Group, where his expertise significantly contributes to advancing sustainable architectural practices.

Dan's remarkable contributions to the field include pioneering the integration of Building Information Modeling (BIM) for lighting analysis with ElumTools, early-stage energy modeling using Autodesk Insight, and immersive experiences through virtual reality (VR) and augmented reality (AR) technologies with platforms like HTC Vive, Oculus Rift, Fuzor, Enscape, and Microsoft HoloLens 2. He has also co-developed the Electrical Productivity Pack for Revit, a tool that enhances electrical design efficiency, marketed by ATG under the CTC Software brand. His innovative work has caught the attention of industry giants Dell and Autodesk, leading to feature videos that showcase his impactful contributions to architecture and design technology.

Dan is an esteemed speaker who has shared his insights on BIM, design technology, and sustainable design at prestigious international forums across the USA, Mexico, Canada, Europe, Asia, and Australia, including Autodesk University, RTC/BILT, and various AIA national and state conferences. His engagement with the global architectural community led to a special invitation from Autodesk to collaborate on future Revit features at their largest R&D facility in Shanghai, China, in 2017.

Dedicated to educating the next generation of architects, Dan holds a teaching position at North Dakota State University, where he mentors graduate architecture students. His academic contributions span lectures and workshops at universities across the United States, fostering growth and innovation in both architecture and interior design education. Dan's affiliations with professional organizations like the American Institute of Architects, Construction Specifications Institute, Illuminating Engineering Society, and the Autodesk Developer Network underscore his commitment to excellence and innovation in the architectural field. Through his teaching and pioneering work, Dan continues to shape the future of architectural design and technology.

Dan writes about design on his blog, BIM Chapters, and in his textbooks published by SDC Publications:

- *Residential Design Using Autodesk Revit 2025*
- *Commercial Design Using Autodesk Revit 2025*
- *Design Integration Using Autodesk Revit 2025 (Architecture, Structure and MEP)*
- *Interior Design Using Autodesk Revit 2025 (with co-author Aaron Hansen)*
- *Autodesk Revit 2021 Architectural Command Reference*
- *Residential Design Using AutoCAD 2025*
- *Commercial Design Using AutoCAD 2023*
- *Chapters in Architectural Drawing (with co-author Steven H. McNeill, AIA, LEED AP)*
- *Interior Design Using Hand Sketching, SketchUp and Photoshop (also with Steven H. McNeill)*
- *SketchUp 2024 for Interior Designers (with co-author Maria Delgado, PhD)*

You may contact the publisher with comments or suggestions at **service@SDCpublications.com**

BIM Chapters (Author's Blog):
http://bimchapters.blogspot.com

X/Twitter
@DanStine_MN

LinkedIn
https://www.linkedin.com/in/danstinemn

Certification Study Guides

Daniel Stine has also prepared the following certification study guides students can use to achieve successful results when taking certification exams which can help make the difference when trying to secure that dream job!

- *Microsoft Office Specialist Excel Associate 365/2019 Exam Preparation*
- *Microsoft Office Specialist Word Associate 365/2019 Introduction & Exam Preparation*
- *Microsoft Office Specialist PowerPoint Associate 365/2019 Introduction & Exam Preparation*
- *Autodesk Revit for Architecture Certified User Exam Preparation (Revit 2025 Edition)*
- *Autodesk AutoCAD Certified User Study Guide (2024 Edition)*

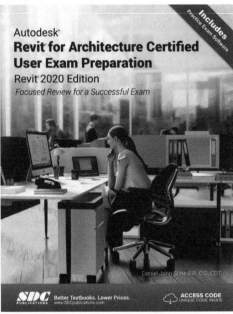

Table of Contents

Exclusive Online Content
See the inside-front cover of this book for download instructions and your unique access code.

Bonus Chapters

Bonus Videos

This book comes with several short videos, approximately 3-5 minutes long, which can be watched in order while working through this book. The full index of video titles can be found on the next page.

These videos are supplementary and not required to complete the exercises in this book. Rather, the videos are intended to cover many of the fundamental Revit topics in a different context to help the reader better understand this important material. A few videos cover intermediate topics, such as the sloped "cloud" ceiling shown in the image to the right.

Bonus Videos Index:

Lesson 1
Getting Started with Autodesk Revit 2025:

This chapter will introduce you to Autodesk® Revit® 2025. You will study the User Interface and learn how to open and exit a project and adjust the view of the drawing on the screen. It is recommended that you spend an ample amount of time learning this material, as it will greatly enhance your ability to progress smoothly through subsequent chapters.

Exercise 1-1:
What is Revit 2025?

What is Autodesk Revit used for?

Autodesk Revit (Architecture, Structure and MEP) is the world's first fully parametric building design software. This revolutionary software, for the first time, truly takes architectural computer aided design beyond simply being a high-tech pencil. Revit is a product of Autodesk, makers of AutoCAD, Civil 3D, Inventor, 3DS Max, Maya and many other popular design programs.

Revit can be thought of as the foundation of a larger process called **Building Information Modeling** (BIM). The BIM process revolves around a virtual, information rich 3D model. In this model all the major building elements are represented and contain information such as manufacturer, model, cost, phase and much more. Once a model has been developed in Revit, third-party add-ins and applications can be used to further leverage the data. Some examples are Facilities Management, Analysis (Energy, Structural, Lighting), Construction Sequencing, Cost Estimating, Code Compliance and much more!

Revit can be an invaluable tool to designers when leveraged to its full potential. The iterative design process can be accomplished using special Revit features such as *Phasing* and *Design Options*. Material selections can be developed and attached to various elements in the model, where one simple change adjusts the wood from oak to maple throughout the project. The power of schedules may be used to determine quantities and document various parameters contained within content (this is the "I" in BIM, which stands for Information). Finally, the three-dimensional nature of a Revit-based model allows the designer to present compelling still images and animations. These graphics help to more clearly communicate the design intent to clients and other interested parties. This book will cover many of these tools and techniques to **assist** in the creative process.

What is a parametric building modeler?

Revit is a program designed from the ground up using state-of-the-art technology. The term parametric describes a process by which an element is modified and an adjacent element(s) is automatically modified to maintain a previously established relationship. For example, if a wall is moved, perpendicular walls will grow, or shrink, in length to remain attached to the

related wall. Additionally, elements attached to the wall will move, such as wall cabinets, doors, windows, air grilles, etc.

Revit stands for **Rev**ise **In**stantly; a change made in one view is automatically updated in all other views and schedules. For example, if you move a door in an interior elevation view, the floor plan will automatically update. Or, if you delete a door, it will be deleted from all other views and schedules.

A major goal of Revit is to eliminate much of the repetitive and mundane tasks traditionally associated with CAD programs allowing more time for design, coordination and visualization. For example, all sheet numbers, elevation tags and reference bubbles are updated automatically when changed anywhere in the project. Therefore, it is difficult to find a mis-referenced detail tag.

The best way to understand how a parametric model works is to describe the Revit project file. A single Revit file contains your entire building project. Even though you mostly draw in 2D views, you are actually drawing in 3D. In fact, the entire building project is a 3D model. From this 3D model you can generate 2D elevations, 2D sections and perspective views. Therefore, when you delete a door in an elevation view you are actually deleting the door from the 3D model from which all 2D views are generated and automatically updated.

Another way in which Revit is a parametric building modeler is that **parameters** can be used to control the size and shape of geometry. For example, a window model can have two *parameters* set up which control the size of the window. Thus, from a window's properties it is possible to control the size of the window without using any of the drawing modify tools such as *Scale* or *Move*. Furthermore, the *parameter* settings (i.e., width and height in this example) can be saved within the window model (called a *Family*). You could have the 2' x 4' settings saved as "Type A" and the 2' x 6' as "Type B." Each saved list of values is called a *Type* within the *Family*. Thus, this one double-hung window *Family* could represent an unlimited number of window sizes! You will learn more about this later in the book.

Window model size controlled by parameters *Width* and *Height*

What Disciplines does Revit Support?

Revit supports **Architecture, Interior Design, Structure**, and **MEP** (which stands for Mechanical, Electrical, and Plumbing). There used to be three discipline-specific versions of Revit and an all-in-one version—now there is just the all-in-one version. You can download the free 30-day trial from autodesk.com. Students may download a free 3-year version of Revit at www.autodesk.com/education.

Revit is not meant to support professional Civil design.

Now is as good a time as any to make sure the reader understands that Revit is not, nor has it ever been, backwards compatible. This means there is no *Save-As* back to a previous version of Revit. Also, an older version of Revit cannot open a file, project or content saved in a newer format. So, make sure you consider what version your school or employer is currently standardized on before upgrading any projects or content. Revit will automatically upgrade an older

Revit model of an existing building, with architecture, structural, mechanical, plumbing and electrical all modeled.
Image courteous of LHB, Inc. www.LHBcorp.com

format when opened in a newer version of the software. This is a one-time process that can take several minutes.

3D model of lunchroom created in *Interior Design Using Autodesk Revit 2025*

Download Autodesk Content

The content provided with Revit, such as doors, windows, toilets, etc., are no longer downloaded to your hard drive during the installation of the software. Instead, the content is downloaded as needed from the **Insert → Load Autodesk Family** tool. Thus, an internet connection will be required to place content. Optionally, all content may be downloaded, see: https://autode.sk/30EJYkX

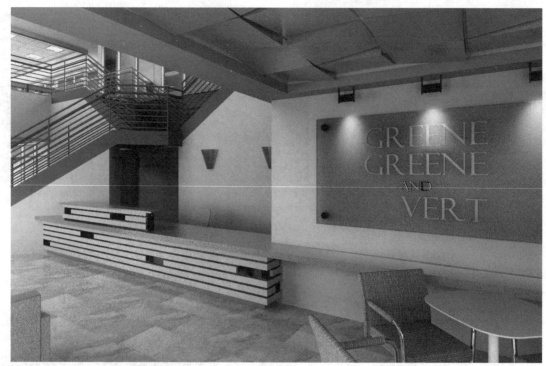

Lobby rendering from *Interior Design Using Autodesk Revit 2025*

Why use Revit?

Many people ask the question, why use Revit versus other programs? The answer can certainly vary depending on the situation and particular needs of an individual or organization.

Generally speaking, this is why most companies use Revit:

- Many designers and drafters are using Revit to streamline repetitive drafting tasks and focus more on designing and detailing a project.
- Revit is a very progressive program and offers many features for designing buildings. Revit is constantly being developed, and Autodesk provides incremental upgrades and patches on a regular basis; Revit 2025 was released about a year after the previous version.
- Revit was designed specifically for architectural design and includes features like:
 - *Autodesk Renderer* Photo Realistic Rendering
 - Phasing; model changes over time
 - Design Options; model changes during the same time period
 - Live schedules
 - Cloud Rendering via *Autodesk cloud services*
 - Conceptual Energy Analysis via *Autodesk cloud services*
 - Daylighting Analysis via *Autodesk cloud services*

A few basic Revit concepts

The following is meant to be a brief overview of the basic organization of Revit as a software application. You should not get too hung up on these concepts and terms as they will make more sense as you work through the tutorials in this book. This discussion is simply laying the groundwork so you have a general frame of reference on how Revit works.

The Revit platform has three fundamental types of elements:
- Model Elements
- Datum Elements
- View-Specific Elements

Model Elements

Think of *Model Elements* as things you can put your hands on once the building has been constructed. They are typically 3D but can sometimes be 2D. There are two types of *Model Elements*:

- **Host Elements**: walls, floors, slabs, roofs, ceilings.

- **Model Components**: Stairs, Doors, Furniture, Casework, Beams, Columns, Pipes, Ducts, Light Fixtures, Model Lines, etc.

 o Some *Model Components* require a host before they can be placed within a project. For example, a window can only be placed in a host, which could be a wall, roof or floor depending on how the element was created. If the host is deleted, all hosted or dependent elements are automatically deleted.

Model Elements
Windows & Doors; require a host

Host Elements
Roof, Wall & Floor

Model Elements
Stair & Washer/Dryer; do not require a host

Datum Elements

Datum Elements are reference planes within the building that graphically and parametrically define the location of various elements within the model. These are the three types of *Datum Elements,* all of which define 3D planes:

- **Grids**
 - ○ Typically laid out in a plan view to locate structural elements such as columns and beams, as well as walls. Grids show up in plan, elevation and section views. Moving a grid in one view moves it in all other views as it is the same element. (See the next page for an example of a grid in plan view.)

- **Levels**
 - ○ Used to define vertical relationships, mainly surfaces that you walk on. They only show up in elevation, section and 3D views. Most elements are placed relative to a *Level;* when the *Level* is moved those elements move with it (e.g., doors, windows, casework, ceilings). **WARNING:** *If a* Level *is deleted, those same "dependent" elements will also be deleted from the project!*

- **Reference Planes**
 - ○ These are similar to grids in that they show up in plan and elevation or sections. They do not have reference bubbles at the end like grids. Revit breaks many tasks down into simple 2D tasks which result in 3D geometry. *Reference Planes* are used to define 2D surfaces on which to work within the 3D model. They can be placed in any view, horizontally, vertically or at an angle.

View-Specific Elements

As the name implies, the items about to be discussed only show up in the specific view in which they are created. For example, notes and dimensions added in the architectural floor plans will not show up in the structural floor plans. These elements are all 2D and are mainly communication tools used to accurately document the building for construction or presentations.

- **Annotation elements** (text, tags, symbols, dimensions)
 - Size automatically set and changed based on selected view scale

- **Details** (detail lines, filled regions, 2D detail components)

File Types (and extensions):

Revit has four primary types of files that you will work with as a Revit user. Each file type, as with any Microsoft Windows based program, has a specific three letter file name extension; that is, after the name of the file on your hard drive you will see a period and three letters:

.RVT	Revit project files; the file most used (also for backup files)
.RFA	Revit family file; loadable content for your project
.RTE	Revit template; a project starter file with office standards preset
.RFT	Revit family template; a family starter file with parameters

The Revit platform has three fundamental ways in which to work with the elements (for display and manipulation):

- Views
- Schedules
- Sheets

The following is a cursory overview of the main ideas you need to know. This is not an exhaustive study on views, schedules and sheets.

Views

Views, accessible from the *Project Browser*, is where most of the work is done while using Revit. Think of views as slices through the building, both horizontal (plans) and vertical (elevations and sections).

- **Plans**
 - A *Plan View* is a horizontal slice through the building. You can specify the location of the **cut plane** which determines if certain windows show up or how much of the stair is seen. A few examples are architectural floor plan, reflected ceiling plan, site plan, structural framing plan, HVAC floor plan, electrical floor plan, lighting [ceiling] plan, etc. The images below show this concept; the image on the left is the 3D BIM. The middle image shows the portion of building above the cut plane removed. Finally, the last image on the right shows the plan view in which you work and place on a sheet.

- **Elevations**
 - An "exterior" elevation is a vertical slice, but where the slice lies outside the floor plan as in the middle image below. Each elevation created is listed in the *Project Browser*. The image on the right is an example of a South exterior elevation view, which is a "live" view of the 3D model. If you select a window here and delete it, the floor plans will update instantly.

 - Similarly, an "interior" elevation is a vertical slice, but inside the building. This view is cropped at the perimeter of a given room.

- **Sections**
 - o Similar to elevations, sections are also vertical slices. However, these slices cut through the building. A section view can be cropped down to become a wall section or even look just like an elevation. The images below show the slice, the portion of building in the foreground removed, and then the actual view created by the slice. A setting exists, for each section view, to control how far into that view you can see. The example on the right is "seeing" deep enough to show the doors on the interior walls.

- **3D and Camera**
 - o In addition to the traditional "flattened" 2D views that you will typically work in, you are able to see your designs more naturally via 3D and Camera views. A 3D view is simply an orthogonal view of the project viewed from the exterior. A Camera view is a perspective view; cameras can be created both in and outside of the building. Like the 2D views, these 3D/Camera views can be placed on a sheet to be printed. Revit provides a number of tools to help explore the 3D view, such as Section Box, Steering Wheel, Temporary Hide and Isolate, and Render.

 The image on the left is a 3D view set to "shade mode" and has shadows turned on. The image on the right is a camera view set up inside the building; the view is set to "hidden line" rather than shaded, and the camera is at eyelevel.

Schedules

Schedules are lists of information generated based on content that has been placed, or modeled, within the project. A schedule can be created, such as the door schedule example shown below, that lists any of the data associated with each door that exists in the project. Revit allows you to work directly in the schedule views. Any change within a schedule view is a change directly to the element being scheduled. Again, if a door were to be deleted from this schedule, that door would be instantly deleted from the project.

| DOOR NUMBER | DOOR | | | | FRAME | | DETAIL | | | GLAZING | FIRE RATING | HDWR GROUP | |
	WIDTH	HEIGHT	MATL	TYPE	MATL	TYPE	HEAD	JAMB	SILL				
1000A	3' - 8"	7' - 2"	WD		HM		11/A8.01	11/A8.01					
1046	3' - 0"	7' - 2"	WD	D10	HM	F10	11/A8.01	11/A8.01 SIM				34	
1047A	6' - 0"	7' - 10"	ALUM	D15	ALUM	SF4	6/A8.01	6/A8.01	1/A8.01 SIM	1" INSUL		2	CARD READER N. LEAF
1047B	8' - 0"	7' - 2"	WD	D10	HM	F13	12/A8.01	11/A8.01 SIM			60 MIN	85	MAG HOLD OPENS
1050	3' - 0"	7' - 2"	WD	D10	HM	F21	8/A8.01	11/A8.01		1/4" TEMP		33	
1051	3' - 0"	7' - 2"	WD	D10	HM	F21	8/A8.01	11/A8.01		1/4" TEMP		33	
1052	3' - 0"	7' - 2"	WD	D10	HM	F21	8/A8.01	11/A8.01		1/4" TEMP		33	
1053	3' - 0"	7' - 2"	WD	D10	HM	F21	8/A8.01	11/A8.01		1/4" TEMP		33	
1054A	3' - 0"	7' - 2"	WD	D10	HM	F10	8/A8.01	11/A8.01		1/4" TEMP	-	34	
1054B	3' - 0"	7' - 2"	WD	D10	HM	F21	8/A8.01	11/A8.01		1/4" TEMP	-	33	
1055	3' - 0"	7' - 2"	WD	D10	HM	F21	8/A8.01	11/A8.01		1/4" TEMP	-	33	
1056A	3' - 0"	7' - 2"	WD	D10	HM	F10	9/A8.01	9/A8.01			20 MIN	33	
1056B	3' - 0"	7' - 2"	WD	D10	HM	F10	11/A8.01	11/A8.01			20 MIN	34	
1056C	3' - 0"	7' - 2"	WD	D10	HM	F10	20/A8.01	20/A8.01			20 MIN	33	
1057A	3' - 0"	7' - 2"	WD	D10	HM	F10	8/A8.01	11/A8.01			20 MIN	34	
1057B	3' - 0"	7' - 2"	WD	D10	HM	F30	9/A8.01	9/A8.01		1/4" TEMP	20 MIN	33	
1058A	3' - 0"	7' - 2"	WD	D10	HM	F10	9/A8.01	9/A8.01			-	33	

DOOR AND FRAME SCHEDULE

Sheets

You can think of sheets as the pieces of paper on which your views and schedules will be printed. Views and schedules are placed on sheets and then arranged. Once a view has been placed on a sheet, its reference bubble is automatically filled out and that view cannot be placed on any other sheet. The setting for each view, called "view scale," controls the size of the drawing on each sheet; view scale also controls the size of the text and dimensions.

Exercise 1-2:
Overview of the Revit User Interface

Revit is a powerful and sophisticated program. Because of its robust feature set, it has a measurable learning curve, though its intuitive design makes it easier to learn than other CAD or BIM-based programs. However, like anything, when broken down into smaller pieces, we can easily learn to harness the power of Revit. That is the goal of this book.

This section will walk through the different aspects of the User Interface (UI). As with any program, understanding the user interface is the key to using the program's features.

File Tab

Quick Access Toolbar (QAT)

RIBBON: *Tabs*

Application Title Bar

Info Center

RIBBON: *Panel*

RIBBON: *Tool*

Options Bar

View Tabs

Type Selector, *on Properties Palette*

Properties Palette

Navigation Bar

Context Menu

Canvas

Project Browser

FIGURE 1-2.1
Revit User Interface

View Control Bar

Status Bar

Selection Toggles

The Revit User Interface

TIP: *See the online videos on the User Interface.*

Application Title Bar

In addition to the *Quick Access Toolbar* and *Info Center*, which are all covered in the next few sections, you are also presented with the product name, version and the current **file-view** in the center. As previously noted already, the version is important as you do not want to upgrade unless you have coordinated with other staff and/or consultants; everyone must be using the same version of Revit. **Note:** the "R" icon on the left is not the same as in previous versions—the *Application Menu* has been replaced with the File Tab covered next.

File Tab

Access to *File* tools such as *Save*, *Plot*, *Export* and *Print* (both hardcopy and electronic printing). You also have access to tools which control the Revit application as a whole, not just the current project, such as *Options* (see the end of this section for more on *Options*).

Recent and Open Documents:

These two icons (from the *File tab*) toggle the entire area on the right to show either the recent documents you have been in (icon on the left) or a list of the documents you currently have open.

In the *Recent Documents* list you click a listed document to open it. This saves time as you do not have to click *Open* → *Project* and browse for the document (*Document* and *Project* mean the same thing here). Finally, clicking the "Pin" keeps that project from getting bumped off the list as additional projects are opened.

In the *Open Documents* list the "active" project you are working in is listed first; clicking another project switches you to that open project.

The list on the left, in the *File Tab* shown above, represents three different types of buttons: *button, drop-down button* and *split button*. Save and Close are simply **buttons**. Save-As and Export are **drop-down buttons**, which means to reveal a group of related tools. If you click or hover your cursor over one of these buttons, you will get a list of tools on the right. Finally, **split buttons** have two actions depending on what part of the button you click on; hovering over the button reveals the two parts (see bottom image to the right). The main area is the most used tool; the arrow reveals additional related options.

Button

Drop-down Button

Split Button

Quick Access Toolbar

Referred to as *QAT* in this book, this single toolbar provides access to often used tools (Home, *Open, Save, Undo, Redo, Measure, Tag*, etc.). It is always visible regardless of what part of the *Ribbon* is active.

The *QAT* can be positioned above or below the *Ribbon* and any command from the *Ribbon* can be placed on it; simply right-click on any tool on the *Ribbon* and select *Add to Quick Access Toolbar*. Moving the *QAT* below the *Ribbon* gives you a lot more room for your favorite commands to be added from the *Ribbon*. Clicking the larger down-arrow to the far right reveals a list of common tools which can be toggled on and off.

The first icon toggles back to the Home screen seen when Revit first opens. Some of the icons on the *QAT* have a down-arrow on the right. Clicking this arrow reveals a list of related tools. In the case of *Undo* and *Redo*, you can undo (or redo) several actions at once.

Ribbon – Architecture Tab

The *Architecture* tab on the *Ribbon* contains most of the tools the architect needs to model a building, essentially the things you can put your hands on when the building is done. The specific discipline versions of Revit omit some of the other discipline tabs.

Each tab starts with the *Modify* tool, i.e., the first button on the left. This tool puts you into "selection mode" so you can select elements to modify. Clicking this tool cancels the current tool and unselects elements. With the *Modify* tool selected you may select elements to view their properties or edit them. Note that the *Modify* tool, which is a button, is different than the *Modify* tab on the *Ribbon*.

The *Ribbon* has three types of buttons: *button*, *drop-down button* and *split*, as covered on the previous page. In the image above you can see the *Wall* tool is a **split button**. Most of the time you would simply click the top part of the button to draw a wall. Clicking the down-arrow part of the button, for the *Wall* tool example, gives you the option to draw a *Wall*, *Structural Wall*, *Wall by Face*, *Wall Sweep*, and a *Reveal*.

TIP: The Model Text tool is only for placing 3D text in your model, not for adding notes!

Ribbon – Annotate Tab

To view this tab, simply click the label "Annotate" near the top of the *Ribbon*. This tab presents a series of tools which allow you to add notes, dimensions and 2D "embellishments" to your model in a specific view, such as a floor plan, elevation, or section. All of these tools are **view specific**, meaning a note added in the first floor plan will not show up anywhere else, not even another first floor plan: for instance, a first floor electrical plan.

Notice, in the image above, that the *Dimension* panel label has a down-arrow next to it. Clicking the down-arrow will reveal an **extended panel** with additional related tools.

Finally, notice the *Component* tool in the image above; it is a **split button** rather than a *drop-down button*. Clicking the top part of this button will initiate the *Detail Component* tool. Clicking the bottom part of the button opens the fly-out menu revealing related tools. Once you select an option from the fly-out, that tool becomes the default for the top part of the split button for the current session of Revit (see image to right).

Ribbon – Modify Tab

Several tools which manipulate and derive information from the current model are available on the *Modify* tab. Additional *Modify* tools are automatically appended to this tab when elements are selected in the model (see *Modify Contextual Tab* on the next page).

> *TIP: Do not confuse the Modify <u>tab</u> with the Modify <u>tool</u> when following instructions in this book.*

Ribbon – View Tab

The tools on the *View* tab allow you to create new views of your 3D model; this includes views that look 2D (e.g., floor plans, elevations and sections) as well as 3D views (e.g., isometric and perspective views).

The *View* tab also gives you tools to control how views look, everything from what types of elements are seen (e.g., Plumbing Fixture, Furniture or Section Marks) as well as line weights.

> *NOTE: Line weights are controlled at a project wide level but may be overridden.*

Finally, notice the little arrow in the lower-right corner of the *Graphics* panel. When you see an arrow like this you can click on it to open a dialog box with settings that relate to the panel's tool set (*Graphics* in this example). Hovering over the arrow reveals a tooltip which will tell you what dialog box will be opened.

Ribbon – Modify Contextual Tab

The *Modify* tab is appended when certain tools are active or elements are selected in the model; this is referred to as a *contextual tab*. The first image below shows the *Place Wall* tab which presents various options while adding walls. The next example shows the *Modify Walls* contextual tab which is accessible when one or more walls are selected.

Place Wall contextual tab – visible when the Wall tool is active.

Modify Walls contextual tab – visible when a wall is selected.

Ribbon – Customization

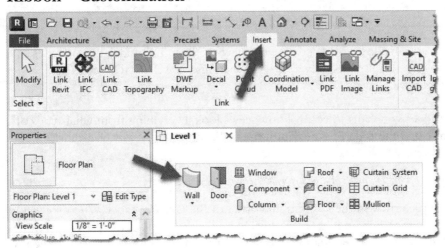

There is not too much customization that can be done to the *Ribbon*. One of the only things you can do is pull a panel off the *Ribbon* by clicking and holding down the left mouse button on the titles listed, e.g. "Link" or "Build," across the bottom. This panel can be placed within the *drawing window* or on another screen if you have a dual monitor setup.

The image above shows the *Build* panel, from the *Architecture* tab, detached from the *Ribbon* and floating within the drawing window. Notice that the *Insert* tab is active. Thus, you have constant access to the *Build* tools while accessing other tools. Note that the *Build* panel is not available on the *Architecture* tab as it is literally moved, not just copied.

When you need to move a detached panel back to the *Ribbon* you do the following: hover over the detached panel until the sidebars show up and then click the "Return panels to ribbon" icon in the upper right (identified in the image above).

FYI: Whenever the resolution of your monitor is too low or you don't have the Revit application maximized on the screen, the buttons may be modified to take up less room on the Ribbon; typically the words are removed. Compare the image to the right with the Build panel above.

If you install an **add-in** for Revit on your computer, you will likely see a new tab appear on the Ribbon. Some add-ins are free, while others require a fee. If an add-in only has one tool, it will likely be added to the catch-all tab called Add-Ins (shown in the image below). Some firms have custom tools they developed, and may have their own tab, like LF Tools.

Tab Visibility

This is not really customizing the User Interface, but in the Options dialog, there are several adjustments one can make – such as turning off tabs and tools not used.

TIP: When Revit is first opened, you are prompted to indicate your role, or what type of work you do. This will turn off tabs here.

Ribbon – States

The *Ribbon* can be displayed in one of four states:

- Full Ribbon (default)
- Minimize to Tabs
- Minimize to Panel Tiles
- Minimize to Panel Buttons

The intent of this feature is to increase the size of the available drawing window. It is recommended, however, that you leave the *Ribbon* fully expanded while learning to use the program. The images in this book show the fully expanded state. The images below show the other three options. When using one of the minimized options you simply hover (or click) your cursor over the Tab or Panel to temporarily reveal the tools.

FYI: Double-clicking on a Ribbon tab will also toggle the states.

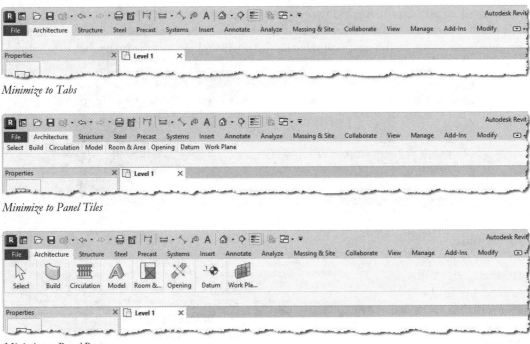

Minimize to Tabs

Minimize to Panel Tiles

Minimize to Panel Buttons

Ribbon – References in this Book

When the exercises make reference to the tools in the *Ribbon* a consistent method is used to eliminate excessive wording and save space. Take a moment to understand this so that things go smoothly for you later.

Throughout the textbook you will see an instruction similar to the following:

> 23. Select **Architecture → Build → Wall**

This means click the ***Architecture*** tab, and within the ***Build*** panel, select the ***Wall*** tool. Note that the *Wall* tool is actually a split button, but a subsequent tool was not listed so you are to click on the primary part of the button. Compare the above example to the one below:

> 23. Select **Architecture → Build → Wall → Structural Wall**

The above example indicates that you should click the down-arrow part of the *Wall* tool in order to select the *Structural Wall* option.

Thus the general pattern is this:

Tab → Panel → Tool → drop-down list item

#1 *Tab:* This will get you to the correct area within the *Ribbon*.

#2 *Panel:* This will narrow down your search for the desired tool.

#3 *Tool:* Specific tool to select and use.

Drop-down list item: This will only be specified for drop-down buttons and sometimes for split buttons.

The image below shows the order in which the instructions are given to select a tool; note that you do not actually click the panel title.

Options Bar

This area dynamically changes to show options that complement the current operation. The *Options Bar* is located directly below the *Ribbon*. When you are learning to use Revit, you should keep your eye on this area and watch for features and options appearing at specific times. The image below shows the *Options Bar* example with the *Wall* tool active.

Properties Palette – Element Type Selector

The *Properties Palette* provides instant access to settings related to the element selected or about to be created. When nothing is selected, it shows information about the current view. When multiple elements are selected, the common parameters are displayed.

The *Element Type Selector* is an important part of the *Properties Palette*. Whenever you are adding elements or have them selected, you can select from this list to determine how a wall to be drawn will look, or how a wall previously drawn should look (see lower left image). If a wall type needs to change, you never delete it and redraw it; you simply select it and pick a new type from the *Type Selector*.

The **Selection Filter** drop-down list below the *Type Selector* lets you know the type and quantity of the elements currently selected. When multiple elements are selected you can narrow down the properties for just one element type, such as *wall*. Notice the image below shows four walls are in the current

Properties Palette; nothing selected

selection set. Selecting **Walls (4)** will cause the *Palette* to only show *Wall* properties even though many other elements are selected (and remain selected).

The width of the *Properties Palette* and the center column position can be adjusted by dragging the cursor over that area. You may need to do this at times to see all the information. However, making the *Palette* too wide will reduce the useable drawing area.

Type Selector; Wall tool active or a Wall is selected

Selection Filter; multiple elements selected

The *Properties Palette* should be left open; if you accidentally close it you can reopen it by **View → Window → User Interface → Properties** or by typing **PP** on the keyboard.

Project Browser

Primary nodes in project browser

The *Project Browser* is the "Grand Central Station" of the Revit project database. All the views, schedules, sheets, and content are accessible through this hierarchical list. The first image to the left shows the seven major categories; any item with a "plus" next to it contains sub-categories or items.

Double-clicking on a View, Legend, Schedule, or Sheet will open it for editing; the current item open for editing is bold (**Level 1** in the example to the left). An empty box indicates the view is not placed on a sheet. Right-clicking will display a pop-up menu with a few options such as *Delete* and *Copy*.

Alternative configuration

Right-click on *Views (all)*, at the top of the *Project Browser*, and you will find a **Search** option in the pop-up menu. This can be used to search for a *View, Family*, etc., a useful tool when working on a large project with 100s of items to sift through.

Like the *Properties Palette*, the width of the *Project Browser* can be adjusted. When the two are stacked above each other, they both move together. You can also stack the two directly on top of each other; in this case you will see a tab for each at the bottom as shown in the second image to the left.

The *Project Browser* should be left open; if you accidentally close it by clicking the "X" in the upper right, you can reopen it by
View → Window → User Interface → Project Browser.

The *Project Browser* and *Properties Palette* can be repositioned on a second monitor, if you have one, when you want more room to work in the drawing window.

Status Bar

This area will display information, on the far left, about the current command or list information about a selected element. The right-hand side of the *Status Bar* shows the number of elements selected. The small funnel icon to the left of the selection number can be clicked to open the *Filter* dialog box, which allows you to reduce your current selection to a specific category; for example, you could select the entire floor plan, and then filter it down to just the doors. This is different than the *Selection Filter* in the *Properties Palette* which keeps everything selected.

On the *Status Bar*, the first five icons on the right, shown in the image below, control how elements are **selected**. From left to right these are

- Select Links
- Select Underlay Elements
- Select Pinned Elements
- Select Elements by Face
- Drag Elements on Selection

Hover your cursor over an icon for the name and for a brief description of what it does. These are toggles that are on or off; **the red 'X' in the upper right of each icon means you cannot select that type of element within the model**. These controls help prevent accidentally moving or deleting things. Keep these toggles in mind if you are having trouble selecting something; you may have accidentally toggled one of these on.

Finally, the two drop-down lists towards the center of the *Status Bar* control **Design Options** and **Worksets** (see image on previous page). Design Options facilitate multiple design solutions in the same area, and *Worksets* relate to the ability for more than one designer to be in the model at a time.

View Control Bar

This is a feature which gives you convenient access to tools which control each view's display settings (i.e., scale, shadows, detail level, graphics style, etc.). The options vary slightly between view types: 2D View, 3D view, Sheet and Schedule. The important thing to know is that these settings only affect the current view, the one listed on the *Application Title Bar*. Most of these settings are available in the *Properties Palette*, but this toolbar cannot be turned off (i.e. hidden) like the *Properties Palette* can.

Context Menu

The *context menu* appears near the cursor whenever you right-click on the mouse (see image at right). The options on that menu will vary depending on what tool is active or what element is selected.

Canvas (aka Drawing Window)

This is where you manipulate the Building Information Model (BIM). Here you will see the current view (plan, elevation or section), schedule or sheet. Any changes made are instantly propagated to the entire database.

Context menu example with a wall selected

Elevation Marker

This item is not really part of the Revit UI but is visible in the drawing window by default via the various templates you can start with, so it is worth mentioning at this point. The four elevation markers point at each side of your project and ultimately indicate the drawing sheet on which you would find an elevation drawing of each side of the building. All you need to know right now is that you should draw your floor plan generally in the middle of the four elevation markers that you will see in each plan view; DO NOT delete them as this will remove the related view from the *Project Browser*.

Revit Options

There are several settings, related to the *User Interface*, which are not tied to the current model. That is, these settings apply to the installation of Revit on your computer, rather than applying to just one model or file on your computer. A few of these settings will be briefly discussed here. It is recommended that you don't make any changes here right now.

Click on **Options** in the *File* tab. The options under File Locations tell Revit where to find templates and content. In the Hardware section you can verify your display driver information and compare with the "Supported Hardware" link. If Revit is crashing or exhibits strange behavior, it often relates to your graphics driver. You can bypass this problem by turning off *Use Hardware Acceleration*, but this will make Revit run slower when panning and zooming.

This concludes your brief overview of the Revit user interface. Many of these tools and operations will be covered in more detail throughout the book.

Efficient Practices

The *Ribbon* and menus are really helpful when learning a program like Revit; however, most experienced users rarely use them! The process of moving the mouse to the edge of the screen to select a command and then back to where you were is very inefficient, especially for those who do this all day long, five days a week. Here are a few ways experienced BIM operators work:

- Use the wheel on the mouse to Zoom (spin the wheel), Pan (press and hold the wheel button while moving the mouse) and Zoom Extents (double-clicking the wheel button). All this can be done while in another command; so, if you are in the middle of drawing walls and need to zoom in to see which point you are about to Snap to, you can do it without canceling the Line command and without losing focus on the area you are designing by having to click an icon near the edge of the screen.

- Revit conforms to many of the Microsoft Windows operating system standards. Most programs, including Revit, have several standard commands that can be accessed via keyboard shortcuts. Here are a few examples (press both keys at the same time):

 - Ctrl + S Save *(saves the current model)*
 - Ctrl + A Select All *(selects everything in text editor)*
 - Ctrl + Z Undo *(undoes the previous action)*
 - Ctrl + X Cut *(Cut to Windows clipboard)*
 - Ctrl + C Copy *(does not replace Revit's Copy tool)*
 - Ctrl + V Paste *(used to copy between models/views/levels)*
 - Ctrl + Tab Change View *(toggles between open views)*
 - Ctrl + P Print *(opens print dialog)*
 - Ctrl + N New *(create new project file)*
 - F7 Spelling *(launch spell check feature)*
 - ENTER Previous Command *(repeat previous command)*

- If you recall, the *Open Documents* area, on the *File Tab,* lists all the views that are currently open on your computer. By clicking one of the names in the list you "switch" to that view. A shortcut is to press **Ctrl + Tab** to quickly cycle through the open drawings.

- Many Revit commands also have keyboard shortcuts. So, with your right hand on the mouse (and not moving from the "design" area), your left hand can type **WA** when you want to draw a Wall, for example. You can see all the

Keyboard shortcuts dialog

preloaded shortcuts and add new ones by clicking *View (tab)* → *User Interface (drop-down)* → *Keyboard Shortcuts.*

It should be noted that any customized keyboard shortcuts are specific to the computer you are working on, not the project. You can use the *Export* button (see image to right) to save the entire keyboard shortcuts list to a file, and then *Import* it into another computer's copy of Revit.

View Tabs

Revit displays a tab for each open view. Clicking a tab is a quick way to switch between open views. Click the "X" in each view tab to close that view. Drag a tab to change its position. Tabs can also be pulled outside of the main Revit application window – even to a second monitor. In the second image below, the schedule could be filling a second computer screen while reviewing the same information in the floor plan.

View Tabs; one for each open view

View Tabs can be pulled outside of the main Revit application window, even to a second screen

This concludes your brief overview of the Revit user interface. Many of these tools and operations will be covered in more detail throughout the book.

Exercise 1-3:
Open, Save and Close a Revit Project

To *Open* **Revit 2025** in Windows 11, select Start and then click Revit 2025 or search for it.

Or double-click
the Revit icon
from your
desktop.

Revit 2025

This may vary slightly on
your computer; see your
instructor or system
administrator if you need
help.

How to Open a Revit Project

By default, Revit will open in the *Home*
screen, which will display thumbnails of
recent projects you have worked on.
Clicking on the preview will open the
project.

1. Click the **Open** button as shown to
 the right.

FYI: If the preview is for a *Central* file (aka
Worksharing) it will automatically open a
Local file. Worksharing allows multiple
people to work in the same file.

FIGURE 1-3.1 Revit Home screen

Next, you will open an existing Revit project file. You will select a sample file provided online at SDCpublications.com; see the inside front cover for more information.

2. Click the drop-down box at the top of the *Open* dialog (Figure 1-3.2). Browse to your downloaded files (provided files need to be downloaded from www.SDCpublications.com; see the backside cover of this book for more info).

> *TIP: If you cannot access the online sample files, you can substitute any Revit file you can find. Some sample files may be found here on your computer's hard drive: C:\Program Files\Autodesk\Revit 2025\Samples, or via* **File** *(tab)* → **Open** *(fly-out)* → **Sample Files**.

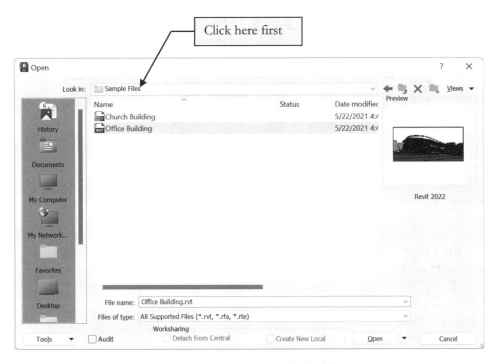

FIGURE 1-3.2 Open dialog; downloaded content from SDCpublications.com

3. Select the file named **Office Building.rvt** and click **Open**.

FYI: Notice the preview of the selected file. This will help you select the correct file before taking the time to open it.

The *Office Building.rvt* file is now open and the last saved view is displayed in the drawing window (Figure 1-3.3).

FIGURE 1-3.3 Sample file "Office Building.rvt"

The *File Tab* lists the projects and views currently open on your computer (Figure 1-3.4).

4. Click **File Tab → Open Documents (icon)** (Figure 1-3.4).

Notice that the *Office Building.rvt* project file is listed. Next to the project name is the name of a view open on your computer (e.g., floor plan, elevation).

FIGURE 1-3.4 File Tab: Open Documents view

Additional views will be added to the list as you open them. Each view has the project name as a prefix. The current view, the view you are working in, is always at the top of the list. You can quickly toggle between opened views from this menu by clicking on them.

You can also use the *Switch Windows* tool on the *View* tab; both do essentially the same thing.

Opening Another Revit Project

Revit allows you to have more than one project open at a time.

5. Click **Open** from the *Quick Access Toolbar*.

6. Per the instructions previously covered, browse the downloaded online files.

7. Select the file named **Church Building.rvt** and click **Open** (Figure 1-3.5).

> ***TIP:*** *If you cannot locate the online files, substitute one of the sample files.*

FIGURE 1-3.5 Sample file "Church Building.rvt"

8. Click **Open Documents** from the *File Tab* (Figure 1-3.6).

Notice that the *Church Building.rvt* project is now listed along with a view: Floor Plan: Level 1.

Try toggling between projects by clicking on *Office Building.rvt – 3D View: 3D View 1*.

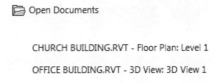

FIGURE 1-3.6 Open Documents

Close a Revit Project

9. Select **File Tab → Close**; click **No** if prompted to save.

This will close the current project/view. If more than one view is open for a project, only the current view will close. The project and the other opened views will not close until you get to the last open view.

10. Repeat the previous step to close the other project file.

If you made changes and have not saved your project yet, you will be prompted to do so before Revit closes the view. **Do not save at this time**. When all open project files are closed, you find yourself back in the *Recent Files* screen – which is where you started.

Saving a Revit Project

At this time we will not actually save a project.

To save a project file, simply select *Save* from the *Quick Access Toolbar*. You can also select *Save* from the *File Tab* or press *Ctrl + S* on the keyboard.

You should get in the habit of saving often to avoid losing work due to a power outage or program crash. Revit offers a reminder to save; this notification is based on the amount of time since your last save. The **Save Time Interval** is set in *Options*.

You can also save a copy of the current project by selecting **Save As** from the File Tab. Once you have used the *Save As* command you are in the new project file and the file you started in is then closed.

Closing the Revit Program

Finally, from the *File Tab* select **Exit Revit**. This will close any open projects/views and shut down Revit. Again you will be prompted to save, if needed, before Revit closes the view. **Do not save at this time**.

You can also click the "X" in the upper right corner of the *Revit Application* window.

Exercise 1-4:
Creating a New Project

Open **Autodesk Revit**.

Creating a New Project File

The steps required to set up a new Revit model project file are very simple. As mentioned earlier, simply opening the Revit program starts you in the *Recent Files* window.

To manually create a new project (maybe you just finished working on a previous assignment and want to start the next one):

1. Select **File tab → New → Project**.

FYI: You can also select the New *(Models) button on the Home screen.*

After clicking *New → Project* you will get the ***New Project*** dialog box (Figure 1-4.1).

FIGURE 1-4.1 New Project dialog box

The *New Project* dialog box lets you specify the template file you want to use, or not use a template at all via the <None> option (which is not recommended). You can also specify whether you want to create a new project or template file.

2. Select the **Imperial Multi-discipline** option in the drop-down list. Leave *Create new* set to **Project** (Figure 1-4.1).

3. Click **OK**. You now have a new unnamed/unsaved project file.

To name an unnamed project file you simply *Save*. The first time an unnamed project file is saved you will be prompted to specify the name and location for the project file.

4. Select **File Tab → Save** from the *Menu Bar*.

5. Specify a **name** and **location** for your new project file.
 Your instructor may specify a location or folder for your files if in a classroom setting.

What is a Template File?

A template file allows you to start your project with specific content and certain settings preset the way you like or need them.

For example, you can have the units set to *Imperial* or *Metric*. You can have the door, window and wall families you use most loaded and eliminate other less often used content. Also, you can have your company's title block preloaded and even have all the sheets for a project set up.

A custom template is a must for design firms and contractors using Revit and will prove useful to the student as he or she becomes more proficient with the program.

Be Aware:
It will be assumed from this point forward that the reader understands how to create, open and save project files. Please refer back to this section as needed. If you still need further assistance, ask your instructor for help.

Exercise 1-5:
Using Revit's Help System

This section of your introductory chapter will provide a quick overview of Revit's *Help System*. This will allow you to study topics in more detail if you want to know how something works beyond the introductory scope of this textbook.

1. Click the **round question mark** icon in the upper-right corner of the screen.

You are now in Revit's Help site (Figure 1-5.1). This is a website which opened in your web browser. The window can be positioned side by side with Revit, which is especially nice if you have a dual-screen computer system. This interface requires a connection to the internet. As a website, Autodesk has the ability to add and revise information at any time, unlike files stored on your hard drive. This also means that the site can change quite a bit, potentially making the following overview out of date. If the site has changed, just follow along as best you can for the next three pages.

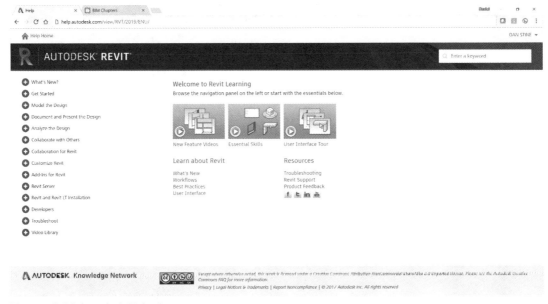

FIGURE 1-5.1 Autodesk Help site

In the upper left you can click the magnifying glass and search the *Help System* for a word or phrase. You may also click any one of the links to learn more about the topic listed. The next few steps will show you how to access the *Help System's* content on the Revit user interface, a topic you have just studied.

2. On the left, click the **plus symbol** next to **Get Started**.

3. Now, click the **plus symbol** next to **User Interface** (Figure 1-5.2).

Notice the tree structure on the left. You can use this to quickly navigate the help site.

4. Finally, click directly on **Ribbon**.

You now see information about the *Ribbon* as shown in the image below. Notice additional links are provided below on the current topic. You can use the browser's *Back* and *Forward* buttons to move around in the *Help System*.

FIGURE 1-5.2 Help window; Ribbon overview

Next, you will try searching the Revit *Help System* for a specific Revit feature. This is a quick way to find information if you have an idea of what it is you are looking for.

5. In the upper right corner of the current *Help System* web page, click in the search text box and then type **gutter**.

6. Press **Enter** on the keyboard.

The search results are shown in Figure 1-5.3. Each item in the list is a link which will take you to information on that topic.

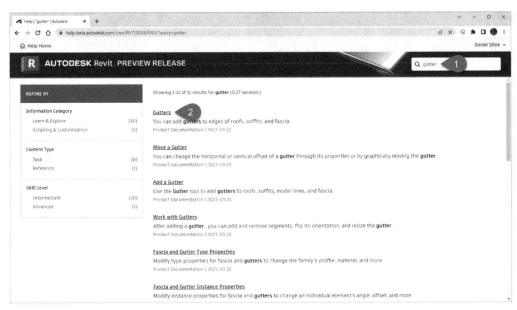

FIGURE 1-5.3 Help search results

The *Help System* can be used to complement this textbook. It is a reference resource, not a tutorial. As you are working through the tutorials in this book, you may want to use the *Help System* to fill in the blanks if you are having trouble or want more information on a topic.

Exercise 1-6:
Introduction to Autodesk Drive

We will finish this chapter with a look at Autodesk Drive, which is an integral foundation for much of Autodesk's **Cloud Services**. You do not necessarily need to use Autodesk's cloud services to complete this book successfully.

Here is how Autodesk describes *Autodesk Drive* on their website:

Autodesk Drive is a cloud storage solution that allows individuals and small teams to organize, preview, and share any type of design or model data.

You can use Autodesk Drive to:
- Upload data to a personal cloud drive.
- Organize and manage your data into folders.
- View 2D and 3D designs and models within the browser on any device.
- Share files & folders with others for viewing, editing, uploading, and managing data.
- You can also access, view, and edit data that others have shared with you.
- With Desktop Connector installed, you can also view and organize files stored in Drive from your desktop and desktop applications.

The Cloud, Defined

Before we discuss *Autodesk Drive* with more specificity, let's define what the *Cloud* is. **The Cloud is a service, or collection of services, which exists partially or completely online.** This is different from the *Internet*, which mostly involves downloading static information, in that you are creating and manipulating data. Most of this happens outside of your laptop or desktop computer.

The cloud gives the average user access to massive amounts of storage space, computing power and software they could not otherwise afford if they had to actually own, install and maintain these resources in their office, school or home. In one sense, this is similar to a *Tool Rental Center*, in that the average person could not afford, nor would it be cost-effective to own, a jackhammer. However, for a fraction of the cost of ownership and maintenance, we can rent it for a few hours. In this case, everyone wins!

Creating an Autodesk Account

The first thing an individual needs to do in order to gain access to *Autodesk Drive* is create a free Autodesk account at <u>drive.autodesk.com</u>. If you are already using an Autodesk product, you likely already have an Autodesk account.

This account is for an individual person, not a computer, not an installation of Revit or AutoCAD, nor does it come from your employer or school. Each person who wishes to access *Autodesk Drive* services must create an account, which will give them a unique username and password.

Generally speaking, there are two ways you can access *Autodesk Drive* cloud services:
- Autodesk Drive website
- Within Revit or AutoCAD; local computer

Autodesk Drive Website

When you have documents stored in the *Cloud* you can access them via your web browser. Here you can manage your files, view them without the full application (some file formats are not supported) and share them. These features use some advanced browser technology, so you need to make sure your browser is up to date; Chrome works well.

Using the website, you can upload files from your computer to store in the *Cloud*. To do this, you create or open a **Folder** and click the **Upload** option (Fig. 1-6.1).

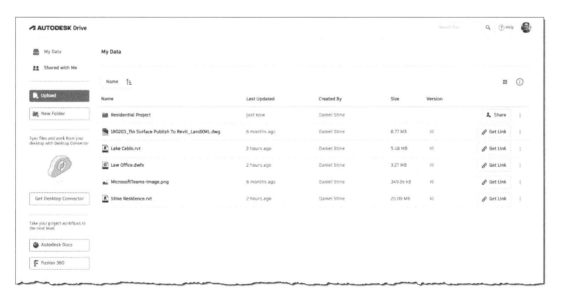

FIGURE 1-6.1 Viewing files stored in the cloud

Tip: If using a modern browser, you can drag and drop documents onto the window. This is a great way to create a secure backup of your documents.

You can share files stored in the *Cloud* with others. Simply hover over a file within *Autodesk DRIVE* and click the **Get Link** icon to see the Share dialog (Figure 1-6.2). Anyone who has this link can view and download (if Allow Downloads is toggled on) the files shared.

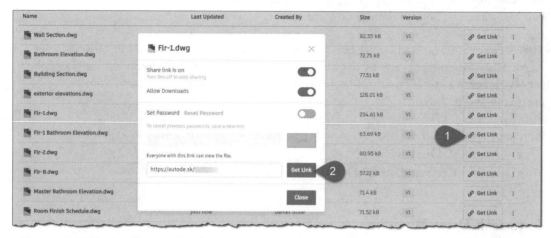

FIGURE 1-6.2 Sharing files stored in the cloud

Installing Autodesk Desktop Connector

To sync the Autodesk Drive files with your local computer and work directly on these files, you must first download Autodesk Desktop Connector. This tool creates an environment similar to OneDrive or Dropbox but with enhanced features supporting Autodesk-specific CAD and BIM workflows.

To install Desktop Connector, first close any open Autodesk products, such as AutoCAD. Then click the **Get Desktop Connector** button (Fig. 1-6.1) to access the download page. Download the latest version and install it.

Files are now synced locally and found via **This PC** within Windows Explorer (Fig. 1-6.3).

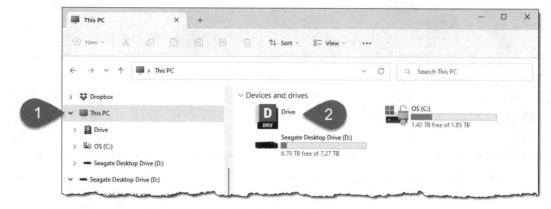

FIGURE 1-6.3 Accessing Drive files via Desktop Connector

The same cloud-based files appear locally, as shown below (Fig. 1-6.4).

FIGURE 1-6.4 Accessing Drive files via Desktop Connector

Accessing Autodesk DRIVE within your CAD/BIM application

Another way in which you can access your data, stored in the cloud, is from within your Autodesk application; for example, Revit or AutoCAD. This is typically the most convenient, as you can open, view and modify your drawings. Once logged in, you will also have access to any *Cloud Services* available to you from within the application, such as rendering services or *Green Building Studio*.

To sign into your *Autodesk Account* within your application, simply click the **Sign In** option in the notification area in the upper right corner of the window. You will need to enter your student email address and password, or work/personal email if you are not a student. When properly logged in, you will see your username or email address listed, as shown in Figure 1-6.5 below. Note: most likely, you are already signed in, as this is required to use most Autodesk products.

FIGURE 1-6.5 Example of user logged into their Autodesk account

The files within Autodesk Drive may be accessed via the **Open** dialog, as shown in Figure 1-6.6 below. Working on the files in this location ensures they are automatically backed up every time you save your work.

FIGURE 1-6.6 Accessing Autodesk Drive within AutoCAD or Revit

Notice that once a file has been modified, you can tell by the **Version** being incremented in Autodesk Drive when viewed in your browser, as shown below (Fig. 1-6.7).

FIGURE 1-6.7 Accessing Autodesk Drive within AutoCAD or Revit

When using AutoCAD, and when sharing files with others via Drive, only one person may modify a DWG file at a time. Similarly, when working with Revit models, an additional paid service called BIM 360 Collaborate Pro is required for multiple users to work within the same BIM.

It is recommended, as you work though this book, that you save all of your work in the *Cloud*, via Autodesk Drive, so you will have a safe and secure location for your files. These files can then be accessed from several locations via the two methods discussed here. It is still important to maintain a separate copy of your files on a flash drive, portable hard drive or in another *Cloud*-type location such as *Dropbox*. This will be important if your main files

ever become corrupt. You should manually back up your files to your backup location, so a corrupt file does not automatically corrupt your backup files.

TIP: If you have a file that will not open try one of the following:

- In AutoCAD: Open AutoCAD and then, from the *Application Menu*, select Drawing Utilities → Recover → Recover. Then browse to your file and open it. AutoCAD will try and recover the drawing file. This may require some things to be deleted but is better than losing the entire file.

- In Revit: Open Revit and then, from the *Application Menu*, select Open, browse to your file, and select it. Click the Audit check box, and then click Open. Revit will attempt to repair any problems with the project database. Some elements may need to be deleted, but this is better than losing the entire file.

Be sure to check out the Autodesk website to learn more about Autodesk BIM 360 Collaborate Pro and the growing number of cloud services Autodesk is offering.

Autodesk User Certification Exam

Be sure to check out Appendix A for an overview of the official Autodesk User Certification Exam and the available study book and software, sold separately, from SDC Publications and this author.

Successfully passing this exam is a great addition to your resume and could help set you apart and help you land a great a job. People who pass the official exam receive a certificate signed by the CEO of Autodesk, the makers of Revit.

Self-Exam:

The following questions can be used as a way to check your knowledge of this lesson. The answers can be found at the bottom of this page.

1. The *View* tab allows you to save your project file. (T/F)

2. You can zoom in and out using the wheel on a wheel mouse. (T/F)

3. Revit is a parametric architectural design program. (T/F)

4. A _____ file allows you to start your project with specific content and certain settings preset the way you like or need them.

5. Autodesk Drive allows you to save your files safely in the _____ .

Review Questions:

The following questions may be assigned by your instructor as a way to assess your knowledge of this section. Your instructor has the answers to the review questions.

1. The *Options Bar* dynamically changes to show options that complement the current operation. (T/F)

2. Revit is strictly a 2D drafting program. (T/F)

3. The Projects/Views listed in the *Open Documents* list allow you to see which Projects/Views are currently open. (T/F)

4. When you use the *Pan* tool you are actually moving the drawing, not just changing what part of the drawing you can see on the screen. (T/F)

5. Revit was not originally created for architecture. (T/F)

6. The icon with the floppy disk picture (🖫) allows you to _____ a project file.

7. Clicking on the _____ next to the *Zoom* icon will list additional zoom tools not currently shown in the *View* toolbar.

8. You do not see the *ViewCube* unless you are in a _____ view.

9. Creating an *Autodesk* user account is free. (T/F)

10. The *Views/Schedules cannot be positioned outside of the Revit application window.* (T/F)

Lesson 2
Model Navigation:

It is important to understand how to visualize the Building Information Model using Autodesk® Revit®. Zooming in, cutting sections and walking through the virtual model are fundamental steps to exploring the architect's or structural engineer's model and your design as it unfolds.

Exercise 2-1:
Using Zoom and Pan to View Your Drawings

Learning to *Pan* and *Zoom* in and out of a drawing is essential for accurate and efficient drafting and visualization. We will review these commands now so you are ready to use them with the first design exercise.

Open **Revit**.

You will select a sample file from the SDC Publications-provided online files.

1. Select **Open** from the *Quick Access Toolbar*.

2. Browse to the **downloaded online files**.

3. Select the file named **Church Building.rvt** and click **Open** (Figure 2-1.1).

> *TIP: If you cannot locate the online files, substitute any of the training files that come with Revit, found at C:\Program Files\Autodesk\Revit 2025\Samples.*

FIGURE 2-1.1 Church Building.rvt project

If the default view that is loaded is not **Floor Plan: Level 1**, double-click on **Level 1** under **Views\Floor Plans** in the *Project Browser*. Level 1 will be bold when it is the active or current view in the drawing window.

Using Zoom and Pan Tools

You can access the zoom tools from the *Navigation Bar*, or the *scroll wheel* on your mouse.

The *Zoom* icon contains several *Zoom* related tools:
- The default (i.e., visible) icon is *Zoom in Region*, which allows you to window an area to zoom into.
- The *Zoom* icon is a **split button**.
- Clicking the down-arrow part of the button reveals a list of related *Zoom* tools.
- You will see the drop-down list on the next page.

Zoom In

4. Select the top portion of the *Zoom* icon (see image to right).

5. Drag a window over your plan view (Figure 2-1.2).

Selection window: Pick two diagonal points, that is, two opposite corners to define the window.

FIGURE 2-1.2 Zoom In window

You should now be zoomed in to the specified area (Figure 2-1.3).

FIGURE 2-1.3 Zoom In results

Zoom Out

6. Click the down-arrow next to the zoom icon (Figure 2-1.4). Select **Previous Pan/Zoom**.

You should now be back where you started. Each time you select this icon you are resorting to a previous view state. Sometimes you have to select this option multiple times if you did some panning and multiple zooms.

7. **Zoom** into a smaller area, and then **Pan**, i.e., adjusting the portion of the view seen, by holding down the scroll wheel button.

The Pan tool just changes the portion of the view you see on the screen; it does not actually move the model.

FIGURE 2-1.4 Zoom Icon drop-down

Take a minute and try the other *Zoom* tools to see how they work. When finished, click **Zoom to Fit** before moving on.

TIP: You can double-click the wheel button on your mouse to Zoom Extents *in the current view.*

Default 3D View

Clicking on the *Default 3D View* icon, on the *QAT*, loads a 3D View. This allows you to quickly switch to a 3D view.

8. Click on the **Default 3D View** icon.

9. Go to the **Open Documents** listing in the *File Tab* and notice the *3D View* and the *Floor Plan* view are both listed at the bottom.

 REMEMBER: *You can toggle between views here.*

10. Click the **Esc** key to close the *File Tab*.

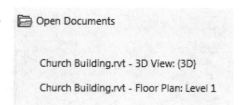

ViewCube

The *ViewCube* gives you convenient view control over the 3D view. This technology has been implemented in many of Autodesk's programs to make the process seamless for the user.

11. You should notice the ***ViewCube*** in the upper right corner of the drawing window (Figure 2-1.5). If not, you can turn it on by clicking *View* → *Windows* → *User Interface* → *ViewCube*.

 TIP: *The ViewCube only shows up in 3D views.*

Hovering your cursor over the *ViewCube* activates it. As you move about the *Cube* you see various areas highlight. If you click, you will be taken to that highlighted area in the drawing window. You can also click and drag your cursor on the *Cube* to "roll" the model in an unconstrained fashion. Clicking and dragging the mouse on the disk below the *Cube* allows you to spin the model without rolling. Finally, you have a few options in a right-click menu, and the **Home** icon, just above the *Cube*, gets you back to where you started if things get disoriented!

FIGURE 2-1.5 ViewCube

12. Give the *ViewCube* a try, then click the **Home** icon when you are done.

 REMEMBER: *The* Home *icon only shows up when your cursor is over the ViewCube.*

13. **Close** the *Church Building* project **without** saving.

Using the Scroll Wheel on the Mouse

The scroll wheel on the mouse is a must for those using BIM software. In Revit you can *Pan* and *Zoom* without even clicking a zoom icon. You simply **scroll the wheel to zoom** and **hold the wheel button down to pan**. This can be done while in another command (e.g., while drawing walls). Another nice feature is that the drawing zooms into the area near your cursor, rather than zooming only at the center of the drawing window. Give this a try before moving on. Once you get the hang of it, you will not want to use the icons. Also, double-clicking the wheel button does a *Zoom to Fit* so everything is visible on the screen.

TIP: Avoid a mouse with a wheel that tilts left and right as this makes using the wheel-button harder to use, making it not ideal for CAD/BIM users.

FIGURE 2-1.6 Furniture and flooring added by the reader later in the book

Exercise 2-2:
Creating 2D Sections

This section covers the simple steps of creating 2D sections in order to survey the vertical detail of the model. This is necessary when you are considering raising the ceiling and want to verify that there is space above for the structural framing and ductwork. You can also use the Section tool to verify model elements, such as furniture, are sitting properly on the floor, or to simply see a wall in elevation.

The *Elevation* tool works very much like the *Section* tool but will be covered later in the book. The *Section* tool is a great catchall tool for quick section and elevation views. These "quick" views are often temporary and not meant to be part of a submittal or used in a drawing set.

In this chapter you are focused simply on viewing the model. But keep in mind that all views, except drafting views and legends, are live views of the 3D model. Therefore, the two-dimensional views can be used to develop the design. You will have a chance to practice these techniques later in the book.

1. From the provided online files, open **Law Office.rvt** in the Book Starter folder; you will not need to save changes to this file while working through this chapter.

2. Make sure your current view is **Level 1 Floor Plan** via the *Project Browser*. Be careful **not** to select the Level 1 Ceiling Plan.

Seeing as sections are used so frequently, Autodesk has placed the *Section* tool on the *Quick Access Toolbar* (*QAT*). You should try to select the tool from this location rather than taking the time to switch to the *View* tab.

3. Select the *Section* tool from the *QAT*.

4. Pick the two points shown in Figure 2-2.1, being careful to create a horizontal line.

You have now created a section! Notice a new item is listed under "Sections" in the *Project Browser*. By default, immediately after creating the section, the section mark is selected. While the section is selected, dashed lines indicate the extent of the model that will be seen in the section view, for both width and depth.

There are three ways to open the new section view:

* Double-click the section view name in the *Project Browser*.
* Double-click the blue section head bubble; section mark cannot be selected for this technique to work (See Figure 2-2.2).
* Right-click on the section mark and select **Go to View** from the pop-up menu.

FIGURE 2-2.1 Creating a section

5. Open the new section view using one of the three options just mentioned.

Below is the anatomy of the *Section* "mark" in plan (Figure 2-2.2). Revit refers to both the section graphics in plan view (shown below) and the section view itself as a section. In this book we will refer to the plan view graphics as a *Section Mark*.

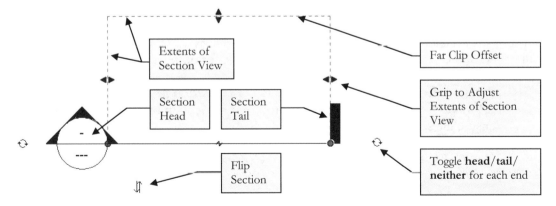

FIGURE 2-2.2 Section "mark" components

You should now see what is often referred to as a longitudinal building section. Revit automatically detects the extents of the model (height, width and depth). The entire height of the building is showing; this would be true even if this was a 40 floor high-rise. The *Far Clip Offset* is such that nothing is cut off; we can see doors and windows on the exterior wall.

By default, the *Crop Region* is visible and the *Detail Level* is set to coarse (Figure 2-2.3). Notice the building elements that the section mark touches (in plan view) are in section and represented with a heavier line. The portion of building that falls within the view extents (i.e., dashed lines when section mark is selected) is seen in elevation or projection.

These Walls, Floors and Roof are shown in Section

Crop Region

FIGURE 2-2.3 Viewing a section

Windows, Doors and Walls shown in Elevation/Projection

In this new view you can see elements not visible in plan view. For example, you can see the building structure and the thickness of the floors and ceilings. Here you can also see the distance between floors; notice the *Level* datum lines (i.e., Level 1, Level 2 and Roof). One could use the *Measure* tool to check how high a window is off the floor; will a desk work next to it? Or how deep a beam is above the room you are designing; how high can the ceiling be here?

Next, you will create another section looking towards the East (right). In this section you will adjust the view extents to limit what is seen in the view, for both width and depth.

6. Return to **Level 1 Floor Plan** view and add the ***Section Mark*** shown in Figure 2-2.4; make sure your first pick is at the top, and the section pick is at the bottom to ensure the section points/looks towards the East.

 TIP: *The* Flip Direction *icon can be used to quickly change the direction of the view, but the view depth is not automatically updated.*

FIGURE 2-2.4 Creating a second section

Looking at Figure 2-2.5, you now have what is often referred to as a building cross section. Again, the grid lines and level datum lines are automatically added. There are a few new aspects we can point out in this view.

First, notice the building longitudinal section mark is showing in this view. Any time a perpendicular section crosses through an elevation or section, it is automatically referenced in that view. If you open your first section view, you will see the new building cross section mark there as well!

Second, in addition to seeing some of the interior of the building in elevation/projection, we also see some of the exterior in elevation/projection.

Finally, having the *Detail Level* set to coarse is more obvious here when looking at the steel bar joists, which support the floor, in elevation. Notice how they are just simple outlines. This may be fine in some situations, but next you will see how to adjust this when more detail is needed.

7. Zoom in on a portion of the exterior wall and steel bar joists – see Fig. 2-2.5.

Steel Bar Joists

Zoom in here
– Step 7

FIGURE 2-2.5 Building cross section view created

8. Switch the *Detail Level* to **Fine** via the *View Control Bar* (Figure 2-2.6).

FIGURE 2-2.6 Adjusting a
view's detail level

Notice the additional detail now shown in the floor, wall
and structure (Figure 2-2.7).

FIGURE 2-2.7 Detail level set to fine; notice
added detail for steel bar joists, floor and wall

FIGURE 2-2.8 Adjusting section mark
far clipping offset – Step #9

Next you will adjust the extents of the view so you only see the part of the building in section; you will adjust the width and depth of what is seen in the section view.

9. In plan view, adjust the depth of the view by pulling the dashed line (extents of view) closer to the section mark (Figure 2-2.8).

 TIP: The section mark must be selected in order to see the dashed lines and the grips to adjust these lines.

10. Switch to the section view; adjust the **Crop Region** so the width is closer to the exterior walls in section (Figure 2-2.9).

Notice how you no longer see the 4'x4' exterior windows (between grids B and D) which are nearly 60 feet away. Also, when the *Crop Region* was adjusted, the levels and grids automatically adjusted, too. In fact, grids A and E totally disappeared because they are no longer intersecting the visible portion of the view.

FIGURE 2-2.9 Adjusted building cross section

11. On the *View Control Bar*, click to turn off the *Crop Region* visibility (Figure 2-2.10).

FIGURE 2-2.10 Crop region visibility

After adjusting the width of the *Crop Region* in the section view, go back to the plan view to see that the *Section Mark* did not change, but the dashed lines did (the section line must be selected to see the dashed lines). The *Section Mark* head and tail could be selected and adjusted if desired; this would not change the *Crop Region*.

Once a section is created, you can quickly move it around in a plan view to inspect other locations in the model. You do not always have to create a brand-new section.

12. Try the following before moving on:

 a. Flip a section to look the other way; use the *Flip* icon when the section mark is selected.

 b. Move a section to cut through the back stairwell; do not create a new one.

 c. Try toggling the visibility of the head and tail graphics. Make a section have two heads (i.e., bubble with arrow).

When you want to delete a section, simply select it in plan view and press the **Delete** key.

13. **Delete** the building cross section from the model.

14. Click **OK** to delete the view (Figure 2-2.11).

FIGURE 2-2.11 Delete view warning

One thing to be careful of when deleting a view (any view, section, elevation, plan view) is if the view has any notes, dimensions or tags. These are view-specific elements that will permanently be deleted with the view. You have not created any of these view-specific elements yet, so this is not a problem. It is also possible to hide a section mark in plan view so it does not need to be deleted; more on this later.

Next you will see how easy it is to list a distance.

15. Switch to the longitudinal section. Move it back into its original location if you moved it in Step 12.

16. Zoom in on grid number "2" between Level 1 and Level 2.

FIGURE 2-2.12 Measure tool on QAT

17. Select the ***Measure*** tool from the *Quick Access Toolbar* (Figure 2-2.12).

18. *Pick your first point.* Pick the bottom of the steel beam (Figure 2-2.13).

19. *Pick your second point.* Pick the top of the concrete slab-on-grade floor.

The distance and angle are now temporarily shown. This information can be used to determine the height of a ceiling (in conjunction with any ductwork passing under it). One more thing: notice the structural steel bar joists are not as deep to the left of the beam as they are to the right. The joists to the right have been designed by the structural engineer to support the additional load for the bookshelves in the law library. If the law library moved due to refined interior development we would have to make sure the structural engineer was informed of this change.

FIGURE 2-2.13 Using the Measure tool

It is good practice to rename a new view as soon as it is created to make it easier to locate it in the future. In this case we are creating temporary, or junk, section views and therefore the name is not as important. However, it would still be a good idea to add your initials as a prefix to the section views. This prefix would let others working on the project know this is your section view and not to delete it.

20. Find the section view in the *Project Browser*, right-click on it and select **Rename**.

21. Add your **initials** and a **dash** per Figure 2-2.14.

FIGURE 2-2.14 Renaming a view

The section head bubble is not filled out (i.e. drawing number and sheet number) by Revit until the section view is placed on a sheet. A **Sheet** is a composition of views (i.e. drawings) and a title block which can be printed and used for presentations or bidding and construction by the contractor. You will learn more about this later in the book. When a section is not placed on a sheet, it is considered **unreferenced**.

When printing, it is possible to hide the tags, elevations and sections for unreferenced views. This only affects printing. You will locate this option next.

22. Click **File Tab → Print,** or just *Ctrl + P.*

23. Select the **Setup…** button.

Notice the check box *Options* in the lower left corner of the *Print Setup* dialog (Figure 2-2.15).

24. **Cancel** the print dialog boxes and close the project without saving.

FIGURE 2-2.15 Print setup dialog

There is still a lot to learn about sections in Revit, but this should be enough to get you started, thus allowing you to explore the model and derive distances.

The main thing to remember with sections is that they are live 2D views of your 3D model. Also, sections can be created anytime you need to look at vertical relationships. The created view does not need to go on a sheet as part of a drawing set. Just name them to maintain order and delete the ones you know you will not need again, or just plan on creating another one when you do.

Exercise 2-3:
Navigating the 3D Model

This exciting section immerses you into the virtual model by way of 3D, perspective camera views and walkthroughs. This knowledge gives the designer and the client a more realistic representation of the proposed space. You will continue to explore the *Law Office* shell model to better understand the context in which you will be working in later chapters.

ViewCube

The *ViewCube* gives you convenient view control over any 3D view. This technology has been implemented in many of Autodesk's programs to make the process seamless for the user.

1. **Open** a copy of the *Law Office* model from the previous section.

2. Click the **Default 3D View** icon on the *Quick Access Toolbar*.

You are now in an isometric 3D view of the model.

3. You should notice the **ViewCube** in the upper right corner of the drawing window (Figure 2-3.1). If not, you can turn it on by clicking *View → Windows → User Interface → ViewCube*.

 TIP: The ViewCube *only shows up in 3D views. There are several related settings via File Tab →* Options → *ViewCube.*

Hovering your cursor over the *ViewCube* activates it. As you move your cursor over the *Cube*, you see various areas highlight. When you click, you will be taken to that highlighted area in the drawing window. You can also click and drag your cursor on the *Cube* to "roll" the model in an unconstrained fashion. Clicking and dragging the mouse on the disk below the *Cube* allows you to spin the model without rolling. You also have a few options in a right-click menu. Finally, the **Home** icon, just above the *Cube*, gets you back to where you started if things get disoriented!

FIGURE 2-3.1 ViewCube

4. Give the *ViewCube* a try, then click the **Home** icon when you are done.

 REMEMBER: The Home *icon only shows up when your cursor is over the* ViewCube.

Keep in mind that the North, East, South, West directions likely do not correspond to True North, but rather Plan North (or Project North).

3D Views

3D views can be duplicated from the *Project Browser*. They can also be **locked** so they don't change. This toggle is found on the *View Control Bar*; it is a 3D House with a padlock over it. When a view is locked the orientation is maintained, with a few exceptions. This allows notes and tags to be added without the concern of the model moving and affecting the annotation. You will not actually lock the 3D view at this time.

Visual Styles

The Revit model can be displayed in one of six ways in any given view. Each view has a toggle on the *View Control Bar* as shown to the right. Most views are typically set to *Hidden Line*, which is basically a black & white drawing where 3D elements hide things behind them. Wireframe is rarely used but is fast and allows you to explore problems. The remaining options have various presentation qualities. *Realistic* is resource

intensive, so panning and zooming can be quite slow in these views on larger models. The image below shows these options from *Wireframe* (left) to *Realistic* (right).

Orient to View; creating a cropped 3D view

The *ViewCube* has an option to align and crop a 3D view to match a specific 2D view. This is great for presentation graphics and troubleshooting.

5. Duplicate the main 3D view; right-click on it in the *Project Browser* and select **Duplicate View → Duplicate**.

You now have a new 3D view called **{3D} Copy 1**; this view is active.

6. Right-click on the *ViewCube* and select **Orient to View** (Figure 2-3.2).

7. Now select the section with your initials from the previous exercise.

Notice you can also create cropped 3D views based on floor plans and elevations. The floor plans will be cropped at the cut plane specified in the 2D plan view (see page 1-8).

The view now looks very similar to the 2D section view – except it is a 3D view (Figure 2-3.3). The view has no connection to the 2D section view. If the *Crop Region* changes, or the *Section Mark* is moved, this 3D view will not change. On the other hand, we are still looking at the same 3D model, so any changes made to the model automatically show up in all views.

The element cropping the 3D view is called a **Section Box**. The *Section Box* can be toggled on and off via a setting in the *Properties Palette*. This setting is only available when nothing in the model is selected; otherwise, the *Properties Palette* displays information about what is selected.

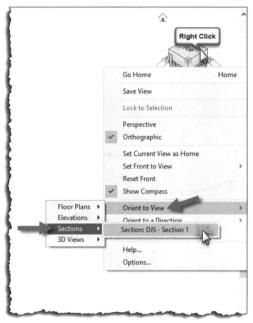

FIGURE 2-3.2 ViewCube – Orient to View

FIGURE 2-3.3 Modified 3D view

Now that this view is "cropped" with the *Section Box*, it will remain that way. The only way it will change is if you manually adjust the *Section Box*, toggle the *Section Box* off or use the *Orient to View* feature again (selecting a different view).

Next, you will adjust the view back to an isometric but not change the *Section Box*.

FIGURE 2-3.4 ViewCube

8. Click the upper left hotspot/link on the *ViewCube* as shown in Figure 2-3.4.

9. Click and drag on the *ViewCube* until your view looks similar to Figure 2-3.5.

FIGURE 2-3.5 Adjusted 3D view

In the view above, Figure 2-3.5, the *Section Box* matches the extents of the 2D section view. The width and height are based on the section view's *Crop Region* size, and the depth is based on the section view's *Far Clip Offset*, when *Far Clipping* is set to one of the two "clip" options. The width and depth also relate to the "Extents of View" dashed lines visible in the plan view when the section mark is selected.

This view has the *Visual Style* set to **Consistent Colors**, which fills the faces with a color. Finally, notice how the site (i.e., earth) is shown in section. Using a *Section Box* is the only way to see the ground with a thickness.

You could leave this view as-is and be able to refer back to it when needed. However, we will instead modify the *Section Box* to see how easy it is to change. And thus, how easy it is to explore the 3D model.

 10. Select the *Section Box*.

Notice how each face of the *Section Box* now has a set of grips. Dragging these grips allows you to modify the *Section Box* on the fly (Figure 2-3.6).

Next you will adjust the view to show the First Floor Plan.

FIGURE 2-3.6 Section Box selected

11. Select the top grip and drag it downward; try to go a foot or two below the top of the doors (Figure 2-3.7). You may have to try adjusting it a few times.

FIGURE 2-3.7 Section Box top lowered

12. Now drag the South face of the *Section Box* out so the entire floor plan is visible (Figure 2-3.8).

You now have a very informative view of the First Floor Plan. You could duplicate this view and set up a cropped view, via *Section Box* adjustments, of the Second Floor. These views can just be used for reference by you, the designer, or they can be placed on sheets and printed out for a presentation; more on sheets later in the book.

FIGURE 2-3.8 Section Box adjusted

The view above has **Shadows** toggled on via the *View Control Bar.*

Camera Views

A *Camera* view is very much like a person standing in a building looking at it through a camera. The camera is a certain distance above the floor; it is looking in a specific direction, and the field of view, or angle, determines how much is seen. A camera view is a perspective view, as compared to the isometric (or parallel line) 3D views previously discussed.

Creating a *Camera* view involves picking the point where you want to be virtually standing and then the direction you wish to be looking. The default *Camera* height, above the floor, is based on the average eye level of 5'-6". You are then automatically switched to that view in the model and a new item shows up in the *Project Browser*.

Next, you will create two *Camera* views, one on each level.

13. Switch to the **Level 2 Floor Plan** via the *Project Browser*.

14. From the *Quick Access Toolbar*, click the down-arrow next to the *3D House* icon.

FIGURE 2-3.9 Using the Camera tool

15. Click the *Camera* tool (Figure 2-3.9).

> *FYI: The* Camera *tool can also be accessed from the* View *tab.*

16. Click the first point approximately as shown in Figure 2-3.11.

17. Click the second point approximately as shown – just outside the building.

You should now have a view similar to Figure 2-3.10. Generally, the first thing you should do after creating a new view is rename it to something more meaningful.

FIGURE 2-3.10 New camera view with shading turned on

FIGURE 2-3.11 Adding a Camera view on Level 2

18. Right-click on **3D View 1** in the *Project Browser* and select ***Rename***.

19. Type **Level 2 Camera View** and press **Enter**.

Now you will adjust the size of the *Crop Region* to increase the field of view. The more you increase the size, the more distorted things become, so be careful not to adjust the *Crop Region* too much. This distortion is similar to what would be seen when using a wide-angle lens on a real camera – the wider the view the more the photo is distorted.

20. Select the *Crop Region*, the view's outline, in the new *Camera* view.

21. Drag the two side grips to make the view wider and the bottom grip downward; try to match Figure 2-3.12.

FIGURE 2-3.12 Adjusting crop region to increase field of view; visual style = consistent colors

22. Toggle off the **Crop View** option as shown in the image below (Fig. 2-3.13).

When a perspective view is un-cropped, you can move within the model by scrolling the wheel on the mouse; normally, this just zooms in or out, but you never moved past a wall. Un-cropped perspective views cannot be placed on sheets.

FIGURE 2-3.13 Crop view toggled off; visual style = hidden lines

You will create one more *Camera* view on Level 1. This will be a view from the main entry, or what a person will see when they walk in the front door!

23. Switch to the **Level 1 Floor Plan** view.

24. Add a *Camera* as shown in Figure 2-3.14.

25. Adjust the *Crop Region* as shown in Figure 2-3.15.

Looking closely at Figure 2-3.15 you should see a void in the far back corner of the model. This relates to a setting for the camera view called *Far Clip Active*, which controls how far into the model you can see.

26. Uncheck **Far Clip Active** in the *Properties Palette* and click **Apply**.

FIGURE 2-3.14 Adding a Level 1 Camera view

27. Rename the new view to **Level 1 Camera View**.

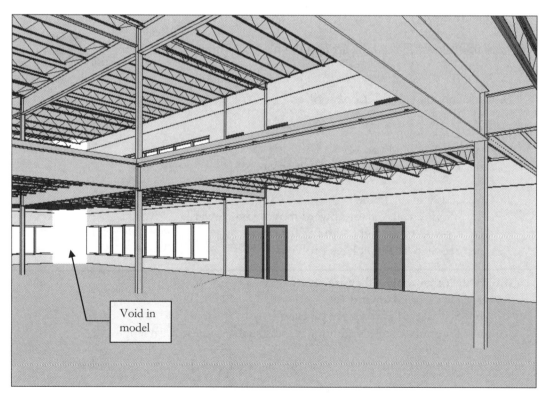

Void in model

FIGURE 2-3.15 New Camera view at Level 1

Steering Wheel

Similar to the *ViewCube*, the *Steering Wheel* aids in navigating your model. With the *Navigation Wheel* you can walk through your model, going down hallways and turning into rooms. Revit has not advanced to the point where the doors will open for you; thus, you walk through closed doors or walls as if you were a ghost!

The *SteeringWheel* is activated by clicking the upper icon on the *Navigation Bar*.

When the *SteeringWheel* (*SW*) is activated, you will see the graphic shown to the left near your cursor. The *SW* will "stick" near your cursor, moving slightly slower than you so you can catch up to the various tools within the *SW*.

Clicking and dragging your mouse on one of these tools activates it. You will try this next.

FIGURE 2-3.16 Navigation Wheel

28. Select the **Full Navigation Wheel** from the *Navigation Bar* (Figure 2-3.17).

The *Look* tool allows you to look around the model without moving your feet.

29. Click and drag on the ***Look*** tool.
 a. Look up.
 b. Look down.
 c. Look behind you.
 d. Look towards the original direction.

FIGURE 2-3.17
Navigation Bar

30. Click and drag on ***Walk***.
 a. Moving your cursor upward moves you forward.
 b. Up and to the right starts to turn right.
 c. See Step 31 if you have problems.

31. Try the other tools – they are pretty easy to figure out.
 a. Do not try **Rewind** yet.
 b. See Step 31 if you have problems.
 c. Right-click reveals additional tools such as adjust focal length.

You can leave the view where it is or use *Rewind* to get back to where you started.

32. Click and drag on the ***Rewind*** tool.
 a. Dragging to the left will rewind the view.
 b. Return all the way back to the beginning.

33. When finished using the *SteeringWheel* tool, click the "X" in the upper right.

34. Close your file **without** saving.

FIGURE 2-3.18 Exploring the Level 1 space using the *Steering Wheel*

Self-Exam:

The following questions can be used as a way to check your knowledge of this lesson. The answers can be found at the bottom of this page.

1. The best way to zoom is by using the wheel on your mouse. (T/F)

2. The North direction on the *ViewCube* always points True North. (T/F)

3. 3D views can be locked. (T/F)

4. This tool allows you to virtually walk through the model: _____ _____.

5. The camera is initially placed on the floor and then needs to be moved up to eye level. (T/F)

Review Questions:

The following questions may be assigned by your instructor as a way to assess your knowledge of this section. Your instructor has the answers to the review questions.

1. When a section mark is deleted from a plan view, the section view is also deleted. (T/F)

2. Double-clicking on the section head opens that view. (T/F)

3. The zoom tools are located on the _____.

4. The detail level does not affect how walls appear. (T/F)

5. You can adjust a 3D view orientation using the *ViewCube*. (T/F)

6. Using the *SteeringWheel's Look* feature actually moves the camera in the model. (T/F)

7. The zoom icon is a split button. (T/F)

8. What cuts off the back of a model in a camera view? _____ _____

9. When multiple people are working in a model, they each have their own default 3D view. (T/F)

10. Using a _____ is the only way to see the ground with a thickness.

Lesson 3
Quick Start: Small Office:

In this lesson, you will get a down-and-dirty overview of the functionality of Autodesk Revit. The very basics of creating the primary components of a floor plan will be covered: Walls, Doors, Windows, Roof, Annotation, and Dimensioning. This lesson will show you the amazing "out-of-the-box" powerful, yet easy-to-use, features in Revit. It should get you very excited about learning this software program. Future lessons will cover these features in more detail while learning other editing tools and such along the way.

Exercise 3-1:
Walls, Grids and Dimensions

In this exercise, you will draw the walls, starting with the exterior. Read the directions carefully; everything you need to do is clearly listed.

Exterior Walls

1. Start a new project named **Small Office** per the following instructions:

 a. **Home Screen → Models → New...**

 b. Click **Browse...** (Figure 3-1.1)

 c. Select the template file named **Commercial-Default.rte**. *(You should be brought to the correct folder automatically.)*

FIGURE 3-1.1 New Project

 d. Click **Open**.

 e. With the template file just selected and *Create new* "Project" selected, click **OK** (Figure 3-1.1).

 See Lesson 1 for more information on creating a new project.

FIGURE 3-1.2 Wall tool

2. Select **Architecture → Build → Wall** on the *Ribbon*. (See Figure 3-1.2.)

Notice that the *Ribbon*, *Options Bar* and *Properties Palette* have changed to show settings related to walls. Next you will modify those settings.

FYI: By default, the bottoms of new walls will be at the current floor level and the tops of the walls are set via the Options Bar as shown in the next step.

FIGURE 3-1.3 Ribbon and Options Bar

3. Modify the *Ribbon*, *Options Bar* and *Type Selector* to the following (Figure 3-1.3):

 a. *Type Selector:* Click in this area and select *Basic Wall:* **Generic – 12″**.

 b. *Height:* Change the height from 14′-0″ to **9′-0″**.

 c. *Location Line:* Set this to **Finish Face: Exterior**.

 d. Click the **Rectangle** icon in the *Draw* panel on the *Ribbon. (This allows you to draw four walls at once [i.e., a rectangle], rather than one wall at a time.)*

You are now ready to draw the exterior walls.

4. In the *Drawing Window*, click in the upper left corner.

 TIP: Make sure to draw within the four elevation markers (see image to right).

5. Start moving the mouse down and to the right. **Click** when the two temporary on-screen listening dimensions are approximately **100′** (East to West) and **60′** (front to back).

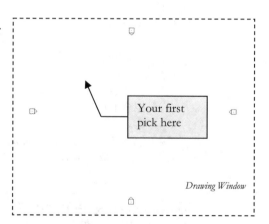

Getting the dimensions exact is not important as they will be revised later on.

Your drawing should look similar to Figure 3-1.4; similar in that the dimensions do not have to be exact right now and the building's location relative to the four elevation tags may vary slightly.

The *Temporary Dimensions* are displayed until the next action is invoked by the user. While the dimensions are displayed, you can click on the dimension text and adjust the wall dimensions. Also, by default the *Temporary Dimensions* reference the center of the wall – you can change this by simply clicking on the grips located on each *Witness Line*; each click toggles the witness line location between center, exterior face and interior face.

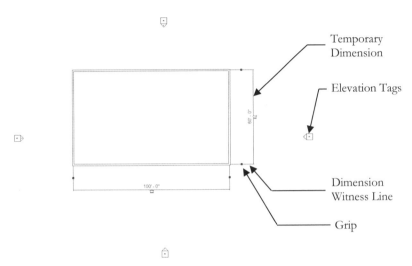

FIGURE 3-1.4 Exterior walls

In the next few steps you will create grid lines and establish a relationship between the walls and the grids such that moving a grid causes the wall to move with it.

Grids:

Grids are used to position structural columns and beams in a building. Adding a grid involves selecting the *Grid* tool and then picking two points in the drawing window.

6. Click **Modify** on the *Ribbon* (to finish using the *Wall* tool) and then select the **Architecture → Datum → Grid**.

 FYI: The same Grid *tool is also found on the* Structure *tab.*

Grid

Next you will draw a vertical grid off to the left of your building. Once you have drawn all the grids you will use a special tool to align the grid with the walls and lock that relationship.

7. [*first pick*] **Click** down and to the left of your building as shown in Figure 2.1-5.

 FYI: 'Click' always means left-click, unless a right-click is specifically called for.

8. [*second pick*] Move the cursor straight up (i.e., vertically) making sure you see a dashed cyan line, which indicates you are snapped to the vertical plane, and the angle dimension reads 90 degrees. Just past the North edge of the building (as shown in Figure 2.1-5), **click**.

FIGURE 3-1.5 Drawing a grid

You have now drawn your first grid line. Next you will quickly draw four more grid lines, two horizontal and two vertical.

> *NOTE: The Grid tool will remain active until you start another tool or select Modify (selecting Modify allows you to select other elements in the drawing window).*

9. Draw another vertical grid approximately centered on your building. BEFORE YOU PICK THE FIRST POINT, make sure you see a dashed horizontal cyan reference line indicating the grid line will align with the previous grid line (you will see this before clicking the mouse at each end of the grid line), then go ahead and pick both points (Figure 3-1.6).

10. Draw the remaining grid lines shown in Figure 3-1.6. Again, do not worry about the exact location of the grid lines, just make sure the ends align with each other.

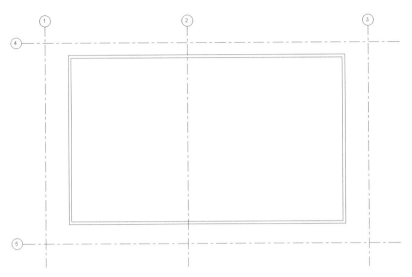

FIGURE 3-1.6 Grids added

Next you will change the two horizontal grid lines to have letters instead of numbers.

11. Zoom in on the grid bubble for the upper horizontal grid line.

12. Click *Modify* and then click on the grid line to select it.

13. With the grid line selected, click on the dark blue text within the bubble.

14. Type **A** and press *Enter* on the keyboard (Figure 3-1.7).

15. Click **Modify** again.

16. Change the other horizontal grid to **B**.

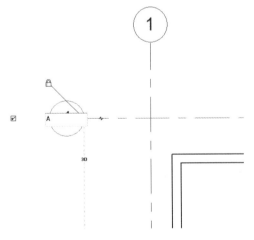

FIGURE 3-1.7 Grid edit

Align

Next you will use the *Align* tool to reposition the grid lines so they "align" with the exterior face of the adjacent walls. The steps are simple: select the *Align* tool from the *Ribbon*; pick the reference line (i.e., the exterior wall face); and then you select the item to move (i.e., the grid line). This tool works on many Revit objects!

17. Select **Modify** → **Modify** → **Align** from the *Ribbon*.

 REMINDER: *Ribbon instructions are* Tab → Panel → Tool

18. [*Align: first pick*] With the *Align* tool active, notice the prompt on the *Status Bar*, select **Wall faces** on the *Ribbon*, next to the label "Prefer", and then select the exterior face of the wall adjacent to grid line 1.

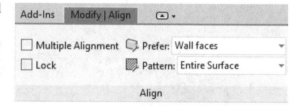

19. [*Align: second pick*] Select **grid line 1**. Be sure to see the next step before doing anything else!

The grid line should now be aligned with the exterior face of the wall. Immediately after using the *Align* tool, you have the option of "locking" that relationship; you will do that next. The ability to lock this relationship is only available until the next tool is activated. After that, you would need to use the *Align* tool again.

20. Click the un-locked **padlock** symbol to "lock" the relationship between the grid line and the wall (see Figure 3-1.8).

21. Use the steps just outlined to **Align** and **Lock** the remaining grid lines with their adjacent walls. Do not worry about the location of grid line 2 (i.e., the vertical grid in the center).

FIGURE 3-1.8 Align "lock"

Dimensions

Next you will add dimensions to the grid lines and use them to drive the location of the walls and grids and learn how to lock them.

22. Select **Modify** and then **right-click** anywhere within the *Drawing Window*; click **Zoom To Fit**.

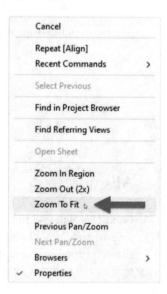

23. Select **Annotate → Dimensions → Aligned** tool.

Aligned

At this point you are in the *Dimension* tool. Notice the various controls available on the *Ribbon* and *Options Bar*. You can set things like the dimension style (via the *Element Type Selector*) and the kind of dimension (linear, angle, radius, etc.) and which portion of the wall to *Prefer* (e.g., face, center, core face).

24. With the *Ribbon*, *Options Bar* and *Properties Palette* as shown above, which should be the default settings, **select grid line 1**.

> *FYI: The grid line will pre-highlight before you select it, which helps make sure you select the correct item (e.g., the grid line versus the wall). Be careful not to select the wall.*

25. Now **select grid line 2** and then **select grid line 3**.

Your last pick point is to decide where the dimension line should be.

26. Click in the location shown in Figure 3-1.9 to position the dimension line.

> *TIP: Do not click near any other objects or Revit will continue the dimension string to that item.*

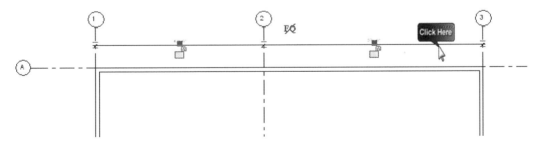

FIGURE 3-1.9 Adding dimensions

Notice that while the dimension string is selected, you see an EQ symbol with a slash through it. This symbol indicates that the individual components of the dimension string are not equal in length. The next step will show you how easy it is to make these dimensions equal!

27. With the dimension string selected, click the EQ symbol located near the middle of the dimension.

The grid lines are now equally spaced (Figure 3-1.10) and this relationship will be maintained until the EQ symbol is selected again to toggle the "dimension equality" feature off.

> *NOTE: When dimension equality is turned off or the dimension is deleted the grid line **will not** move back to its original location; Revit does not remember where the grid was.*

Typically, you would not want to click the padlock icons here because that would lock the current dimension and make it so the grid lines could not be moved at all.

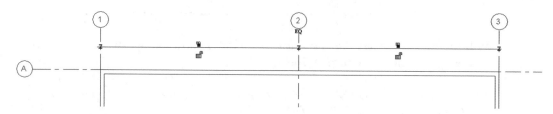

FIGURE 3-1.10 Toggling dimension equality

Next you will add an "overall building" dimension from grid line 1 to grid line 3. This dimension can be used to drive the overall size of your building (all the time keeping grid line 2 equally spaced).

28. Using the **Aligned** *Dimension* tool, add a dimension from grid line 1 to grid line 3 and then pick to position the dimension line (Figure 3-1.11).

FIGURE 3-1.11 Overall building dimension added

When using a dimension to drive the location of geometry, you need to select the item you want to move and then select the dimension text to enter the new value. You cannot just select the dimension because Revit does not know whether you want the left, right or both grid lines to move. The only thing you can do, graphically, by selecting the dimension directly is "lock" that dimension by clicking on the padlock symbol and then click the blue dimension text to add a suffix if desired. Next you will adjust the overall building size.

29. Click **Modify** (or press the *Esc* key twice) to make sure you cancel or finish the *Dimension* tool and that nothing is selected.

30. **Select grid line 3**.

31. With grid line 3 selected, click the dimension text and type **101** and then press *Enter*.

 FYI: Notice that Revit assumes feet if you do not provide a foot or inch symbol.

32. Repeating the previous steps, add a dimension between grid lines A and B, and then adjust the model so the dimension reads **68′-0″**.

FIGURE 3-1.12 Building size established

Your project should now look similar to Figure 3-1.12. You should notice that dimensions must "touch" two or more items (the grid lines in this case). Also, because the walls were aligned and locked to the grids, moving the grids caused the walls to move.

The last thing you will do before moving on to the interior walls is to swap out the generic walls with a more specific wall. This would be a common situation in a design firm; a generic wall is added as a "place holder" until the design is refined to the point where the exterior wall system is selected.

The process for swapping a wall is very simple: select the wall and pick a different type from the *Type Selector*. The next steps will do this but will also show you how to quickly select all the exterior walls so you can change them all at once!

33. Click **Modify** and then hover your cursor over one of the exterior walls so it highlights. (Do not click yet.)

34. *With an exterior wall highlighted,* i.e., pre-selected, take your hand off the mouse and tap the **Tab** key until all four walls pre-highlight.

 FYI: The Tab *key cycles through the various items below your cursor. The current options should include one wall, a chain of walls, and a grid line.*

35. *With all four walls highlighted,* **click** to select them.

36. *With all four walls selected,* pick *Basic Wall:* **Exterior – Brick on CMU** from the *Type Selector* on the *Properties Palette.*

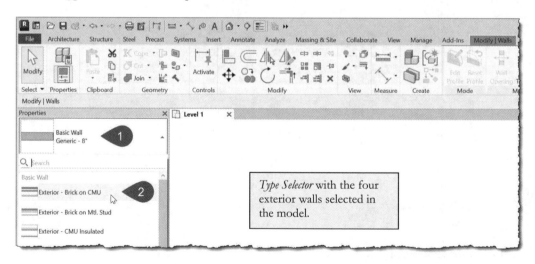

Type Selector with the four exterior walls selected in the model.

Detail Level

Revit allows you to control how much detail is shown in the walls.

37. On the *View Control Bar*, located in the lower left corner of the *Drawing Window*, set the *Detail Level* to **Fine**.

As you can see in the two images below, *Coarse* simply shows the outline of a wall type and *Fine* shows the individual components of the wall (i.e., brick, insulation, concrete block, etc.).

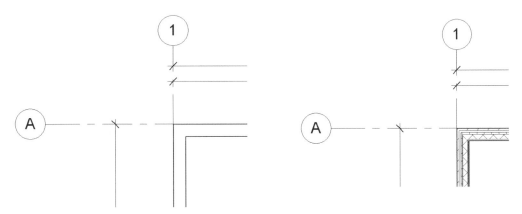

FIGURE 3-1.13 Detail level – coarse (left) vs. fine (right)

Now that you have the exterior walls established, the grid lines properly placed, and their relationships locked in the project, you can now proceed with the layout of the interior spaces.

Interior Walls

38. With the **Wall** tool selected, modify the *Ribbon*, *Options Bar* and *Properties Palette* to the following (also, see image on the following page):

 a. *Type Selector:* set to <u>Basic Wall</u>: **Interior – 4 7/8″ Partition (1-hr)**.

 b. *Height:* **Roof**

 c. *Location Line:* Set this to **Wall Centerline.**

 d. *Turn off Chain.*

39. Draw a wall from the West wall (i.e., "vertical" wall on the left) to the East wall (on the right). See Figure 3-1.14.

 a. Make sure your cursor "snaps" to the wall before clicking.

 b. Before clicking the second point of the wall, make sure the dashed cyan line is visible so you know the wall will be truly horizontal.

 c. The exact position of the wall is not important at this point as you will adjust it in the next step.

 d. With the temporary dimensions still active, proceed to the next step.

FIGURE 3-1.14 Adding interior walls – North indicator added for reference only

40. Click the **witness line grips** (see Figure 3-1.14) until the "clear" space of the room is listed (see Figure 3-1.15).

41. Now click the blue text of the temporary dimension, type **22**, and then press *Enter* (Figure 3-1.15).

42. Click **Modify** to finish the current task.

FIGURE 3-1.15 Repositioning interior wall via temporary dimensions

The clear space between the interior wall and the north wall is now 22'-0". Next you will add additional interior walls to create equally spaced rooms in this area.

43. Using the same settings as the interior wall just added, draw five (5) vertical walls as shown in Figure 3-1.16.

 a. Make sure they are orthogonal (i.e., the dashed cyan line is visible before picking the walls endpoint).

 b. Make sure you "snap" to the perpendicular walls (at the start and endpoint of the walls you are adding).

 c. Do not worry about the exact position of the walls.

 TIP: Uncheck Chain on the Options Bar.

FIGURE 3-1.16 Adding additional interior walls

In the next step you will use a dimension string to reposition the walls so they are equally spaced. This process is similar to what you did to reposition grid line 2. However, you have to specify which part of the wall you want to dimension to (center, face, core center, core face).

FYI: The "core" portion of a wall system typically consists of the structural element(s) such as the concrete block (in your exterior walls) or the metal studs (in your interior walls).

44. Select **Annotate → Dimension → Aligned** tool on the R*ibbon*.

45. On the *Options Bar*, select **Wall Faces**.

This setting will force Revit to only look for the face of a wall system. You can select either face depending on which side of the wall you favor with your cursor. This feature lets you confidently pick specific references without needing to continually zoom in and out all over the floor plan.

46. Select the interior face of the West wall to start your dimension string.

The next several picks will need to reference the wall centerlines. Revit allows you to toggle the wall position option on the fly via the *Options Bar*.

47. Change the setting to **Wall Centerlines**.

48. The next five picks will be on the five "vertical" interior walls. Make sure you see the dashed reference line centered on the wall to let you know you are about to select the correct reference plane.

49. Change the wall location setting back to **Wall Faces** and select the interior face of the East wall (the wall at grid line 3).

50. Your last pick should be away from any elements to position the dimension string somewhere within the rooms (Figure 3-1.17).

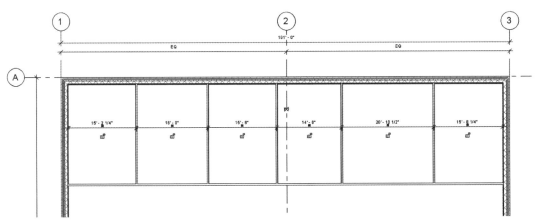

FIGURE 3-1.17 Adding a dimension string

51. Click the EQ symbol to reposition the interior walls.

52. Click **Modify**.

The interior walls are now equally spaced (Figure 3-1.18)!

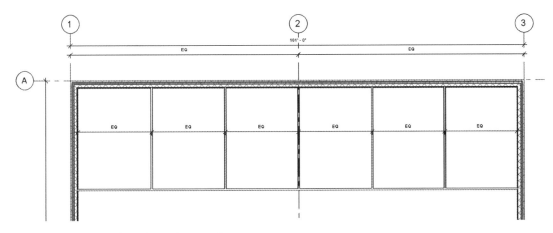

FIGURE 3-1.18 Enabling dimension equality

53. Add a vertical "clear" dimension to indicate the depth of the rooms. Set *Prefer* to **Wall Faces** for both ends of the dimension line. See Figure 3-1.19.

54. Click the **padlock** symbol (🔒) to tell Revit this dimension should not change (Figure 3-1.19).

55. Click **Modify**.

Next you will adjust the overall building dimensions and notice how the various parametric relationships you established cause the model to update!

FIGURE 3-1.19
Locking dimensions

FIGURE 3-1.20 Adjusting dimensions

56. Click grid line 3 and change the overall dimension from 101 to **40**, by clicking on the dimension text and then pressing *Enter*.

TIP: When adjusting the building footprint via the dimensions, you need to select the grid line, not the east wall, because the dimension references the grid line.

Notice the interior walls have adjusted to remain equal, and grid line 2 is still centered between grids 1 and 3 (Figure 3-1.20).

57. Change the 40'-0" dimension to **110'-0"**.

58. Select grid line A and change the 68'-0" dimension to **38'-0"**.

Your model should now look similar to Figure 3-1.21. Notice the interior wall maintained its 22'-0" clear dimension because the interior wall has a dimension which is locked to the exterior wall, and the exterior wall has an alignment which is locked to grid line A.

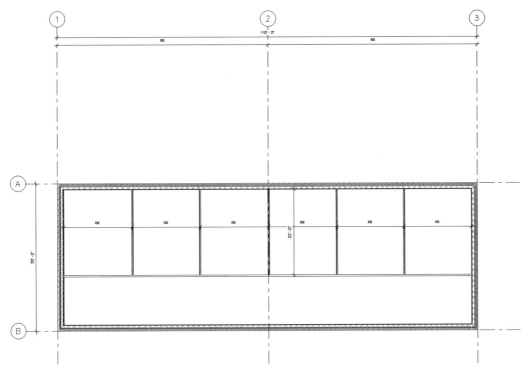

FIGURE 3-1.21 Adjusting dimensions

59. Click **Undo** icon on the *QAT* to restore the 68' dimension.

Your building should now be 110'-0" x 68'-0".

60. **Save** your project (*Small Office.rvt*).

> *TIP: You can use the Measure tool to list the distance between two points. This is helpful when you want to quickly verify the clear dimension between walls. Simply click the icon and snap to two points and Revit will temporarily display the distance. You can also click "chain" on the Options Bar and have Revit add up the total length of several picks.*

Exercise 3-2:
Doors

In this exercise you will add doors to your small office building.

1. Open **Small Office.rvt** created in Exercise 3-1.

Placing Doors

2. Select **Architecture → Build → Door** tool on the *Ribbon* (Figure 3-2.1).

Notice that the *Ribbon, Options Bar & Properties Palette* have changed to show options related to Doors. Next you will modify those settings.

The *Type Selector* indicates the door style, width, and height. Clicking the down arrow to the right lists all the doors pre-loaded into the current project.

FIGURE 3-2.1 Door tool

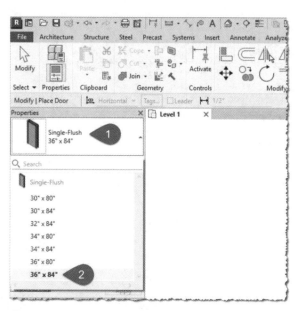

FIGURE 3-2.2 Type selector: Doors

The default template project that you started from has several sizes for a single flush door. Notice, in Figure 3-2.2, that there are two standard heights in the list. The 80″ (6′-8″) doors are the standard residential height and the 84″ (7′-0″) doors are the standard commercial door height.

3. Change the *Element Type Selector* to *Single-Flush:* **36″ x 84″**, and click **Tag on Placement** on the *Ribbon* (it should have a blue background when active).

4. Move your cursor over a wall and position the door as shown in **Figure 3-2.3**. (Do <u>not</u> click yet.)

Notice that the swing of the door changes depending on what side of the wall your cursor is favoring.

Also notice Revit displays listening dimensions to help you place the door in the correct location.

5. Click to place the door. Revit automatically trims the wall and adds a door tag.

 TIP: Press the spacebar before clicking to flip the door swing if needed.

6. While the door is still selected, click on the *change swing (control arrows)* symbol to make the door swing against the wall if it is not already (Figure 3-2.4).

FIGURE 3-2.3 Adding door

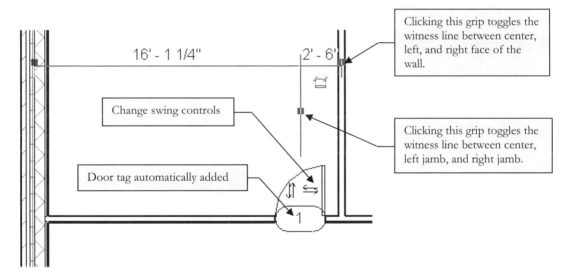

FIGURE 3-2.4 Door just placed

Next you will reposition the door relative to the adjacent wall.

7. Click **Modify** to finish the *Door* tool.

8. Click the door (not the door tag) you just placed to select it.

9. Click the **witness line grips** so the temporary dimension references the right door jamb and the wall face as shown in Figure 3-2.5.

TIP: You can also click and drag the witness line grip to another wall or line if the default location was not what you are concerned with.

10. Click on the dimension text, type **4″** and press *Enter*. Make sure you add the inch symbol or you will get feet rather than inches (Figure 3-2.5).

 Tip: This will be the typical dimension for all doors directly adjacent to a wall as shown here.

FIGURE 3-2.5 Edit door location

Unfortunately, the door *Families* loaded in the commercial template do not have frames. So the 4″ dimension just entered provides for a 2″ frame and 2″ of wall. The library installed on your hard drive, along with the Revit "web library," do provide some doors with frames. It is possible to create just about any door and frame combination via the *Family Editor.* The *Family Editor* is a special mode within Autodesk Revit that allows custom parametric content to be created, including doors with sidelights, transoms and more!

Mirroring Doors

The *Mirror* command will now be used to quickly create another door opposite the adjacent perpendicular wall.

11. With the door selected, click the **Mirror → Pick Mirror Axis** on the *Contextual Tab*.

12. On the *Options Bar*, click **Copy**. If copy was not selected, then the door would be relocated rather than copied.

13. With the door selected and the *Mirror* command active, hover the cursor over the adjacent wall until the dashed reference line appears centered on the wall, keep moving the mouse until you see this, and then click. (Figure 3-2.6)

Walls : Basic Wall : Interior - 4 7/8" Partition (1-hr) : Reference

Dashed cyan reference line

FIGURE 3-2.6 Mirroring a door

As you can see in Figure 3-2.7, the door has been mirrored into the correct location.

Revit does not automatically add door tags to mirrored or copied doors. These will be added later.

TIP: The size of the door tag is controlled by the view's scale.

FIGURE 3-2.7 Door mirrored

Copying Doors

Now you will copy the two doors so the other rooms have access.

14. Click to select the first door (not the door tag) and then press and hold the **Ctrl** key. *While holding the Ctrl key,* click to add the second door to the selection set.

15. With the two doors selected, click the **Copy** tool.

16. On the *Options Bar*, select **Multiple** (if not already checked).

At this point, you need to pick two points: a "copy from here" point and a "copy to there" point. The first point does not have to be directly on the elements(s) to be copied. The next step will demonstrate this; you will pick the midpoint of the wall adjacent to the two doors (first pick), and then you will pick the midpoint of the wall where you want a set of doors (second pick). With "multiple" checked, you can continue picking "second points" until you are finished making copies (pressing *Esc* or *Modify* to end the command).

17. Pick three points:

 a. *First pick:* midpoint/centerline of wall (see Figure 3-2.8);

 b. *Second pick:* midpoint/centerline of wall shown in Figure 3-2.9;

 c. *Third pick:* midpoint/centerline of wall shown in Figure 3-2.9.

18. Pick **Modify** to end *Copy*.

The doors are now copied.

FIGURE 3-2.8 Copy – first point with midpoint symbol visible

FIGURE 3-2.9 Numbers indicate pick-points listed in step #17

You will now add two exterior doors using the same door type.

19. Using the **Door** tool, add two exterior doors approximately located per Figure 3-2.10. Match the swing and hand shown.

FIGURE 3-2.10 Adding exterior doors

Tag All (Not Tagged)

Revit provides a command to quickly add a tag (e.g., a door tag) to any door that does not currently have one in the current view. The tag might have to be moved or rotated once placed, but this still saves time and the possibility of missing a door tag.

20. Select **Annotate → Tag → Tag All**.

21. In the *Tag All Not Tagged* dialog box, check the box for **Door Tags** under *Category* and set *Tag Orientation* to **Vertical**. Click **OK** (see image on next page).

All the doors should now be tagged in your floor plan.

FYI: Door tags can be deleted at any time and added again later at any time. Tags simply display information in the element being tagged – thus, no information about the element is being deleted; the building information integrity remains intact.

Deleting Doors

Next you will learn how to delete a door when needed. This process will work for most elements (i.e., walls, windows, text, etc.) in Revit.

22. Click **Modify**.

23. Click on door number 7 (the exterior door on the left, not the door tag) and press the **Delete key** on your keyboard.

As you can see, the door is deleted and the wall is automatically filled back in. Also, a door tag can only exist by being attached to a door; therefore, the door tag was also deleted.

One last thing to observe: Revit numbers the doors in the order in which they have been placed (regardless of level). Doors are not automatically renumbered when one is deleted. Also, doors can be renumbered to just about anything you want.

24. **Save** your project *(Small Office.rvt)*.

Exercise 3-3:
Windows

In this exercise you will add windows to your small office building.

1. Open **Small Office.rvt** created in Exercise 3-2.

Placing Windows

Window

2. Select **Architecture → Build → Window**.

Notice that the *Ribbon*, *Options Bar* and *Properties Palette* have changed to show options related to windows. Next you will modify those settings.

The *Type Selector* indicates the window style, width and height. Clicking the down arrow to the right lists all the windows loaded in the current project.

3. With the *Window* tool active, do the following (Figure 3-3.1):

 a. Change the *Type Selector* to *Fixed:* **36″ x 48″**.

 b. On the *Ribbon*, verify *Tag on Placement* is toggled off.

 c. Note the **Sill Height** value in the *Properties Palette*.

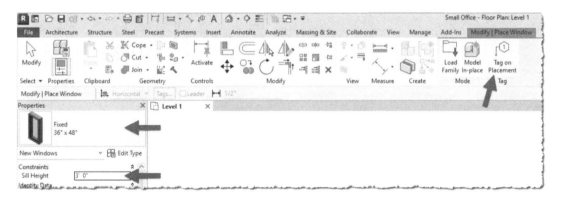

FIGURE 3-3.1 Ribbon and Options Bar: Window tool active

4. Move your cursor over a wall and place **two windows** as shown in **Figure 3-3.2**. *Notice that the position of the window changes depending on what side of the wall your cursor favors.*

FYI: *The window sill height is controlled by Properties Palette which you will study later in this book. For now, the default dimension was used.*

5. Adjust the **temporary dimensions** per the following:

 a. Dimensions per Figure 3-3.2.

 b. Use the witness line grip to adjust the witness line position.

 REMEMBER: The selected item moves when temporary dimensions are adjusted. Pick the left window to set the 6'-0" dimension and the right for the 8'-0" dimension.

6. Using the **Copy** command, in a way similar to copying the doors in the previous exercise, copy the two windows into each office as shown in Figure 3-3.3. (Do not worry about exact dimensions.)

7. **Save** your project *(Small Office.rvt)*.

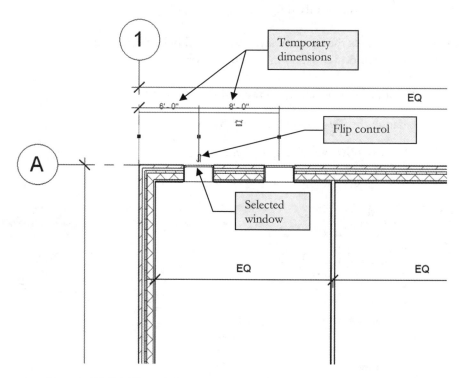

FIGURE 3-3.2 Adding windows – temporary dimensions still active

FIGURE 3-3.3 Windows added to north wall

The windows will not adjust with the grid lines and interior walls; it is possible to add dimensions and get this to work. If you tried to change the overall dimension from 110′ to 40′ again, Revit would let you know it needed to delete some windows before the change could be made. Like doors, windows need their host to exist.

The windows can all be adjusted via the temporary dimensions. The window selected is the largest width available in the project (based on the template file from which the project was started), but it is not a masonry dimension. However, additional window sizes can be added on the fly at any time. Additionally, you can create your own template file that has the doors, windows, walls, etc. that you typically need for the kind of design work you do.

In addition to the preloaded windows, several window styles are available via the family library loaded on your hard drive and the *Autodesk Web Library* (e.g., dbl-hung, casement, etc.). It is also possible to create just about any window design in the *Family Editor*.

Object Snap Symbols:
By now you should be well aware of the snaps that Revit suggests as you move your cursor about the drawing window.

If you hold your cursor still for a moment while a snap symbol is displayed, a tooltip will appear on the screen. However, when you become familiar with the snap symbols you can pick sooner (Figure 3-3.4).

The TAB key cycles through the available snaps near your cursor.

The keyboard shortcut turns off the other snaps for one pick. For example, if you type SE on the keyboard while in the Wall command, Revit will only look for an endpoint for the next pick.

Finally, typing SO (snaps off) turns all snaps off for one pick.

Symbol	Position	Keyboard Shortcut
✕	Intersection	SI
☐	Endpoint	SE
△	Midpoint	SM
○	Center	SC
✕	Nearest	SN
└	Perpendicular	SP
Ω	Tangent	ST

FIGURE 3-3.4 Snap Reference Chart

Exercise 3-4:
Roof

You will now add a simple roof to your building.

1. Open **Small Office.rvt** created in Exercise 3-3.

The first thing you will do is take a quick look at a 3D view of your building and notice an adjustment that needs to be made to the exterior walls.

2. Click the **Default 3D View** icon on the *QAT*.

The 3D icon switches you to the default 3D view in the current project. Your view should look similar to Figure 3-4.1. Notice the exterior walls are not high enough, which is due to a previous decision to set the wall height to 9'-0". Next you will change this, which can be done in the plan view or the current 3D view.

FIGURE 3-4.1 3D view of small office

3. In the 3D view, hover your cursor over one of the exterior walls to pre-highlight it, then (before clicking) press the **Tab** key to pre-highlight a "chain of walls" (i.e., all the exterior walls), and then click to select them.

Next you will access the properties of the selected walls so you can adjust the wall height. In Revit, most any design decisions that are made can be adjusted at any time.

4. Change the following in the *Properties Palette*:

 a. *Top Constraint:* **Up to level: Roof**

 b. *Top Offset:* **2'-0"**

 c. Click **Apply** (Figure 3-4.2B).

FIGURE 3-4.2A Exterior wall heights adjusted

FIGURE 3-4.2B Selected wall properties

Setting the top of wall to be associated with a *Level* establishes a parametric relationship that causes the wall height to automatically adjust if the level datum is adjusted (e.g., from 12'-0" to 14'-0").

Plus, the *Top Offset* at 2'-0" creates a 2'-0" parapet, which will always be 2'-0" high no matter what the roof elevation is set to. There are instances when you would want the height to be fixed.

All of the settings related to the selected wall show up here; these are called instance parameters.

Sketching a Roof

FIGURE 3-4.3 Roof tool

Now that the exterior walls are the correct height, you will now add the roof. This building will have a flat roof located at the roof level.

5. Double-click **Level 1** (floor plan) in the *Project Browser* to switch back to that view.

6. Click **Architecture → Build → Roof** (Figure 3-4.3).

The fly-out prompts you to choose the method you want to use to create the roof.

7. Click **Roof by Footprint**.

 FYI: If you get a prompt to move the completed roof to "Level 2" level, that means you started this chapter/exercise with the wrong template.

At this point you have entered *Sketch Mode* where the Revit model is grayed out so the perimeter you are about to sketch stands out.

Also notice the *Ribbon, Options Bar* and *Properties Palette* have temporarily been replaced with Sketch options relative to the roof (Figure 3-4.4).

8. Click **Extend to Core** on the *Options Bar*, and make sure **Defines Slope** is not checked (Figure 3-4.4).

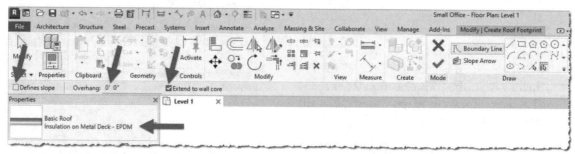

FIGURE 3-4.4 Roof sketch tools

9. Select the exterior walls:

 a. Hover your cursor over one of the exterior walls to pre-highlight the wall.

 b. Press **Tab** to select a "Chain of Walls" (i.e., all the exterior walls).

 c. **Click** to select the exterior walls, which will add the sketch lines.

At this point you should have four magenta sketch lines, one on each wall, which represent the perimeter of the roof you are creating. When sketching a roof footprint, you need to make sure that lines do not overlap and corners are cleaned up with the *Trim* command if required. Your sketch lines require no additional edits because of the way you added them (i.e., Pick walls and Tab select).

Notice for the roof sketch you need to adjust the level on which the roof will be created. By default, the top surface of the roof element will be parametrically aligned with the current level (i.e., Level 1 in this case). You will change this to the roof level.

FIGURE 3-4.5 Roof instance properties

10. Set *Base Level* to **Roof** (Figure 3-4.5).

Now you are ready to finish the roof and exit sketch mode.

11. Click the **green check mark** on the *Ribbon*.

12. Click **Yes** to the join geometry prompt (Figure 3-4.6).

The join geometry option will make the line work look correct in sections. If you clicked "No," the wall and floor lines would just overlap each other and look messy.

FIGURE 3-4.6 Join geometry prompt

The roof is now created and, in section, will extend through the finishes to the concrete block because "extend to core" was selected when the sketch lines were added. Also, because you used the "Pick Walls" option on the *Ribbon* (which was the default), the roof edge will move with the exterior walls.

13. To see the roof, click the **Default 3D View** icon.

14. To adjust the 3D view, press and hold the **Shift** key while pressing the **wheel button** and dragging the mouse around.

FIGURE 3-4.7 Default 3D View

15. Click the **X** on the {3D} drawing tab to close that view. This will close the 3D view but not the project or the Level 1 view.

 REMEMBER: *Clicking the **X** in the upper right of the application title bar will close Revit, but it will prompt you to save first if needed.*

16. **Save** your project.

Exercise 3-5:
Annotation, Room Tags & Schedules

Adding text is very simple in Revit. In this exercise you will add a title below the floor plan. You will also place room tags.

Placing Text

1. Open **Small Office.rvt** created in Exercise 3-4.

2. Make sure your current view is **Level 1**. The word "Level 1" will be bold under the *Floor Plans* heading in your *Project Browser*. If Level 1 is not current, simply double-click on the Level 1 text in the *Project Browser*.

3. Select **Annotate → Text → Text** tool on the *Ribbon*.

Once again, notice the *Ribbon* has changed to display some options related to the active tool (Figure 3-5.1).

FIGURE 3-5.1 Ribbon with Text tool active

The *Type Selector* indicates the text style (which determines the font style, height and more); users can create additional text styles. From this *Contextual Tab*, on the Ribbon, your alignment (i.e., Left justified, Centered or Right justified) can also be set.

4. Set the *Ribbon* settings to match those shown above, and **Click** below the floor plan to place the text (Figure 3-5.2).

5. Type **OFFICE BUILDING – Option A**, then click somewhere in the plan view to finish the text (do not press *Enter*).

The text height, in the *Type Selector*, refers to the size of the text on a printed piece of paper. For example, if you print your plan you should be able to place a ruler on the text and read ¼″ when the text is set to ¼″ in the *Type Selector*.

Text size can be a complicated process in CAD programs; Revit makes it very simple. All you need to do is change the **view scale** for **Level 1** and Revit automatically adjusts the text and annotation to match that scale – so it always prints ¼″ tall on the paper.

Text just placed – move and rotate icons still visible

•OFFICE BUILDING - Option A•

FIGURE 3-5.2 Placing text

You will not change the scale now, but it can be done via the *View Control Bar* (Figure 3-5.3). If you want to try changing it, just make sure it is set back to ⅛″ = 1′-0″ when done.

You should now notice that your text and even your door and window symbols are half the size they used to be when changing from ⅛″ to ¼″.

You should understand that this scale adjustment will only affect the current view (i.e., Level 1). If you switched to Level 2 (if you had one) you would notice it is still set to ⅛″=1′-0″. This is nice because you may, on occasion, want one plan at a larger scale to show more detail.

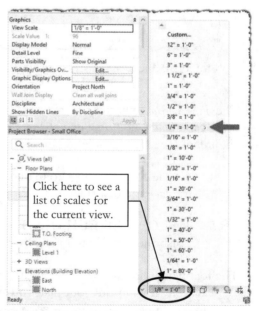

Click here to see a list of scales for the current view.

FIGURE 3-5.3 Set View Scale

Placing Room and Room Tags

Placing *Room Tags* must be preceded by placing a *Room*. A *Room* element is used to define a space and hold information about a space (e.g., floor finish, area, department, etc.).

The *Room* feature searches for areas enclosed by walls; a valid area is pre-highlighted before you click to create it.

By default, Revit will automatically place a *Room Tag* at the cursor location when you click to add the *Room* element.

6. Select **Architecture → Room & Area → Room**.

7. Set the *Type Selector* to **Room Tag: Room Tag With Area** and make sure **Tag on placement** is selected on the *Ribbon*.

8. Click within each room in the order shown in Figure 3-5.4; watch for the dashed reference line to align the tags.

FIGURE 3-5.4 Placing rooms and room tags

While the *Room* tool is active, the placed rooms in the model are shaded light blue so you can see which spaces already have rooms placed. The large "X" is also part of the room. When the *Room* tool is not active, you can hover the cursor over the approximate location of the "X" until it pre-highlights, then you can click to select the room. With the *Room* object selected you can add information or delete it via the *Properties Palette*.

9. Click **Modify** to end the current tool.

Notice the rooms are not visible and the "X" is gone. Also notice the *Room Tag* selected shows the following information stored within the *Room* object: Name, Number and Area.

> *FYI:* The area updates automatically when the walls move. Next you will change the room names.

10. Click on the *Room Tag* for **room number 1** to select it.

When a *Room Tag* is selected the "dark blue" text is editable and the "lighter blue" text is not. An example of text that cannot be edited would be the actual text "Sheet Number" next to the sheet number on a sheet border.

11. Click on the room name text, type **OFFICE**, and then press **Enter** on the keyboard.

12. Change rooms 3-4 to also be named **OFFICE**.

13. Change the large room name to **LOBBY**.

14. Leave two rooms (5 and 6) as "Room" for now.

Schedules

The template you started your project from had room and door schedules set up. So from the first door and room you placed, these schedules started filling themselves out! You will take a quick look at this to finish out this section.

15. In the *Project Browser*, click the "+" symbol next to *Schedules/Quantities* to expand that section (if required) and then double-click on **Room Schedule** to open that view. *FYI:* The blue filled box indicates this item is placed on a sheet (in this case, due to the template the project was started from).

The room schedule is a tabular view of the Revit model. This information is "live" and can be changed (Figure 3-5.5).

FIGURE 3-5.5 Room Schedule

Next you will change the two rooms named "room," and see that the floor plan is automatically updated!

16. Click in the *Room Name* column for room number 5 and change the text to read **MEN'S TOILET RM**.

17. Click in the *Room Name* column for room number 6 and change the text to read **WOMEN'S TOILET RM**.

Room Number	Room Name
1	OFFICE
2	OFFICE
3	OFFICE
4	OFFICE
5	MEN'S TOILET RM.
6	WOMEN'S TOILET RM
7	LOBBY

18. Click the lower "X" in the upper right of the drawing window to close the Room Schedule view.

19. Switch to *Level 1* (if required) and **zoom in** on rooms 5 and 6 (Figure 3-5.6).

Notice that the room names have been updated because the two views (floor plan and schedule) are listing information from the same "parameter value" in the project database.

TIP: You can zoom in and out in a schedule by holding down the **CTRL** key and **scrolling the wheel** on your mouse. Use **CTRL + 0** to reset the view.

FIGURE 3-5.6 Room names updated

20. Open and Close the door schedule to view its current status.

21. **Save** your project.

Exercise 3-6:
Printing

The last thing you need to know to round off your basic knowledge of Revit is how to print the current view.

Printing the current view:

1. In *Level 1* view, right-click anywhere and select **Zoom to Fit**.

2. Select **File Tab → Print**.

3. Adjust your settings to match those shown in **Figure 3-6.1**.

 - Select a printer, from the list, that you have access to.
 - Set *Print Range* to **Visible portion of current window**.

FIGURE 3-6.1 Print dialog

4. Click on the **Setup** button to adjust additional print settings.

5. Adjust your settings to match those shown in **Figure 3-6.2**.

 - *Set Zoom to:* **Fit to page**

FIGURE 3-6.2 Print Setup dialog

6. Click **OK** to close the *Print Setup* dialog and return to *Print*.

7. Click the **Preview** button in the lower left corner. This will save paper and time by verifying the drawing will be correctly positioned on the page (Figure 3-6.3).

8. Click the **Print** button at the top of the preview window.

9. Click **OK** to print to the selected printer.

FYI: Notice you do not have the option to set the scale (i.e., ⅛″ = 1′-0″). If you recall from our previous exercise, the scale is set in the properties for each view. If you want a quick half-scale print you can change the zoom factor to 50%. You could also select "Fit to page" to get the largest image possible but not to scale.

FIGURE 3-6.3 Print Preview

Printer versus Plotter?

Revit can print to any printer or plotter installed on your computer.

A <u>Printer</u> is an output device that uses smaller paper (e.g., 8½"x11" or 11"x17"). A <u>Plotter</u> is an output device that uses larger paper; plotters typically have one or more rolls of paper ranging in size from 18" wide to 36" wide. A roll feed plotter has a built-in cutter that can – for example – cut paper from a 36" wide roll to make a 24"x36" sheet.

Plotter with three paper rolls

Color **printer** / copier

In addition to physically printing on paper, it is also possible to export a PDF file using Revit's built-in tool. This is accessed from **File →** **Export → PDF** or via the *Quick Access Toolbar* (see image to right). The PDF file can be emailed to clients, contractors, consultants, and more. The PDF file is also a good way to save progress versions of your drawings.

Export sheets and/or views to a PDF file

Self-Exam:

The following questions can be used as a way to check your knowledge of this lesson. The answers can be found at the bottom of this page.

1. The *Measure* tool is used to dimension drawings. (T/F)

2. Revit will automatically trim the wall lines when you place a door. (T/F)

3. Snap will help you to draw faster and more accurately. (T/F)

4. A 6'-8″ door is a standard door height in _____ construction.

5. While using the wall tool, the height can be quickly adjusted on the

 _____ Bar.

Review Questions:

The following questions may be assigned by your instructor as a way to assess your knowledge of this section. Your instructor has the answers to the review questions.

1. The *View Scale* for a view is set by clicking the scale listed on the *View Control Bar*. (T/F)

2. Dimensions are placed with only two clicks of the mouse. (T/F)

3. The relative size of text in a drawing is controlled by the *View Scale*. (T/F)

4. You can quickly switch to a different view by clicking on a View tab. (T/F)

5. You cannot select which side of the wall a window is offset to. (T/F)

6. The _____ key cycles through the available snaps near your cursor.

7. The _____ tool can be used to list the distance between two walls without drawing a dimension.

8. While in the *Door* tool you can change the door style and size via the

 _____ within the *Properties Palette*.

Notes:

Lesson 4
Revit Materials:

Organizing and documenting materials is an important task for the interior design professional. This involves developing material collections/systems for spaces, understanding products being specified (and material options available for them), building code constraints (e.g., flame spread ratings), cost, availability and much more. Autodesk® Revit® allows much of this to be managed within the BIM database. This chapter will cover the basics and be applied to our *Law Office* project.

Exercise 4-1:
Introduction to Using Materials in Revit

In Revit, a *Material* is more than just something you can touch or hold in your hand. It is a "container" which helps to keep track of the following:

- **Identity** settings
 - o Descriptive Information
 - o Product Information
 - o Annotation Information
- **Graphics** settings
 - o Shading
 - o Surface Pattern and color
 - o Cut Pattern and color
- **Appearance** asset
 - o Photo-realistic render appearance
 - ▪ Texture
 - ▪ Reflectivity
 - ▪ Transparency
 - ▪ Self-Illumination
 - ▪ Etc.
- **Physical** asset
 - o Structural engineering data
 - ▪ Concrete
 - ▪ Steel
 - ▪ Aluminum
 - ▪ Wood
- **Thermal** asset
 - o Energy modeling data
 - ▪ Thermal Conductivity
 - ▪ Specific Heat
 - ▪ Etc.

In Revit, a single named item called a *Material* holds all of this information. We will discuss each of the items listed above in more detail, but first let's talk about how a *Material* is used.

> **FYI:** *Anytime a Revit term or tool is mentioned in this book it is title-case and italicized. This helps the reader to distinguish a software term, such as* Material, *from a design industry term such as* **material**.

There are three primary ways in which a *Material* is used, or applied, in a model:
- *By Category*
- *By Element*
- *By Face*

Each of these will be discussed in detail in a moment, but first notice the three walls below. The left wall segment has a light color and 4″ tile pattern applied *By Category*. The middle wall has a light color with a brick pattern *By Element*. The last wall has a solid darker color painted on a single face (*By Face*).

By Category By Element By Face

By Category

Object Styles

When a 3D element does not have a specific *Material* applied, the project-wide *Object Style* is used, accessed by this dialog: **Manage → Settings → Object Styles**.

In Figure 4-1.1, notice the various categories listed: Furniture, Casework, Doors, etc. These are hard-wired categories Revit uses to organize elements, control visibility, and in the case of this discussion, controls how elements look. Each category may be assigned a default *Material*. When a 3D element has its *Material* parameter set to **<By Category>**, the *Object Style Material* is used.

It is often not too important to set the *Material* here because each 3D element often has a *Material* assigned to it, which takes precedence over *Object Styles*. For example, the pedestal sink shown to the right has a "Sink – Porcelain" and a "Sink – Faucet" material assigned directly to

the various 3D elements contained within the *Family* (i.e., component). Therefore, the Plumbing Fixture *Material* in *Object Styles* is ignored. It might be best to set the *Object Styles* materials to an unusual color to draw attention to any element which does not have a specific *Material* selected in your project.

FIGURE 4-1.1 Object Styles dialog

By Element

The most common way a *Material* is applied to something is *By Element*. This is true for *System Families* and *Loadable Families*. Note the following:

- **System Families** (walls, ceilings, floors, etc.):
 These elements cannot exist outside of a project, therefore the *Material* adjustment option is built into the dialog box which is used to develop the assembly, the *Edit Assembly* dialog box; an example is shown on the next page.

- **Loadable Families** (furniture, casework, specialty equipment, doors, etc.):
 These are *Families* which are developed in the *Family Editor* environment. Every 3D element in a *Family* has a *Material* parameter. When the item is first created, the *Material* parameter is set to **By Category**. This parameter is usually changed in one of two ways:

- o **Picking a Material**
 When a *Material* is selected directly, it is basically hard-wired and cannot be changed in the project.

- o **Mapping to a Material Parameter**
 When a *Material Parameter* is employed, the *Material* can easily be changed in the project; this can be an *Instance* or *Type* parameter.

We will expand on this to help clarify the concept. Some of this information might be difficult to understand when one first begins using Revit. However, it is best to quickly read through this information now and then plan on referring back here to help fill in the gaps as needed.

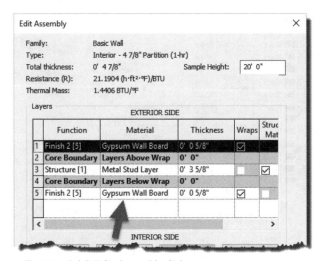

The ***Edit Assembly*** dialog, shown to the left (Figure 4-1.2), highlights how the *Material* selection is built into the workflow for developing an assembly (i.e., *System Family*), a wall in this case. This is all done directly in the project environment.

The *Edit Assembly* dialog is similar for the other *System Families* such as Floor, Roof and Ceiling.

FIGURE 4-1.2 Edit Assembly dialog

Next we will look at a simple chair family (Figure 4-1.3) which comes with the Revit software. We will see how three *Materials* have been utilized in this *Family*, but only two are selectable in the project.

The *Family* is **Chair-Tablet Arm.rfa** and can be found via **Insert ->> Load Autodesk Family** and searching for "Chair-Tablet Arm".

If we load this chair into a project, we can observe a few things about it. You may try this in Revit while reading, but it is not required. First, if the chair is selected, you can click the ***Edit Type*** button, on the *Properties Palette*, to view the *Type Properties*.

FIGURE 4-1.3 Chair-Tablet Arm family

Notice the two *Material Parameters*: **Seat/Back Material** and **Tablet Material** (Figure 4-1.4). These can both be changed to another *Material* by clicking in the *Value* field and selecting the small button that appears to the right. This opens the *Materials* dialog, from which another *Material* can be selected or created.

> *FYI: By changing a* Type Parameter*, every chair of this type within the project will be instantly updated.*

FIGURE 4-1.4 Type Properties for tablet arm chair

However, unlike the seat/back rest and tablet surface, the chair legs and back support *Materials* are not directly changeable in the project. There is a specific *Material* in use, and it appears in the *Material* dialog, but the *Material* used for the legs/back rest cannot be changed in the project – only in the *Family Editor*.

By Face

The third way in which a *Material* is typically applied is by using the ***Paint*** tool, which adds a *Material* to a single face. This feature has some limitations in the context of a pure BIM project.

The ***Paint*** tool is used on *System Families*, such as Walls, Floors, Ceilings, etc. The following description of how the tool is used is meant to be superficial and does not need to be done in a Revit file at this time. The *Paint* tool is selected on the *Modify* tab, a *Material* is selected from a dialog that appears, and then individual faces are selected.

The image to the right (Figure 4-1.5) shows a wall with three of the six faces visible from this vantage point. Thus, using the *Paint* tool you can only change the surface *Material* of these three faces. You would need to change the view of the wall to override the other faces.

When a surface is painted, Revit paints the entire face, not just the portion visible in the current

FIGURE 4-1.5
Three faces of wall visible in this view

view. For example, an interior elevation might be of a wall common to a corridor or hallway. Revit will paint the entire side of the corridor wall, not just the portion visible in the specific room you are in (Figure 4-1.6A & 16B).

FIGURE 4-1.6A Original corridor wall

FIGURE 4-1.6B
Entire wall receives the painted material, not just a single room

Revit does provide a way to constrain the area, or extents, of a *Material* to be painted on a face. This is accomplished using the **Split Face** tool, found adjacent to the *Paint* tool on the *Modify* tab. When the *Split Face* tool is selected, you select a surface and then Revit enters sketch mode; here you draw lines, which are constrained to the plane of the surface. These lines must define an enclosed area.

The image below (Figure 4-1.7) shows the *Split Face* tool used to define the entire wall in room number one, between the demising walls and up to the ceiling. The *Split Face* was also used to define the wainscot finish in room number 2. Note that the same corridor wall has two separately defined areas using the *Split Face* tool.

FIGURE 4-1.7
Split Face tool used to constrain painted areas

In the image below, the *Split Face* and *Paint* tools were used to define an area for the accent band and wall base (Figure 4-1.8). Everything in this image is created by the reader, which consists of Shaw flooring, Toto plumbing fixtures, and Dal Tile.

FIGURE 4-1.8 Toilet room created by reader later in this textbook

This section will take a closer look at the Revit *Material* itself. Here you will gain a better understanding of what information, both textual and graphical, is contained within any given *Material*.

Exercise 4-2:
A Detailed Look at Revit Materials

As mentioned in the previous section, a *Material* in Revit is a "container" which holds information about the various materials used in the project. Some of the information contained in a *Material* is used prior to construction for visualization and documentation and some can be used during Construction Administration and after for Facilities Management.

To explore the *Materials* in a project, most of which come from the project template, click the **Materials** button on the *Manage* tab.

Materials

You will see the *Materials* dialog shown below (Figure 4-2.1).

FIGURE 4-2.1 Materials dialog

ARROW #1: *Clicking the drop-down list, pointed out with the #1 arrow in Figure 4-2.1, allows you to switch the material list to just be text rather than a preview (see also Figure 4-2.2). This adjustment allows you to see more items in the limited space.*

ARROW #2: *Clicking this icon will toggle the visibility of the external materials library. The advantage of not having the external library visible is that you can see more of the materials in your project.*

ARROW #3: *If the Material Browser dialog is not fully expanded you need to click the icon pointed out with the #3 arrow to see the selected material's properties.*

We will now take a close look at the various aspects of the *Material* dialog. It is a good idea to review this information closely now and refer back when needed. Having a good understanding of this fundamental feature will benefit the interior designer substantially. It is not necessary to follow along yet; if you do, it is recommended you do not save. The next exercise will have several steps which will be done in a project to apply what is presented in this section.

In the *Materials* dialogs (Figure 4-2.1), notice the list of *Materials* on the left side of the screen.

This list can be sorted in three ways:
- Material usage state
- Search words
- Classification

FIGURE 4-2.2 Sort by Material usage state

Selecting by the drop-down list (#1, Figure 4-2.2) allows you to narrow the list down to just *Materials* used in the project, or not used. The default is *Show All*, so both used and not used appear in the list; you would normally want to see both.

Word search allows you to enter a term by which the *Materials* list is filtered. The image to the right (Figure 4-2.3) filters the *Materials* list down to any *Material* or *Category* containing the word "wood."

A combination of the two can be used to further filter the list down, and quickly get to the information you need. Notice in Figure 4-2.4A, the list is sorted by the word "tile" and for unused materials in the current project.

FIGURE 4-2.3 Sort by word search

It is also possible to sort the materials by a parameter called **Class**. Clicking the drop-down arrow pointed out in Figure 4-2.4B gives you the list to choose from. All of the Autodesk materials have a value entered for *Class* (e.g. Glass, Paint, Wood). When creating a material, you can select from a *Class* value previously entered in the project or enter a value of your own.

FIGURE 4-2.4A Sort by word search and material usage

To wrap up the discussion on the list of *Materials*, when you right-click on a Material name, you can ***Duplicate***, ***Rename*** and ***Delete*** materials as seen in Figure 4-2.5. Also, at the bottom of the *Material Browser* dialog, a small icon allows one to create a new *Material* in addition to duplicating an existing one.

FIGURE 4-2.4B Right-click options

FIGURE 4-2.5 Right-click options

Now that you understand the mechanics of the *Materials* list itself, you will take a look at the information contained within each *Material*. **The information seen on the right half of the *Material Browser* dialog, under the tabs, is based on which material is selected in the *Project Materials* list on the left**. If the material browser is not fully expanded, click the small icon in the lower right corner of the *Material Browser*; it looks like two arrows pointing to the right.

- **Identity tab**

 With the first *Material* selected in the *Project Materials* list on the left, **Acoustic Ceiling Tile 24" x 24"**, notice the *Identity* options available in (Figure 4-2.6).

 o Descriptive Information
 This is an area to expound upon the material name to help the design team understand what this material is or is used for.

 o Product Information
 Who makes this material, for how much and where can I find more information? You can enter this information here. For example, you might have a *Material* which represents a specific fabric, which is applied to a Haworth or Steelcase furniture *Family*. The parameters for the furniture *Family* will list the furniture manufacturer: Haworth or Steelcase in this example. But the *Material* applied within the furniture *Family* is tracking the fabric information.

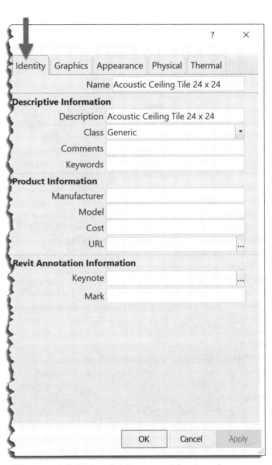

FIGURE 4-2.6 Material Browser; Identity tab

 o Revit Annotation Information
 It is possible to add *Smart* text with optional arrows, which is called a *Tag* (see example on page 10-2), in a drawing view to report information about a specific *Material*. If the information changes, the *Tag* will update automatically. You will learn more about keynoting later in this book.

- **Graphics tab**

 The information found on the *Graphics* tab primarily relates to how the various *Materials* appear on screen and how they print on paper or in a PDF. This closely relates to the development of the *Construction Document (CD)* deliverables. The image shown below still relates to the Acoustic Ceiling Tile (ACT) *Material* (Figure 4-2.7).

 o <u>Shading</u>
 This is the solid color used on-screen when the ***Visual Style*** is set to *Shaded* or *Consistent Colors* (see Page 2-15 for *Visual Styles* examples).

 The ***Transparency*** setting is used to make certain objects see-through. This applies to all of the *Visual Styles*. The most common application of this feature is on glass.

 The color set for *Shading* is not used in any way for photo-realistic renderings. This color can be derived from the information specified for renderings by checking the *Use Render Appearance* option; more on this soon.

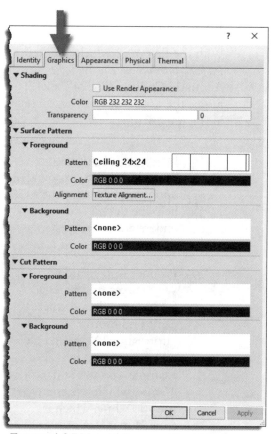

FIGURE 4-2.7 Material Browser; Graphics tab

 o <u>Surface Pattern</u> *(and Color)*
 When a *Surface Pattern* is selected, Revit will display this pattern anytime an object's surface is visible in a view. For example, if the ACT *Material* is applied to a ceiling element, the 24″ x 24″ pattern will be applied to it in the *Ceiling Plan View* and any 3D views. See the next page for an example.

 o <u>Cut Pattern</u> *(and Color)*
 When a *Cut Pattern* is selected, Revit will display this pattern within the boundary of an object, if that object is being cut within the view. Objects will typically be cut in section and detail views. A *Floor Plan* has a horizontal cut plane at about 3′-6″ above the level. So, most elements which pass through this cut plane have the *Cut Pattern* shown. The most common example would be walls. See the next page for an example.

Both the *Surface* and *Cut* patterns have a **Foreground** and **Background** option. This allows two patterns to be used to further define special conditions.

Overview of how Surface and Cut patterns are used in a Revit model:

PLAN VIEW

The three views shown on this page are used to illustrate how surface and cut patterns, as defined within a *Material*, are applied to model elements. See the next page for more detailed notes on these images.

ELEVATION / SECTION VIEW

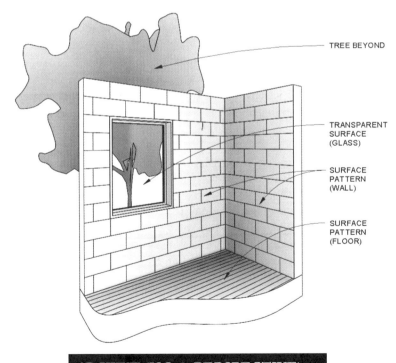

3D VIEW (ISO & PERSPECTIVE)

FLOOR PLAN EXAMPLE:

Looking at the previous page, the floor plan example shows a *Cut Pattern* within the walls and a *Surface Pattern* on the floor. The 8'-0" tall walls extend up through the 3'-6" cut plane. Therefore, for the *Material* applied to the wall, the *Cut Pattern* specified is shown. The floor exists entirely below the 3'-6" cut plane, so we are seeing its surface; therefore, the material's *Surface Pattern* is applied. Refer back to Exercise 1-1 (*Views* discussion starting on pg. 1-8) for an overview on how the *Cut Plane* interacts with the model.

ELEVATION/SECTION EXAMPLE:

In this example, the floor and the wall on the right are in section (see the section line in Figure 4-2.8). The wall with the window is not being cut, rather is seen in elevation, or projection. Note, the two walls are the same type (i.e. 8" Masonry) and use the same *Material*. Therefore, the wall's *Material: Cut Pattern* is shown in the portion of wall in section and its *Material: Surface Pattern* is shown for the wall in elevation/projection. The floor is made up of multiple layers, and

FIGURE 4-2.8 Section cut through building

the larger part which represents the structure does not have a *Cut Pattern* selected in the *Material* being used; it is set to "<none>". Thus, the floor does not have a *Fill Pattern* shown when in section. Final note, even though the glass has a transparent material, it does not show up transparent in elevation and section views, only 3D views as you will see in the next example. This is because we typically don't want extra "clutter" showing up within the windows and doors. As you can see, no part of the tree appears within the window.

3D VIEW EXAMPLE:

In this view everything is in elevation/projection, so the surface pattern is used for the applied *Material*. Note: the walls and floor appear to be cut, but that is just the way they were modeled for this example. In 3D views, the transparency is applied to the element so the tree is visible through the window.

Here are a few more points about patterns used in Materials:

- **Surface and Cut Patterns Optional:**
 Keep in mind that *Surface* and *Cut patterns* are optional. In large commercial projects it is often undesirable to see a floor pattern as the drawing gets too cluttered with notes, dimensions and tags. In building sections the ceiling is not really thick enough for a pattern to show, so the model is more efficient when these extra lines are not all being generated in a given view.

- **Pattern Scale:**
 This topic will be covered in more detail later but is worth pointing out at this juncture. There are two types of *Fill Patterns* in Revit: **Model** and **Drafting**. *Model Patterns* typically represent actual materials such as ceiling tile, concrete block (CMU) and brick to name a few. *Drafting Patterns* symbolically represent a material, thus the density of the pattern changes with the **View Scale** to help improve clarity (in a

similar way that text and dimensions adjust with the view's scale). The example below (Figure 4-2.9) shows the same 12" concrete wall in section. Notice how the density of the *Drafting Pattern* changes when the *View Scale* is adjusted.

FYI: *Notice that the line weights also change when the View's* ***View Scale*** *changes. Larger scales have lighter line weights because they are viewed closer.*

FIGURE 4-2.9 Drafting patterns automatically adjust size with view scale

- **Appearance Asset**
 The settings in this tab only come into play in one of two ways:

 1) *Visual Style* is set to **Realistic** for the current view

 2) When doing a **Rendering**

 Otherwise, the way an element appears in Revit is controlled by the settings on the *Graphics* tab, unless otherwise overridden.

 o Photo-realistic render appearance
 Different types of appearance assets have unique options. For example, paint has an *Application* setting where you can select *Roller*, *Brush*, or *Spray*; compare Figures 4-2.10 and Figure 4-2.11.

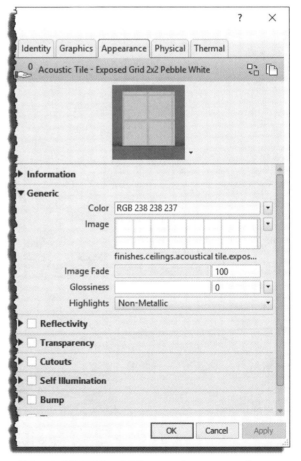

FIGURE 4-2.10 Material Browser: Appearance tab

When creating materials, it is important to use the correct asset-type. Notice in the image to the right (Figure 4-2.11a) that it specifically indicates "**Wall Paint**" and the previous image for the ACT indicated "**Generic**." The use of *Generic* should be used as little as possible. When Revit knows what the material is, e.g. it is not *Generic*, the renderings are more realistic. Revit's new 'advanced materials' will say "**Parameters**" as shown in Figure 4-2.11b; use these materials whenever possible for higher quality renderings.

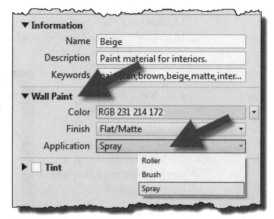

FIGURE 4-2.11A Material Browser: Appearance tab

With many *Appearance Assets* you can specify a raster image file (i.e., texture) to be used to fill or cover a surface during a rendering. This can be an actual photo of a material or a simulation. The main thing is the image needs to be *tileable*. This means when the image is copied over and up, to fill a surface, the edges of the image will not be apparent. See the two examples below (Figures 4-2.12 and 4-2.13).

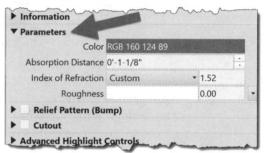

FIGURE 4-2.11B Material Browser: Appearance tab

FIGURE 4-2.12
Tileable image;
ideal for rendering

Image file

Image file applied to surface in rendering (edges not visible)

FIGURE 4-2.13
Non-Tileable image;
bad for rendering

Image file

Image file applied to surface in rendering (edges visible)

When the image of the material is used by a given *Appearance Asset*, clicking on it opens another dialog box, the *Texture Editor*, where you can adjust how the image is applied to the surface and align it with the fill pattern (Figure 4-2.14). In the *Texture Editor*, if a setting does not have a check in the box, it is not considered in the appearance. Many of the settings are self-explanatory and can be quickly tested by adjusting the values to see how it impacts the preview and rendering.

It is important to pay attention to the dimensions listed. Notice the 16'-0" by 16'-0" dimensions in the example. Dividing this dimension by the number of tiles visible in the image tells us each tile will be 2'-0", which is correct. However, if we download an image file from a manufacturer's website, we will want to make sure to verify the number of tiles has not changed. If it did, we would have to adjust the scale values.

FIGURE 4-2.14 Texture Editor Dialog

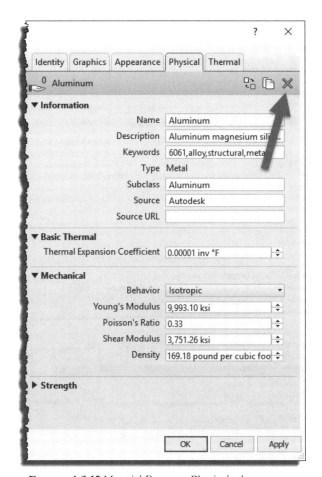

FIGURE 4-2.15 Material Browser: Physical tab

- **Physical Asset**
 The *Physical* tab is primarily used by the structural engineer to manage the structural properties for concrete, steel, aluminum and wood. This information can be used in structural analysis programs. When this information is not applicable, the *Asset* can be deleted from the *Material*.

 The example shown to the left is for the **Aluminum** *Material*. The *Strength* section can be expanded to reveal additional information (Figure 4-2.15). Notice the *Replace, Duplicate* and *Delete* icons in the upper right.

- **Thermal Asset**

 The *Thermal* aspect of a *Material* is used in energy modeling calculations (Figure 4-2.16). The basics are computed directly within Revit, such as the R-Value and Thermal Mass of a wall or floor (Figure 4-2.17). More advanced calculations can be derived in external applications.

FYI: Deleting all the Thermal assets from your *Materials* will cause the calculated values, shown in Figure 4-2.17, to be blank.

It should be pointed out that the values shown are provided by Revit for many standard real-world materials. The average software user is not expected to know this information. However, once a *Physical* or *Thermal* asset is loaded into a project, a person knowledgeable in the area of structural design or energy modeling can make adjustments to these values if needed. An example might be a special new type of rigid

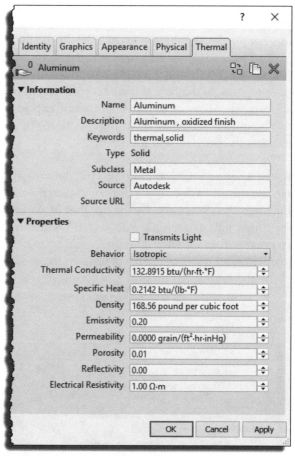

FIGURE 4-2.16 Material Browser: Thermal tab

insulation which has a higher R value. Therefore, the energy modeling expert can adjust the values for this known condition.

Related BIM Chapters article:
Creating a custom Thermal Asset for a Revit Material - early energy modeling with Insight
https://bimchapters.blogspot.com /2021/05/creating-custom-thermal-asset-for-revit.html

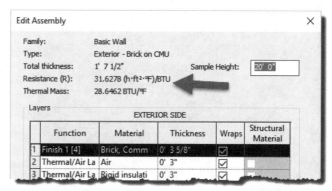

FIGURE 4-2.17 Thermal calculations in wall assembly

Duplicating a Material _and_ its assets:

It is important to know how to properly duplicate a _Material_ in your model, so you do not unintentionally affect another _Material_.

If you use **Duplicate Using Material and Assets** to duplicate a _Material_ in your model, the **Appearance Asset** will be associated with the new _Material_ AND the _Material_ you copied it from! For example, in Figure 4-2.18, we will right-click on Carpet (1) and duplicate it. Before we do so, notice the _Appearance Asset_ named "RED" is not shared (arrow #3 Fig 4-2.18).

FIGURE 4-2.18 Duplicating a material

Once you have duplicated a _Material_ with this command, notice the two carpet materials, in this example, now indicate they both share the same _Appearance Asset_. Changing one will affect the other. Click the **Duplicate this asset** icon in the upper right (# 3 in Figure 4-2.19).

FIGURE 4-2.19 Duplicating an asset

The duplicated _Appearance Asset_ can be renamed (Figure 4-2.20). You can now make changes to this material without affecting other materials.

Using **Duplicate Material and Asset** saves most of the steps just covered.

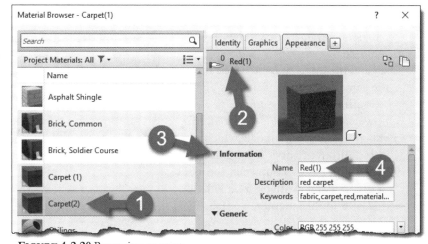

FIGURE 4-2.20 Renaming an asset

Adding or Replacing Assets

All of the *Assets*, except *Graphics* and *Identity* (which are not assets), can be replaced with another option from the provided library, called the *Asset Browser*. Additionally, the *Physical* and *Thermal* assets can be deleted.

FIGURE 4-2.21 Replacing an asset

If the asset already exists, but you want to change it, click the "Replace" icon shown in Figure 4-2.21. This will open the **Asset Browser**, from which you can choose a different *Asset* (Figure 4-2.22). The *Asset's* properties are hard-wired in the libraries and not directly editable until loaded into a project.

Clicking the icon pointed out in Figure 4-2.22 (item # 2) will push the Asset into the currently selected *Material*.

If your *Material* does not have a *Physical* or *Thermal* asset, in the *Material Browser* you use the **Add Asset** icon option shown in Figure 4-2.23. This too will open the *Asset Browser*.

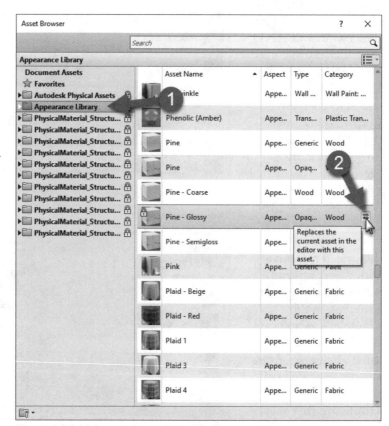

FIGURE 4-2.22 Asset browser dialog; choosing a physical asset

An asset can be associated with multiple *Materials*. For example, several materials may be based on the same *Physical* asset such as "Pine – Southern." Keep in mind, however, if the asset is adjusted, this will affect all the materials which use it. If a different version were needed, you would have to duplicate the asset before making changes. You will do this later in this chapter.

FIGURE 4-2.23 Accessing the asset browser

Materials Library

If the *Material* you want is not available in your project, in the *Project Materials* area of the *Material Browser*, you can either create one or load it from the **AEC Materials** library. If the *Material* you want exists in the library this is much faster than creating your own as it has all the assets, including *Physical* and *Thermal*, preloaded. *FYI:* To see the external materials below, you may have to click the icon shown as step 1 in Figure 4-2.24.

The search box at the top of the *Material Browser* actually searches both the materials in your project and the library (Figure 4-2.24). Notice there are six materials within the project which contain the word tile in the name AND there are several results, within several categories, below in the library.

It is possible to create a custom library here. The library can be stored on a server, allowing multiple users to access it. Use the icon in the lower left to get started with a custom library.

FIGURE 4-2.24 Material Browser; search for material

To load a Material into your project from the library, simply double click on it.

Custom Information

The **Custom Parameters** button, at the bottom of the *Material Browser*, gives you access to a list of *Material*-based parameters in the project. These are user-created parameters used to track pretty much anything you want. One example might be the source distance to better manage which products are locally sourced for LEED accreditation. Clicking this button opens a separate dialog to view the custom parameters (Figure 4-2.25). It is too bad one cannot initiate the creation of a *Material* parameter from this dialog.

FIGURE 4-2.25 Material parameters

Conclusion

This has been a fairly thorough overview of *Materials* in Revit. In the next exercise you will apply some of these concepts to help reinforce the ideas recently put forth.

Exercise 4-3:
Revit Materials Concepts Applied – Part 1

This section will apply many of the concepts covered thus far. You will open the **architectural starter project** provided online and save it to your hard drive. Be sure to back up this file often as it is the file you will work on throughout this book.

Before we open Revit, we will cover a few technical points related to materials. Many materials are noted in drawings using abbreviations. The abbreviations used may vary slightly from firm to firm. A few standards exist and are generally good to follow. However, as long as the abbreviations used are defined within the *Contract Documents*, drawings or project manual, it is "legal" to use whatever you want. Below is an example of an Interior Design group's abbreviations list (Figure 4-3.1). Keep in mind that other disciplines might have their own abbreviations list, which would be located in another section of the project. Using duplicate abbreviations between disciplines should be avoided. For example, CC means Cubical Curtain in the list below, so the Civil design group should not use CC to mean Concrete Curb on their drawings if at all possible.

ABBREVIATIONS

ACP	ACOUSTIC CEILING PANEL	EQ	EQUAL	LCW	LINEAR WOOD CEILING	SQ FT	SQUARE FEET (FOOT)
ACT	ACOUSTIC CEILING TILE	EXP	EXPANSION / EXPOSED	MET	METAL	SS	STAINLESS STEEL
ADJ	ADJACENT	EX	EXISTING	MTL	METAL	SSURF	SOLID SURFACE
AF	ATHLETIC FLOORING	EXIST	EXISTING	MISC	MISCELLANEOUS	ST	STAIN
AFF	ABOVE FINISHED FLOOR	FAF	FLUID APPLIED FLOORING	NA	NOT APPLICABLE	STL	STEEL
ALUM	ALUMINUM	FB	FABRIC	NIC	NOT IN CONTRACT	STN	STONE
AP	ACOUSTIC PANEL	FIN	FINISH	PERF	PERFORATED	SV	SHEET VINYL
AWT	ACOUSTIC WALL TREATMENT	FPR	FIBERGLASS REINFORCED PANEL	PHP	PHENOLIC PANEL	TH	THRESHOLD
BMU	BURNISHED MASONRY UNIT	GB	GRAB BAR	PLAS	PLASTER	TS	TRANSITION STRIP
BR	BUMPER RAIL	GL	GLASS	PLAM	PLASTIC LAMINATE	TYP	TYPICAL
CAB	CABINET	GRT	GROUT	PP	POLYMER PANEL	TZ	TERRAZZO
CC	CUBICAL CURTAIN	GT	GLASS TILE	PREFIN	PREFINISHED	TXT	TEXTILE
CJ	CONTROL JOINT	GYP	GYPSUM	PRT	PORCELAIN TILE	UNFIN	UNFINISHED
CT	CERAMIC TILE	GWB	GYPSUM WALL BOARD	PT	PAINT	UNO	UNLESS NOTED OTHERWISE
CG	CORNER GUARD	HM	HOLLOW METAL	QT	QUARRY TILE	VCT	VINYL COMPOSITION TILE
CONC	CONCRETE	HPC	HIGH PERFORMANCE COATING	QS	QUARTZ SURFACING	VIF	VERIFY IN FIELD
CPT	CARPET (BROADLOOM/TILE)	HPL	HIGH PRESSURE LAMINATE	RB	RESILIENT BASE	WCV	WALL COVERING
CR	CRASH RAIL	HR	HAND RAIL	RES	EPOXY RESIN	WD	WOOD
EJ	EXPANSION JOINT	HT	HEIGHT	RF	RESILIENT FLOORING	WP	WALL PROTECTION
EM	ENTRY MAT	LF	LINEAR FEET (FOOT)	SCONC	SEALED CONCRETE	WT	WINDOW TREATMENT
EP	EPOXY	LN	LINOLEUM	SCT	SPECIALTY CEILING TILE		

FIGURE 4-3.1 Abbreviation list for interior design drawings

Generally speaking, generic terms should be used rather than trademarked manufacturer terms. For example, drawings noting the construction of an interior wall or soffit should say "GYPSUM BOARD" rather than something manufacturer-specific such as DRYWALL or SHEETROCK. This helps to avoid confusion during bidding and construction, especially on publicly funded projects which require competitive bidding. That is, two or more manufacturers must be specified, not just the designer's favorite choice, for better or worse. Having said that, sometimes the drawings list a specific product to be used as the benchmark for other manufacturers or approved equals.

Next you will be creating a number of *Materials* in Revit and entering some of the descriptive data about them.

Creating Materials

You will select a starter file from the available online files for this book.

1. Open Revit and then select **Open…**, under Models, on the *Home* screen.

2. Browse to the downloaded **Law Office.rvt** model.

 a. See the inside front cover for more information on downloading files for this book from www.SDCpublications.com.

3. Select **Manage** *tab* ➔ **Settings** *panel* ➔ **Materials**.

Next you will create a duplicate of the Carpet(1) *Material.* The easiest thing to do is first select a similar *Material* to the one you wish to create. In this case you are about to create a new *Material* to represent carpet, so you will be instructed to select an existing carpet material, existing in that it is already in the project, not in the sense that it is already installed in the actual building.

4. In the *Project Materials* list, select ***Carpet (1)***.

5. Right-click *Carpet (1)* and then click the ***Duplicate Material and Assets*** in the pop-up
 menu (Figure 4-3.2).

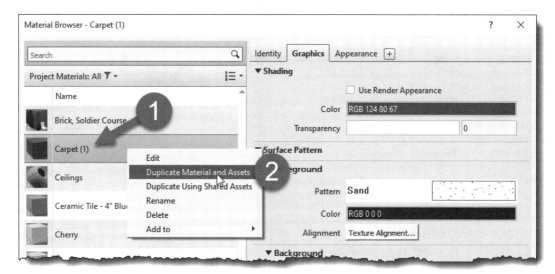

FIGURE 4-3.2 Duplicating a material

A new Material is created with a name similar to the original: Carpet (2). You will rename it.

6. **Rename** *Carpet(2)* to **CPT1**; you should be able to just select the text and change it. If you clicked off of the new material, you will need to right-click and select *Rename*.

7. Switch to the **Identity** tab for the new *Material – CPT1*; information shown will be entered in Step 13 (Figure 4-3.3).

Next you will go online and find the information needed to complete the *Material* definition for CPT1. Finding information such as this at the beginning of a project is not usually necessary. This could be information already saved in your Revit template file, if this is a specific product you regularly use in your designs.

Please refer to the following list for the next several steps. This list is similar to what one might find in a set of construction documents.

09 6800 - TILE CARPETING

CPT1 CARPET TILE 1
MFG:	INTERFACE FLOR
GROUP:	THE POND COLLECTION
STYLE:	BERLIN (12422)
SIZE:	20"X20" NOMINAL (50 CM X 50 CM)
COLOR:	6711 BARK
LOC:	PUBLIC SPACES

CPT2 CARPET TILE 2
MFG:	INTERFACE FLOR
GROUP:	THE AERIAL COLLECTION
STYLE:	AE310 (138820)
SIZE:	20"X20" NOMINAL (50 CM X 50 CM)
COLOR:	106574 MUSHROOM
LOC:	OPEN OFFICE, PUBLIC ACCENT

CPT3 CARPET TILE 3
MFG:	INTERFACE FLOR
GROUP:	THE AERIAL COLLECTION
STYLE:	AE310 (138820)
SIZE:	20"X20" NOMINAL (50 CM X 50 CM)
COLOR:	106575 TOPAZ
LOC:	ACCENT

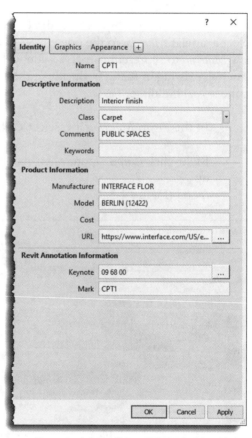

FIGURE 4-3.3 Editing CPT1 Identity data

First you will focus on the carpet material CPT1, which is manufactured by Interface Flor.

8. Open your web browser; browse to http://www.interfaceflor.ca

9. Select your region, e.g. *Americas\United States*, and then click the **Search** link (Figure 4-3.4).

 FYI: The *website design may change; follow along as best you can.*

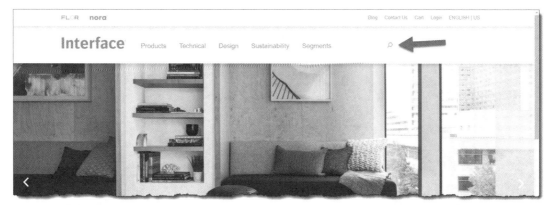

FIGURE 4-3.4 Interface Flor website

10. In the search area type *Berlin* and then click **See all results**.

11. Find color **6711 (Bark)** as shown in Figure 4-3.5; click on it to see more details.

12. Click **Specifications** near the top of the page.

You now have access to all the information you need to define the Revit *Material* (Figure 4-3.6).

13. In Revit, enter the information shown in Figure 4-3.3.

 a. Copy/Paste the website address for the current product page to the URL field in Revit.

 b. Click the small button to the right of the **Keynote** field and then select **09 68 00 Carpet** as shown in Figure 4-3.7.

 NOTE: You will learn more about keynoting later in the book.

Berlin
The Pond Collection
Bark 6711

FIGURE 4-3.5
Product preview data

As you can see on the webpage for this product, there is much more information available than Revit has built-in parameters for. That is why you are adding a web address (URL) to this information. However, as you will learn later, it is possible to add additional custom material parameters in Revit.

14. Switch back to the **InterfaceFlor** website.

15. Near the top of the screen, click **High-res Images** (Figure 4-3.6).

16. Click to **download a zip archive of all high-res images**.

The ZIP should now be saved to your computer's *Downloads* folder.

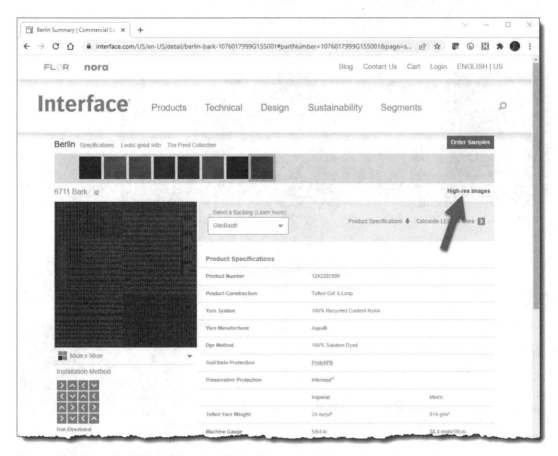

FIGURE 4-3.6 Interface Flor website - specific product page

Next you will switch to the *Appearance* tab and make some adjustments, and then come back to the *Graphics* tab.

Notice, on the *Appearance* tab (Figure 4-3.8), that a PNG raster image is assigned to this material. You will use an image you just downloaded from the carpet manufacturer's website and then edit your *Material* to use this new image. Keep in mind that anyone who might work on the project needs access to the image file location. So, in an office environment, you would not want to save this file to your local hard drive.

TIP: You can set the path to the raster images via File Tab → Options → Rendering. Do this on each user's computer.

FIGURE 4-3.7 Keynotes selection dialog

Next you will specify a name and location for the raster image you are about to save. Pay close attention to the drive and folders in which the file is saved. You will need to point Revit to this file location in a moment.

17. Double-click on the Zip file in File Explorer (Windows) and Copy/Paste **berlin_6711_Bark_50cmx50cm_NonDirectional.jpg** to your project folder.

You now have an image file which represents the specified carpet style available on your computer or network. You will assign this image to the *Appearance* asset.

18. Back in Revit, in the material browser, make sure material CPT1 is selected.

19. Expand the **Information** area and change the *Name* to **CPT1** (see Figure. 4-3.8).

 a. You should also edit the *Description* and keywords.

FIGURE 4-3.8 Editing an appearance asset

You now have an *Appearance* asset in your project named CPT1. This can be changed without affecting any other material, because you used Duplicate Material and Assets to create this material.

20. In the *Generic* section, click on the file name link pointed out in Figure 4-3.9.

21. Browse to the image file saved in Step 17 above, select it and click **Open**.

FIGURE 4-3.9 Selecting a new image file for appearance asset

Now, when using the *Realistic* display style OR doing a rendering, this image will be mapped to any surface which has been assigned the CPT1 *Material*. However, right now the carpet tiles will not be the correct size. You will adjust this next.

22. *To edit the image size:* on the *Appearance* tab, click directly on the image swatch located above the image name (i.e., the link you clicked on to change the image).

 FYI: *Not the preview image at the top.*

You are now in the *Texture Editor* dialog (Figure 4-3.10). Here you may adjust several settings related to the raster image.

At the moment, the edge dimensions are not equal, thus the image is being stretched. You will need to deactivate the "lock aspect ratio" feature so the two edge dimensions can be changed independently.

23. Click the **chain icon** pointed out in Figure 4-3.10.

Next you will adjust the *Scale* dimensions. Even though the project is set up for feet and inches, it is possible to enter metric values and Revit will automatically convert it. This is true anywhere you enter a length in Revit, other than the plain *Text* command. Interface happens to manufacture their carpet tile to a metric size, so you will try this now.

24. Highlight the value listed next to *Width* and type: **200cm**. FYI: the image we selected shows (4x4) 50cm tiles.

25. Make the same change for *Height*.

26. Click **Done**.

27. Uncheck **Bump** on the *Appearance* tab.

The *Bump* mapping helps materials look more three-dimensional when rendered. However, the map is typically a grayscale image based on the main image file. This requires working on a copy of the image in another program such as Adobe Photoshop, which will not be done at this time.

28. Switch to the *Graphics* tab and check **Use Render Appearance** under *Shading*.

Notice how Revit automatically adjusts to the color to match the color of the raster image. This shading color is what is used for *Shaded* and *Consistent Colors* Display Styles.

You will leave the surface and cut patterns as they are for now. Because the manufacturer provided a 4x4 sample area (16 tiles) the result is a more realistic representation of an installation. In some cases, individual tiles may need to be combined and rotated in Photoshop to achieve similar results.

FIGURE 4-3.10 Texture Editor

29. Using the steps just covered, create two more *Materials* for **CPT2** and **CPT3** as defined on page 4-24. Be sure to use the *Duplicate Material and Assets* option.

> ***FYI:*** *When you duplicate* CPT1 *you save some typing (i.e. data entry) as all the values for the selected* Material *are also duplicated. However, you need to be careful and review everything, so you do not leave incorrect information assigned to the new* Material.

The process just covered will usually be the same for any carpet manufacturer or any product manufacturer in general. The only challenge is some websites are easier to navigate than others, and raster images may not always be available or of good quality (i.e., resolution and size).

Next you will look at creating several paint sample materials. This will involve locating the specific paint color's RGB color rather than a raster image. The RGB color can then be entered for shaded views as well as for rendered views.

Each paint manufacturer organizes their color information a little differently. We will look at the Sherwin-Williams website specifically. Often, the other paint manufacturer's sites will be similar. Be sure to check out the various color tools provided, such as the smart phone and Adobe Photoshop options. Below is the list of colors you will be working with:

09 91 00 - PAINTING AND COATING

PT1 PAINT 1 (WALLS) *R242 G224 B195*
 COLOR: "BEIGE"
 MFG: SHERWIN-WILLIAMS
 Model: SW0051 CLASSIC IVORY
 Revit Material Settings: *Application*: Roller, *Finish*: Eggshell

PT2 PAINT 2 (INTERIOR HM DOORS & FRAMES / STEEL) *R99 G101 B98*
 COLOR: "GRAY"
 MFG: SHERWIN-WILLIAMS
 Model: SW7068 GRIZZLE GRAY
 Revit Material Settings: *Application*: Brush, *Finish*: Semi-gloss

PT3 PAINT 3 (CEILINGS & SOFFITS, EXPOSED STRUCTURE) *R233 G235 B231*
 COLOR: "WHITE"
 MFG: SHERWIN-WILLIAMS
 Model: SW7007 CEILING BRIGHT WHITE
 Revit Material Settings: *Application*: Roller, *Finish*: Flat

PT4 PAINT 4 (ACCENT WALL COLOR) *R117 G103 B97*
 COLOR: "BROWN"
 MFG: SHERWIN-WILLIAMS
 Model: SW6040 Nutshell
 Revit Material Settings: *Application*: Roller, *Finish*: Eggshell

PT5 PAINT 5 (TOILET ROOMS) *R202 G179 B160*
 COLOR: "BROWN"
 MFG: SHERWIN-WILLIAMS
 Model: SW6093 FAMILIAR BEIGE
 Revit Material Settings: *Application*: Roller, *Finish*: Eggshell

30. Browse to the Sherwin-Williams website: http://www.sherwin-williams.com.

31. Choose your region if prompted.

32. Select **Explore Color >**.

33. In the search box at the top, type **SW0051**.

34. Press **Enter**.

35. Click on the preview swatch for **Classic Ivory** and then click **Details** to see more information (Figure 4-3.11).

You should now see the information shown below (Figure 4-3.11). Notice the RGB Value listed: 242,224,195. These are the Red, Green, and Blue colors which make up the Classic Ivory color. This information can be entered into the Revit material to create the proper color representation. Next you will create the PT1 *Material* in Revit.

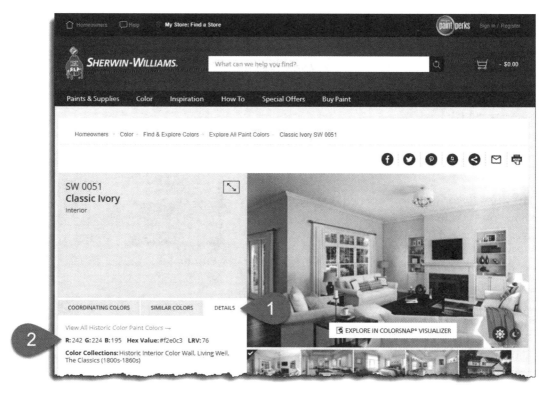

FIGURE **4-3.11** Sherwin Williams website

The paint material is a little more complicated than carpet in that the Revit *Material* will also represent the building material being painted. In Revit, the paint finish is applied to the face of the concrete block (CMU) or gypsum board (GWB). Therefore, you have to create a PT1 paint *Material* for both CMU and for GWB. This is due to the fact that each material has a different cut and surface pattern. This might make more sense after the next few steps.

36. Back in your *Law Office* Revit model, open the ***Materials*** dialog.

37. Select the *Material* **Gypsum Wall Board**.

38. Use **Duplicate Material and Asset** to make a copy of the *Material* (via right-click).

39. Rename the new material to **Gypsum Wall Board – PT1**.

40. Fill out the **Identity** tab using the information from the website as shown in Figure 4-3.12.

FIGURE **4-3.12** Identity info for PT1

41. Switch to the **Appearance** tab.

Next you will need to replace the *Appearance* asset with one specifically meant for paint. Currently it is set to *Generic* and not *Wall Paint* as pointed out in Figure 4-3.13.

42. Click the **Replace this asset** icon (Figure 4-3.13).

The *Asset Browser* has now opened (Figure 4-3.14). Here you need to select a *Wall Paint* asset and overwrite the *Generic* asset in the law office model.

All the *Wall Paint* assets provided are for specific colors, so we essentially need to duplicate one and rename it. However, the assets in the Autodesk libraries are not editable. So, we need to load one into the project and then rename it. We will do this next.

FIGURE 4-3.13 Replace asset icon

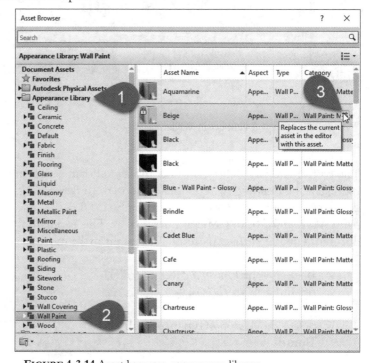

43. In the *Appearance Library* list on the left, select **Wall Paint** (Figure 4-3.14).

44. Click the "load into project" icon for the **Beige** asset, or double-click the asset name.

The appearance asset should now be updated in your project. Next you will rename it. This can be done in the *Asset Browser*.

FIGURE 4-3.14 Asset browser, appearance library

45. **Close** the *Asset Browser* dialog.

46. Expand the **Information** area in the *Appearance* tab (Figure 4-3.15).

47. Change the *Name* **SW 0051 Classic Ivory**.

Now you can adjust the specific paint information for your *Material* (Figure 4-3.16). Keep in mind that this *Appearance* asset may be applied to other *Materials* in your project. For example, the asset could be applied to a CMU wall or metal railing. Any adjustments made to the asset are instantly applied to all *Materials* which use it.

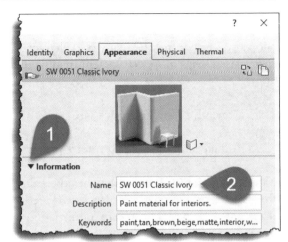

FIGURE 4-3.15 Rename new assert

48. Click on the **RGB color numbers** to adjust them.

49. Change the RGB numbers as follows (Figure 4-3.16):

 b. R: 242

 c. G: 224

 d. B: 195

As you recall, the RGB value came from the paint manufacturer's website. The color swatch in Revit should look very similar to that shown on the website. Keep in mind that individual monitors and color printers can vary significantly in how they represent a color. It is always best to get a physical color sample before making final selections. But this will get you close!

FIGURE 4-3.16 Final appearance settings

50. Adjust the *Finish* option to **Eggshell**.

51. Adjust the *Application* option to **Roller**.

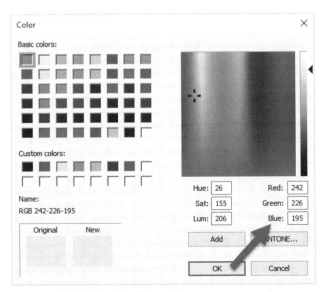

FIGURE 4-3.17 Materials dialog; Adjust appearance color

Notice how the *Appearance* options available vary from one material type to another. This allows you to control the real-world characteristics more accurately for a number of types or materials. This is more convenient than having all possible options when they do not really apply.

52. Switch to the **Graphics** tab.

53. Make sure the **Use Render Appearance** under *Shading* is checked (Figure 4-3.18).

Notice how the RGB color matches the value you entered for the render *Appearance*. Again, the *Appearance* information is for *Realistic* & *Ray Trace* mode and used when creating photo-realistic renderings. The *Shading* information is used in plan, elevation and 3D views when the *Visual Style* is set to *Shaded* or *Consistent colors*.

Next you will set the cut pattern to *None*. It is not typically necessary to hatch this material (i.e. gypsum board) because it is not very thick (1/2″ – 5/8″ typically), so the pattern does not read well.

54. Set the *Foreground Cut Pattern* to **None** (Figure 4-3.18).

55. Repeating the steps just covered, create a material for **PT2** through **PT5** as shown on page 4-30.

 a. **PT2** should be duplicated from *Steel, Paint Finish, Ivory, Matte* in AEC Materials library.

 b. **PT3** should be duplicated from *Gypsum Ceiling Board*.

 c. **PT4** & **PT5** should be duplicated from **PT1**.

56. Remember to **Save** your project often to avoid losing work.

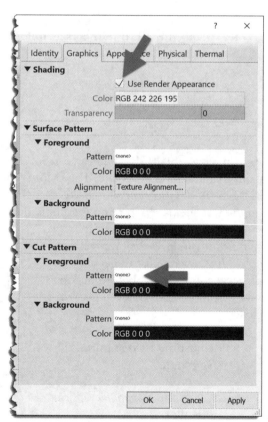

FIGURE 4-3.18
Material Browser; Graphics tab

Exercise 4-4:
Revit Materials Concepts Applied – Part 2

We will continue to develop several real-world materials in Revit.

```
09 30 00 - TILING
    CT1  CERAMIC TILE 1 (SHOWER WALL TILE)
        MFG:        DALTILE
        STYLE:      MATTE WALL TILE
        SIZE:       4-1/4" X 4-1/4" NOMINAL
        COLOR:      X735 ALMOND
        GROUT:      944 LIGHT CHOCOLATE (RGB = 99, 77, 61)

    CT2  CERAMIC TILE 2 (SHOWER CEILING TILE)
        MFG:        DALTILE
        STYLE:      GLOSSY WALL TILE
        SIZE:       4-1/4" X 4-1/4" NOMINAL
        COLOR:      0190 ARCTIC WHITE
        PATTERN:    INSTALLED AT 45 DEGREES
                    ON SHOWER CEILINGS
        GROUT:      44 BRIGHT WHITE (RGB = 248, 248, 248)

    PRT1 PORCELAIN TILE 1 (FLOOR TILE)
        MFG:        DALTILE
        STYLE:      CITY VIEW
        SIZE:       18" X 18" NOMINAL
        COLOR:      CY04 DOWNTOWN NITE
        GROUT:      TEC 927 LIGHT PEWTER (RGB = 126, 128, 130)

    PRT2 PORCELAIN TILE 2 (WALL BASE TILE)
        MFG:        DALTILE
        STYLE:      CITY VIEW
        SIZE:       6" X 12" NOMINAL COVE BASE
        COLOR:      CY04 DOWNTOWN NITE
        GROUT:      TEC 927 LIGHT PEWTER (RGB = 126, 128, 130)
```

The third material you will create is tile, both ceramic and porcelain. This is similar to the previous examples, except you will set up a surface pattern for the size tile to be used and also deal with entering a color when the RGB color is not directly available. The materials you will create are listed above.

1. Open the *Material Browser* dialog in Revit.

2. Filter the material list by *Class* **Ceramic**; click the down arrow as shown in Figure 4-4.1.

FIGURE 4-4.1 Filter materials by class

You should see two materials: "Ceramic Tile – 4″ Blue" and "Porcelain, Ivory." FYI: The filter list is based on materials present in the project, and not the library.

First you will create a ceramic tile *Material* called CT1. Similar to the floor finish you created for carpet, you will not worry about the floor construction. The floor finishes will be applied differently than how the wall finishes are applied. The main reason for this is to show two ways in which materials may be applied. We will discuss this more in the chapters dealing with modeling.

3. Select **Ceramic Tile – 4″ Blue**, right-click and Duplicate Material and Assets.

4. Rename the new *Material* to **CT1**.

5. With the new *Material* selected, switch to the **Identity** tab.

Next you will go to the Daltile website and search for the specific tile under consideration. You will find on this specific website that each tile does not have its own page. Thus, the URL will have to either point to the main website or the page for the series of the selected tile.

6. Go to www.daltile.com using your web browser.

7. In the search box enter **X735**.

8. Click the **Matte Almond - Square** option.

9. Use the current web page address for the *Material* **URL** in Revit.

10. Find the X735 tile sample (click on it).

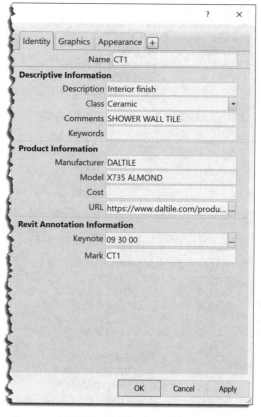

FIGURE 4-4.2 Material Browser; Identity tab

Notice the information available for this product, such as it not being appropriate for use as a floor tile. Also, the RGB color is not given. One technique to derive the RGB color is to right-click on the image on the website, select **Save Picture As** and then save the preview image to a file on your computer's hard drive. Next you could open this image in an image editing program such as Adobe Photoshop and use the *Color Picker* tool to list the RGB values of the image. You will be provided the RGB colors, so you do not need to use another program at this time.

11. Finish entering the information on the *Identity* tab as shown in Figure 4-4.2.

Note that the *Class* being set to *Ceramic* on the *Identity* tab (Figure 4-4.2) only relates to which category the *Material* is listed under in the *Material Bowser* and does not directly relate to the actual material assigned on the *Appearance* tab. However, you should make sure they match so there is no contradiction.

12. In Revit, switch to the **Graphics** tab.

13. Make sure **Use Render Appearance** under *Shading* is **not** checked.

14. Click the color swatch to adjust the RGB color.

15. Set the RGB color to
 R=231, G=220, B=202
 (similar to Figure 4-3.17).

On the website, we can find the actual tile size 4.287" x 4.287". We will update the surface pattern to a nominal 4.25" square.

16. Click the *Foreground Surface Pattern* preview, select the **Model** option, and then select **4.25″ Tile**. Click **OK** to close the *Fill Patterns* dialog.

17. Set the *Cut Pattern* for the tile to **None** (Figure 4-4.3); click the preview and then click the *No Pattern* button.

With these settings one could begin to design using the CT1 material. However, the *Appearance* tab does not have a proper image file selected to represent the selected tile. It is still set to the blue tile.

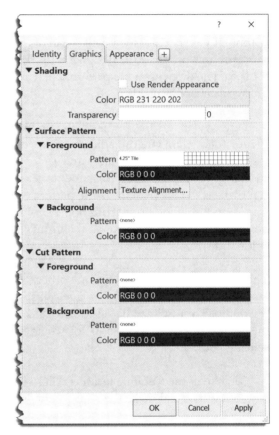

FIGURE 4-4.3 Material Browser; Graphics tab

Because we do not have a representative image with a repeating grout pattern to download, you must either find one in the *Autodesk Library*, modify a material, take a picture of the tile installed somewhere or find something online.

18. Switch to the **Appearance** tab.

19. Rename it **CT1** (Steps 1 & 2, Figure 4-4.4).

Notice this material is currently using an image file to represent the tile (Figure 4-4.4). This would be like taking a photo of an installed tile and then cropping it so that it repeats nicely. In the next few steps, you will take a look at another option for representing tile.

20. Click the down-arrow to the right of the image preview (Step 3, Figure 4-4.4).

21. Select **Tiles** from the list.

Revit still understands this material to be "ceramic" but rather than using an image, it is now using a "synthetic" tile pattern. The appearance asset is synthetic in that it is computer generated. As you will see in a moment, this provides the designer with the tools to quickly create a tile pattern when an image file is not available.

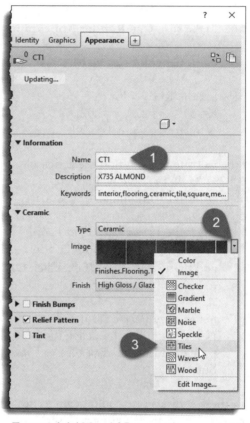

You are now in the *Texture Editor* shown in Figure 4-4.5. Here you can adjust all sorts of settings related to how the tile pattern should appear. One of the adjustments one can make is the tile pattern. Is the pattern stack bond, running bond, common Flemish bond, etc.? You can even adjust the grout color and size.

This method can be used for many other materials such as sidewalks. For sidewalks, the number of rows can be set to zero and the column set to be about 5'-0" or whatever spacing you want.

FIGURE 4-4.4 Material Browser; Appearance tab

22. Under the Ceramic heading, click **Edit...** and make the adjustments shown in Figure 4-4.5.

23. Click **Done** to exit the Texture Editor.

FYI: The grout types listed are manufactured by TEC. If interested, you can visit their website at www.tecspecialty.com. Look for AccuColor® Premium Unsanded Grout (Cementitious Grout).

FIGURE 4-4.5 Texture editor

The *Material Browser* dialog below shows one of the other options shown in Figure 4-4.4. This is the **Checker** option where you can select two separate colors. This can be used for an entire wall or in an accent area (Figure 4-4.6). Be sure to explore the other options, such as *Wood*, which can be used to represent wood or other things such as art glass.

FIGURE 4-4.6 Material Browser (alternate option)

A *Relief Pattern* would be like embossing a piece of paper. In the raster image currently being used (TileSquareBump.png in this case) the non-white areas are debossed on the otherwise flat/infinite ceramic material. This *Relief Pattern* gives the general appearance of a tile pattern.

24. Switch the *Relief Pattern* to a **Tiles** pattern, similar to steps 20 and 21.

25. Adjust the settings to match Figure 4-4.7.

Notice the other options, including the ability to rotate the pattern.

26. Click **Done**.

27. Switch back to the *Graphics* tab.

28. Click the **Texture Alignment** button.

This can be used to make sure the surface pattern and the appearance pattern align.

FIGURE 4-4.8 Texture alignment

29. Click the arrows if necessary to make the fill pattern align with the render appearance pattern (Figure 4-4.8).

30. Click **OK** to close the *Material Browser*.

FIGURE 4-4.7 Texture Editor for relief pattern

The following two images in Figure 4-4.9 highlight the difference in a rendered image when the *Relief Pattern* is used, as seen in the example on the left, and when one is not, on the right. The *Relief Pattern* actually gives the material a three-dimensional look. The lighting in the space is actually in sync with the *Relief Patterns* as well. Notice how the shade and highlights correspond with the lighting primarily coming from the top right in the image on the left. The image on the right is flat and void of any shading or highlights at the edge of the tiles.

FIGURE 4-4.9 Rendered material with *Relief Pattern* on left, and without on right

Now you will create another ceramic tile pattern using CT1 as a starting point.

31. Create *Material* **CT2**:

 a. *Duplicate* CT1 via a right-click using *Duplicate Material and Asset*

 b. *Rename* to CT2

32. Switch to the **Graphics** tab.

33. Use the following RGB values for shading: R=244, G=240, B=237.

Next you will adjust the *Surface Pattern* so it is rotated 45 degrees.

34. On the **Graphics** tab, click the pattern preview next to *Foreground Surface Pattern*.

35. Set the *Pattern Type* to **Model** and then click the **New fill pattern** button.

36. Adjust the settings to match Figure 4-4.10; enter the *Name* **4.25″ Tile at 45 Degrees**.

FIGURE 4-4.10
Add Surface Pattern dialog

37. Click **OK** to create the new model pattern. Remember, model patterns do not change size when the view scale is changed for a given view.

This new *Fill Pattern* is now available to any tool or feature which utilizes the fill patterns.

38. In the Fill Pattern dialog, click **OK** to accept the new pattern for the material's *Foreground Surface Pattern*.

39. Switch to the *Appearance* tab.

Before making any changes to the *Appearance* asset you need to make sure the appearance asset is not linked to the material you copied it from, CT1 in this case. Revit reminds you of this by the icon pointed out in Figure 4-4.11; it indicates that this asset is shared with one other material. So, if you rotate the *Appearance* asset for CT2, it will also change CT1. Since you duplicated the material using Duplicate Material and Asset, you should not have this problem. If you did, simply click the *Duplicate this Asset* icon to the far right.

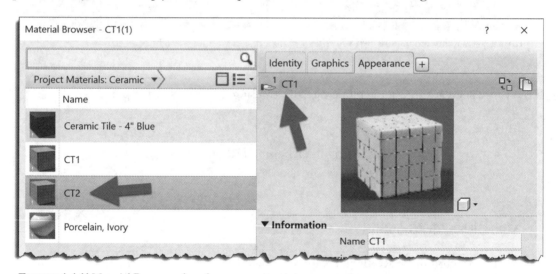

FIGURE 4-4.11 Material Browser; shared appearance asset icon example

40. Expand the *Information* section, on the *Appearance* tab, and change the *Name* to **CT2**.

41. Adjust the rotation of the *Appearance Asset* to **45** degrees; click **Edit...** in both the *Ceramic* and *Relief Pattern* sections (Figure 4-4.12).

FIGURE 4-4.12 Rotate texture 45 degrees

42. Adjust the RGB color for the ceramic to R=244, G=240, B=237.

43. Switch to the **Graphics** tab.

44. In the *Surface Pattern* area, click the **Texture Alignment** button.

45. Notice that the *Surface Pattern* and the render texture, i.e., appearance, both align. Try clicking the arrows, but click **Cancel** to close without making changes.

46. Click **OK** to close the *Material* dialogs.

47. **Save** your project.

Your *Material* preview should look like that shown in Figure 4-4.13 to the right.

FIGURE 4-4.13 Completed CT2

Next you will create the last two tile materials.

48. In the *Materials* dialog, select CT1 and then **Duplicate** it; name the new material **PRT1**. Tip: Be sure to use *Duplicate Material and Assets*.

49. Adjust the **Identity** information per the steps previously covered to create the ceramic materials; see page 4-35.

50. Switch to the **Graphics** tab.

51. Check **Use Render Appearance for Shading**.

52. Click the *Foreground Surface Pattern* preview area.

53. Set the *Pattern Type* to **Model** and then click **New**.

54. Create a new pattern named **18″ Tile**; adjust the settings to match Figure 4-4.14.

55. Click **OK** twice to get back to the *Material Browser*.

FIGURE 4-4.14 Add surface pattern

Next you will modify the *Appearance* tab. But first you might want to review the *Texture Alignment* to see how things do not currently align.

56. Switch to the **Appearance** tab.

You will notice from the manufacturer's website that the selected porcelain tile is not a simple solid color. It has a natural irregular pattern to it. So, you might **not** want to use the same appearance previously applied to CT1 and CT2. Next you will download an image from the manufacturer's website.

57. Go to www.daltile.com and search for tile number **EQ10**.

 a. Select EQ10 – Ivory – Square

 b. The required JPG file (*Daltile - EQ10 - Ivory - Square.png*) can be accessed from the download files available with this book; see the inside front cover for instructions and your unique access code.

58. In the gallery view, right-click on large preview image.

59. Select **Save Image As** and save it to your project folder.

The only problem with this image is it does not have a grout joint around the perimeter. This will not look great in a rendering; due to the irregular pattern a line will still be apparent. It is possible to use an image editing program, such as Adobe Photoshop, to correct the problem. To add the joint, you would want half a joint so it tiles, or repeats, properly on a surface in the model. You will not be required to make these types of edits at this time. The edited file is provided.

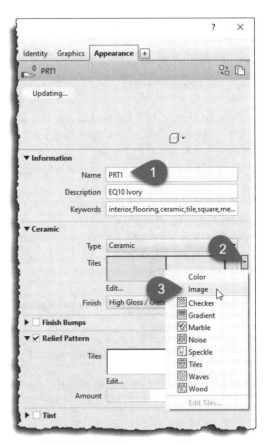

FIGURE 4-4.15 Appearance tab

60. Rename the *Appearance* asset to **PRT1** (similar to step 19).

61. Back in the *Material Browser*, with PRT1 selected, adjust the *Description* as shown in Figure 4-4.15.

Next you will change the *Appearance* asset to use an image rather than the "synthetic" *Tiles*.

62. Select the down-arrow to the right of the *Tiles* preview and select **Image** (Figure 4-4.15).

You now have a raster image-based *Appearance*. You will change it to the image you just downloaded.

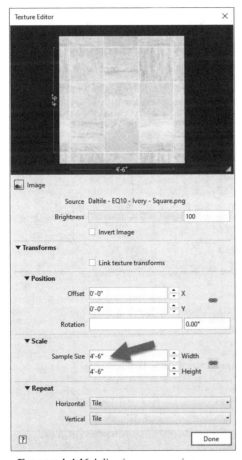

FIGURE 4-4.16 Adjusting texture size

63. Browse to the raster image of the EQ10 tile provided with this book.

64. The *Texture Editor* opens automatically; if you need to open it again later, click on the image preview in the *Material Browser*.

65. In the *Texture Editor*, change the *Sample Size* to **4'-6"x4'-6"**; this is for three 18" tiles in each direction (Figure 4-4.16).

66. Click **Done** to close the *Texture Editor*.

Now you need to adjust the *Relief Pattern*. This will remain the "synthetic" Tiles option. You will adjust the spacing to align with the grout joints in the image. You will also make an adjustment so the joint does not look too deep.

67. Edit the ***Relief Pattern*** as follows:

 a. Tile count: 3.00

 b. Set the *Sample Size* to **4'-6"**

 c. Set the *Grout Color* to R= 125, G=125, B=125

Note that the raster image and the relief pattern are both set up to show full tiles. If this were not the case you would need to offset the relief pattern to align with the image.

To better visualize the changes you are making, it is possible to adjust the preview image to be larger and to represent a wall.

68. Click the down-arrow next to the material preview and select **Walls** (Figure 4-4.17).

69. Adjust the preview size by dragging the width of the dialog box (to adjust the width) and the horizontal line just below the image (to adjust the height).

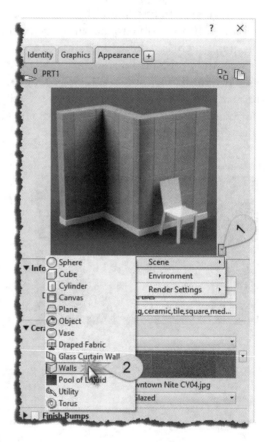

FIGURE 4-4.17 Changing preview image

You now have a larger image which looks like a wall to better visualize your material. Notice how the grout joint does not appear to be as deep. In the *Relief Pattern*, the darker the graphics, the deeper the relief appears in those areas in the material. Therefore, when we make the grout lines a lighter grey, the appearance becomes less recessed.

Next you will duplicate PRT1 to create PRT2. Both are based on the same tile. One is for the floor and the other is meant to be used as wall base. The wall tile is 6″x12″. This can be set up as 12″x12″ as the tile will be cropped by the surface it is applied to in the model. That is, PRT2 will only be used as a wall base and only applied to an area that is six inches tall.

70. Duplicate PRT1 as **PRT2**.

71. Adjust the *Identity* tab.

 a. *Comments*: PORCELAIN WALL BASE TILE

 b. *Mark*: PRT2 *Remember to always set this value when duplicating a material*

 c. Everything else is the same as PRT1.

72. On the *Graphics* tab, set the *Foreground Surface Pattern* to **12″ Tile**.

73. On the *Appearance* tab, use Duplicate this asset (upper right) to duplicate the asset and then adjust the texture size to be **3′x3′**; this is for three 12″ Tiles.

 a. Do this for both the image and the relief pattern.

74. Close the *Materials* dialog; be sure to click the **OK** button and not the red "X" in the upper right as this will cancel and discard your changes.

75. **Save** your project file. You will use this model in subsequent chapters.

You should save all the image files you create or download into a single folder if possible. If working in an office, the folder should be on a network, accessible to everyone on the design team. In Revit, everyone can go to **File Tab → Options → Rendering** and add "additional render appearance paths" so Revit knows where the images are located and can access them when needed.

This wraps up a fairly thorough overview of using and managing *Materials* in Revit. You should hang onto this book so you can refer back to this information as needed. Eventually, you will know how to do these steps by memory as there are lots of materials on projects. Keep in mind that your favorite materials can be added to a custom template file or custom material library, so they are available immediately for all your projects.

Exercise 4-5:
Revit Materials and Custom Parameters

This last section will look at an intermediate concept related to Revit materials. Often there is additional information you might wish to keep track of for a given material. Revit allows you to create custom parameters which are able to hold additional information about the material. It is not essential that you know how to do this now, but it is easy so why not give it a try!

You may have noticed some of the websites you visited in this chapter list data like **Post-Consumer Content**. You may also want to keep track of the distance in miles to the main source of the material. These metrics are used to rate the sustainability of a building and factor into programs such as *Leadership in Environmental and Energy Design* (LEED).

1. In Revit, switch to the **Manage** tab on the *Ribbon*.

2. Select **Project Parameters**.

3. Click the **New Parameter** icon.

4. Enter the information as shown in Figure 4-5.1 below.

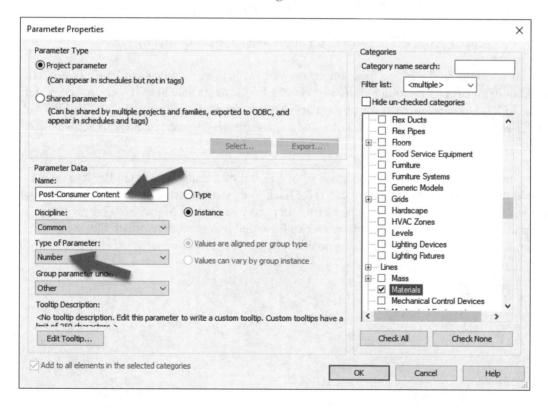

FIGURE 4-5.1 Creating a project parameter

5. Click **OK** to create the parameter.

6. Add another project parameter called **Source (miles)** with all other settings the same as the previous steps.

7. Click **OK** and then click **OK** again to close the *Project Parameters* dialog.

8. Open the *Materials* dialog, and open the *Material Browser*.

9. Click the **Custom Parameters** button in the lower left (Figure 4-5.2).

10. Add the information shown in Figure 4-5.2.

Notice the custom parameters area associated with every *Material* – be sure the correct *Material* is selected before editing the *Custom Parameters*. You are able to add as many custom parameters as you need. This information can also appear in a schedule; you will learn more about schedules later in the book.

FIGURE 4-5.2 Custom material parameters in the *Materials* dialog

Self-Exam:

The following questions can be used as a way to check your knowledge of this lesson. The answers can be found at the bottom of this page.

1. To control the extents of the *Paint* tool on a face, use the *Split Face* tool. (T/F)

2. A material cannot store information about the manufacturer. (T/F)

3. It is not possible to rotate a surface pattern. (T/F)

4. Not all render appearance materials tile properly. (T/F)

5. You can store custom information in a Revit material. (T/F)

Review Questions:

The following questions may be assigned by your instructor as a way to assess your knowledge of this section. Your instructor has the answers to the review questions.

1. To change a material "*By Face*" use the *Paint* tool. (T/F)

2. You can only duplicate materials, not start from scratch. (T/F)

3. A material's surface pattern is what would be seen in an interior elevation. (T/F)

4. List three primary ways materials are added: by _____, by _____, and by

 _____.

5. For a render appearance, a relief pattern would be like debossing a piece of paper. (T/F)

6. The glass material is transparent in 3D but not in elevation. (T/F)

7. The *Identity* tab, in the *Materials* dialog, has a section for shading adjustments. (T/F)

8. Revit does not come with any *Materials*. (T/F)

9. Model *Fill Patterns* adjust size when the view scale changes. (T/F)

10. When a 3D element has its material parameter set to <by category>, the

 _____ material is used.

Lesson 5
Programming:

Before modeling the first wall or space it is important to work with the client and develop a **Space Program**. This is a list of required spaces and what size they need to be. The design professional assists in allocating room sizes based on rule-of-thumb and experience. They also help to make sure uncommon rooms are not overlooked, such as the janitor's closet, mechanical room, recycling, etc. The *Program* establishes the "puzzle pieces" the designer uses to begin the graphical design process. This chapter will look at how Autodesk® Revit® can be used to develop and manage the *Space Program*.

Exercise 5-1:
Introduction to Schedules

Traditionally, *Space Programs* have been managed in a computer application not directly connected to the design software, such as Microsoft Excel or Access. With Revit, much of this can be done directly within the program and be tied directly to the 3D model. In some cases, it may not be possible to completely eliminate these other applications, but for the average project, Revit should suffice.

Before we get into the specifics regarding our *Law Office* project, we will first look at an introduction to *Schedules* in Revit. This overview will cover the fundamental schedule concepts in order to understand the task at hand: space programming. Later in the book you will receive a broader study on the topic of *Schedules*.

The image below is a partial view of a schedule in Revit (Figure 5-1.1), which is a tabular view of the 3D model. Schedules can list the <u>values</u> associated with any element-type in the model. Each column in a schedule represents a parameter; specific <u>values</u> are stored in parameters. This schedule is listing the *Rooms* in the model.

				ROOM FINISH SCHEDULE			
				WALL FINISHES			
ROOM NO.	ROOM NAME	FLOOR FINISH	BASE FINISH	NORTH	EAST	SOUTH	WE
100	LOBBY	CT1	CT1	PT1	PT1	PT1	PT1
101	CLOSET	CT1	RB	P1	P1	P1	P1
102	HALL	CT1	CT1	PT1	PT1	PT1	PT1
103	TOILET	CT3	CT2	CT3 / PT2	CT3 / PT2	CT3 / PT2	CT3 / F
104	TOILET	CT3	CT2	CT3 / PT2	CT3 / PT2	CT3 / PT2	CT3 / F
105	OFFICE	C1	RB1	PT1	PT1	PT1	PT1
106	OFFICE	C1	RB1	PT1	PT1	PT1	PT1
107	CONFERENCE	C1	RB1	PT1	PT1	PT1	PT1

FIGURE 5-1.1 Room Finish Schedule

Let's look at how parameters are associated with elements in the model. When something is selected, its *Instance Properties* are seen in the *Properties Palette* (Figure 5-1.2). Notice how the values for Room 103 correspond to those appearing in the *Schedule* (Figure 5-1.1). It is possible to have multiple schedules listing any combination of the available parameters.

FYI: The 'wall finish' values are manually entered in a room finish schedule, as the room element does not have the ability to read the actual material assigned to a wall element.

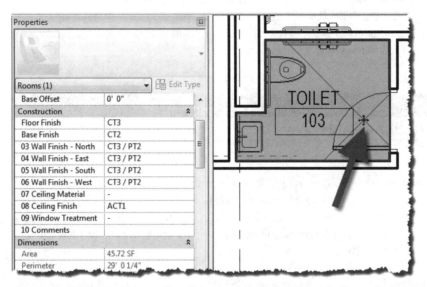

FIGURE 5-1.2 Room element selected, instance properties shown in Properties Palette

Another example of an element's parameters is shown in Figure 5-1.3. Here we see a door selected. The *Properties Palette* shows the door's instance parameters, that is, information unique to this specific door. If you select the *Type Properties* button, within the *Properties Palette* you see the door's *Type Properties* (Figure 5-1.4). These values remain the same for every door of the selected type (i.e., 36″ x 84″ in the image below). If one of these values change, it affects every instance in the model.

FIGURE 5-1.3 Door element selected, instance properties shown in Properties Palette

FIGURE 5-1.4 Type Properties shown for selected door

Revit schedules can list both instance and type parameters. Each parameter appears as a column in a schedule. Model elements make up the rows.

Creating a Schedule

Now that you have a basic understanding of Revit schedules, let's take a look at how they are created. You will get more step-by-step instructions later. Right now we are just laying the ground work. There are a number of people on the design team who might work in schedules and not really do any modeling. For example, a project manager might develop and manage sheet schedules, equipment schedules and more. The point is learning to use schedules this early in your Revit training is not a bad thing, as this is the only thing some individuals on the design team use on a daily basis, that and maybe printing.

When the *Schedule/Quantities* tool is selected, which is used to create a new schedule, the *New Schedule* dialog appears (Figure 5-1.5).

- Tool location: **View → Create → Schedules → Schedules/Quantities**

The first step is to select the *Category*. The *Name* and *Phase* can be changed later, but the *Category* option is a one-time deal.

Every element in Revit goes in a specific *Category*, which is defined when the content is initially created.

The list of Categories is predefined by Revit and cannot be changed.

FIGURE 5-1.5 New Schedule dialog

Most of the *Category* names are pretty straightforward; walls go in the *Walls* category, doors in the *Doors* category, and so on. Below are a few examples that might not be as easy to deduce.

Casework: Base cabinets and wall cabinets, reception desk, built-in bookshelves

Furniture Systems: Cubicles and other built-in office furniture

Specialty Equipment: Toilet room accessories, lockers, ladders, etc.

Topography: Site or ground surface

When a *Category* is selected for a new schedule, Revit knows to look for all the elements that have been placed in the model which are in that *Category*. A **<Multi-Category>** schedule has the ability to look at every element in the model and report on parameters they all have in common, such as *Cost*.

Schedule Properties

Once a schedule is created, you are immediately brought to the ***Schedule Properties*** dialog (Figure 5-1.6). This is the place you select what information you want listed in the schedule and what it should look like. We will take a few minutes to review the various options. You will not use all of these features and settings right now, but you can refer back here later as needed.

You will notice the *Schedule Properties* dialog is divided into five separate areas; each area has its own tab near the top. In a moment, we will take a look at each tab's contents.

Before Revit will create the schedule you have to select one or more parameters to be scheduled. To do this you add ***Available Fields***, list on the left, to the ***Scheduled Fields (in order)*** list, on the right. Simply select a parameter on the left and click the **Add →** button, or just double-click the parameter name.

FIGURE 5-1.6 Schedule Properties dialog

If you simply selected the five parameters shown on the right (Figure 5-1.6, #2) and clicked **OK**, the schedule would be created. A new item would appear in the *Project Browser*, under *Schedules/Quantities*. If you are in a new or empty project, the schedule would be empty (Figure 5-1.8). On the other hand, if the model has 30 items in the selected category, you should see 30 lines/rows, one for each instance in the model. Additionally, if those 30 elements have information entered into any of the scheduled parameters, this information will appear in the schedule automatically.

If you close the *Schedule Properties* box, this is how you get back to it: open the schedule view and then select one of the **Edit** buttons in the *Properties Palette* (Figure 5-1.7). You do not click the *Schedules/ Quantities* button as this is only to create a new one.

FIGURE 5-1.7 View Properties

		<Furniture Schedule>		
A	B	C	D	E
Type Mark	Model	Manufacturer	Description	Cost

FIGURE 5-1.8 New schedule created in an empty model

Schedule Properties: Fields Tab

Looking at the *Fields* tab (Figure 5-1.6) we will take a look at the four areas highlighted.

1). **Available Fields:** Each element type, or *Category*, has a number of fields, or parameters, associated with them. Some of these parameters are hard-wired into the software, some have been added initially in the content, and some are created in the project as needed (see item 3 below).

 In addition to common parameters such as *Cost* and *Manufacturer*, each category has a unique list of options. Here are a few selective examples:

 Casework: Construction Type, Depth, Finish
 Lighting Fixture: Circuit Number, Ballast Loss, Wattage
 Windows: Width, Height, Sill Height

2). **Scheduled fields (in order):** This list determines what information appears in the schedule. Each parameter listed here equals a column in the schedule. The top item is the first column on the left in the schedule. Compare the order of parameters in Figure 5-1.6 with the schedule in Figure 5-1.8.

 These parameters can be reordered using the **Move Up** and **Move Down** buttons below; you cannot drag the parameters. It is also possible to **Edit** and **Delete** from parameters; however, most built-in parameters cannot be deleted nor edited.

 FYI: Edit *allows you to change the parameter* Name, *select a different* Group *in which the parameter should fall, and select which* Categories *this parameter should be available in.*

3). **Custom and Calculated Parameters, plus Combine Parameters:** When you have information that needs to be stored in an element but there is no appropriate parameter, you can create one. You will see two examples, custom and calculated, on the next two pages. A newer option is to combine multiple parameters into a single column – for example, a table width and depth could be listed as 4'-0" x 6'-0".

4). **Select available fields from:** In the example shown, we are looking at a **Furniture Schedule**. Thus, we are able to select *Available Fields* from the *Furniture* category. You may also change this drop-down list to *Room*. This allows you to list which room number and room name each piece of furniture falls in.

 FYI: Changing the drop-down from Furniture *to* Room *also changes the fields that appear in the* Available fields *list above.*

Creating a Parameter:

When you click the **Add Parameter...** button (Figure 5-1.6), the *Parameter Properties* dialog opens (Figure 5-1.9). Here you specify the following:

Project or Shared parameter: The main thing you need to know is if you want the information stored in this parameter to appear in a tag, it has to be a *Shared Parameter*.

Instance or Type parameter: Do you want the value entered to be unique for each instance, or the same? You might want the "Chair Fabric" to be an instance because the same model number can have different colors but the "Recycled Content" to be the same value per model number. To state it another way: if you have four chairs in the model, two may have a red fabric and two blue, but they all have the same **.35** value for recycled content. Changing the recycled content value for one chair automatically updates the other three.

Parameter Name: Be sure to use a simple name everyone can understand. Do not use dashes as they are seen as a minus sign in formulas.

Discipline: *Common*, *Structural*, and *Electrical* are the options in Revit. This determines what is listed in the next drop-down list: *Type of Parameter*.

FIGURE 5-1.9 Parameter Properties; creating a new parameter

Type of Parameter: This option requires you to know what type of information you will be storing in the parameter. Will it be Currency, Volume, Slope, URL, Material, etc.? **This is the one thing in this dialog you cannot change once you click OK.** If it needs to change, you have to delete the parameter and recreate it. Selecting *Text* is the most flexible as it can hold words and numbers. But, selecting something specific, such as *Integer*, will prevent someone from entering bad data (e.g., a word or a decimal value).

Group Parameter Under: This simply determines which heading the parameter falls under in the *Properties* window. Looking back at Figure 5-1.4, you see three group headings: *Construction*, *Dimensions* and *Identity Data*.

Creating parameters is somewhat like "graphical" computer programming. This is part of what makes up the "I" in BIM - **B**uilding **I**nformation **M**odeling.

Creating a Calculated Value:

Whereas *Parameters* show up in the *Properties Palette* and schedules, also in tags when using shared parameters, **Calculated Values only appear in schedules.**

A *Calculated Value* can be used for many situations. Here are just a few examples:

- Calculate sales tax
- List percentage difference between actual and programmed square footage
- Determine occupant load for a room based on building code

FIGURE 5-1.10 Creating a calculated value

Parameters used in a formula are *case sensitive*. That means they need to be typed exactly as they appear in the project. To help avoid making a mistake, you can click the "..." button to the right of *Formula* to select from a list of available parameters, which can include other *Calculated Values*.

Basic mathematical formulas may be used. Here are a few examples:

Parameter A * Parameter B
Parameter A + Parameter B
Parameter A - Parameter B
(Parameter A + Parameter B + Parameter C) * Parameter D

It is also possible to use constant values:

Parameter A * .1 *or* Source_Distance - 500

Sometimes you need to strip the feet and inches from a parameter so it can be used in a formula. For example, you cannot multiply 5'-6" x $5 because Revit is confused by the feet and inches, but 5.5 x $5 works. Here is how that is done:

Parameter X = 10'-6" (Parameter X is a *Length* parameter type)
Parameter Y = Parameter X / 1' = 10.5 (Parameter Y is a calculated value set to *Number*)

Here is a more complicated set of *Calculated Values* used in a schedule to calculate the occupant load for each space in the building; focus on the mathematical expressions rather than the specific example at this time:

Control = if(OLF_Calc > OLF_Integer, 1, 0)
Occ_Load_Calc = if(Control = 1, OLF_Calc + 1, OLF_Integer)
OLF_Integer = (Area / SF_Per_Occupant) / 1 SF
OLF_Calc = ((Area / SF_Per_Occupant) / 1 SF)

To see an example of what we have covered thus far, notice the drawing below (Figure 5-1.11). The drawing depicts two desks and two chairs in a model. For this example, this is all the furniture in the entire model at this time.

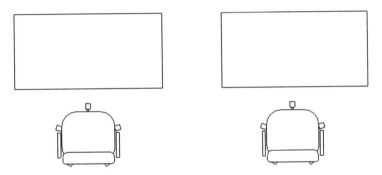

FIGURE 5-1.11 Four furniture elements in a Revit file

<Furniture Schedule>						
A	B	C	D	E	F	G
Type Mark	Model	Manufacturer	Description	Cost	Recycled Content	Non Recycled Con
D1	SC2401F	STEELCASE	OFFICE DESK	2550.00	0.35	0.65
D1	SC2401F	STEELCASE	OFFICE DESK	2550.00	0.35	0.65
C1	LEAPV2	STEELCASE	OFFICE CHAIR	1325.00	0.1	0.9
C1	LEAPV2	STEELCASE	OFFICE CHAIR	1325.00	0.1	0.9

FIGURE 5-1.12 Furniture elements automatically appear in the furniture schedule

The schedule we talked about creating, and is shown in Figure 5-1.8, now has four rows which relate to the elements placed in the model (Figure 5-1.12). Additionally, a new *Parameter* was created to track **Recycled Content**. A *Calculated Value* was created to determine the **Non Recycled Content**.

The width of a column may need to be adjusted to see the title (as in the example above) or the contents of the cells. Simply place your cursor between the columns and drag. Later we will see how the column heading in the schedule can be different from the actual parameter name (e.g., Manufacturer can be abbreviated to MFR.).

One last comment about the *Fields* tab: notice the **Include elements in linked files option**. When you have other Revit models linked into your model, structural, mechanical and electrical, etc., you can check the box in the lower left to be able to schedule elements located in the linked models.

Schedule Properties: Filter Tab

Schedules may be filtered by one or more parameter criteria. In the example below (Figure 5-1.13), the furniture schedule is being filtered by any *Type Mark* parameter which begins with the letter "C." In our two chairs and two desks example drawing, the desks are now hidden because their *Type Mark* begins with "D." This filter assumes the *Type Mark* for all chairs will begin with the letter "C." Obviously, someone could key in the wrong value for *Type Mark* and create a problem. See the filtered result in Figure 5-1.14.

It should be pointed out that each additional filter condition added makes the final result more restrictive. Note the "and" next to the 2nd, 3rd and 4th filters (not an "or").

Revit will only filter by a parameter that is added to the schedule. Looking back at Figure 5-1.6 we cannot filter by furniture located on Level 1 because the parameter *Level* is not in the schedule, even though it is "available" in the model.

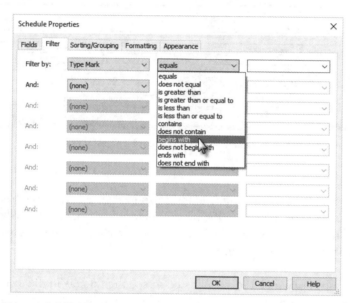

FIGURE 5-1.13 Schedule Properties: Filter tab

	A	B	C	D	E	F	G
	Type Mark	Model	Manufacturer	Description	Cost	Recycled Content	Non Recycled Con
	C1	LEAPV2	STEELCASE	OFFICE CHAIR	1325.00	0.1	0.9
	C1	LEAPV2	STEELCASE	OFFICE CHAIR	1325.00	0.1	0.9

<Furniture Schedule>

FIGURE 5-1.14 Furniture schedule filtered for chairs

Having a schedule, such as this example, which just reports chairs, does not mean there is no way to now schedule desks. It is possible to have multiple "furniture" based schedules: one schedule for each level, one for chairs, one for desks, one for estimating, etc. Each furniture based schedule would just be reporting a different set of parameters from those available.

Schedule Properties: Sorting/Grouping Tab

There is a lot more to the *Sorting/Grouping* tab than meets the eye. As the name implies, we can control sorting. Looking back at Figure 5-1.12, we can see the *Type Marks* are not sorted alphabetically. In Figure 5-1.15, we have told Revit to sort by *Type Mark* in ascending order. The result is seen in Figure 5-1.16; notice the *Type Marks* are in ascending order now.

> *FYI:* Note that we have turned off the Filtering applied on the previous page.

FIGURE 5-1.15 Schedule Properties: Sorting/Grouping tab

A	B	C	D	E	F	G
Type Mark	Model	Manufacturer	Description	Cost	Recycled Content	Non Recycled Con
C1	LEAPV2	STEELCASE	OFFICE CHAIR	1325.00	0.1	0.9
C1	LEAPV2	STEELCASE	OFFICE CHAIR	1325.00	0.1	0.9
D1	SC2401F	STEELCASE	OFFICE DESK	2550.00	0.35	0.65
D1	SC2401F	STEELCASE	OFFICE DESK	2550.00	0.35	0.65
Grand total: 4						

<Furniture Schedule>

FIGURE 5-1.16 Furniture schedule sorted by parameter "Type Mark"

In a sense, when a list is sorted it is grouped, but what Revit means by *Grouping* relates to the **Header**, **Footer** and **Blank Line** check boxes. The two images below show how we can *Group* by the floor level and then *Sort* by the *Type Mark*. The next page has more information.

FIGURE 5-1.17 Schedule Properties: Sorting/Grouping tab revised

	A	B	C	D	E	F	G
	Type Mark	Model	Manufacturer	Description	Cost	Recycled Content	Non Recycled Con
Level 1							
	C1	LEAPV2	STEELCASE	OFFICE CHAIR	1325.00	0.1	0.9
	C1	LEAPV2	STEELCASE	OFFICE CHAIR	1325.00	0.1	0.9
	D1	SC2401F	STEELCASE	OFFICE DESK	2550.00	0.35	0.65
	D1	SC2401F	STEELCASE	OFFICE DESK	2550.00	0.35	0.65
Level 1: 4							
Level 2							
	C1	LEAPV2	STEELCASE	OFFICE CHAIR	1325.00	0.1	0.9
	C1	LEAPV2	STEELCASE	OFFICE CHAIR	1325.00	0.1	0.9
	C1	LEAPV2	STEELCASE	OFFICE CHAIR	1325.00	0.1	0.9
	D1	SC2401F	STEELCASE	OFFICE DESK	2550.00	0.35	0.65
	D1	SC2401F	STEELCASE	OFFICE DESK	2550.00	0.35	0.65
	D1	SC2401F	STEELCASE	OFFICE DESK	2550.00	0.35	0.65
Level 2: 6							
Grand total: 10							

\<Furniture Schedule - Group by Level\>

FIGURE 5-1.18 Furniture schedule grouped by parameter "level"

Of course, for this example, three more chairs and three more desks have been added to Level 2 in our Revit model. Remember, you don't need to be following along in Revit at this time. In order to sort or group by level, the *Level* parameter had to be added to the **Scheduled Fields (in order)** list on the *Fields* tab. Normally, this would create a new column in the schedule. You will see how this column was hidden when you review the *Formatting* tab.

When *Header* was checked in Figure 5-1.17, a header space was added with the word **Level 1** listed, and another above the Level 2 furniture items. The footer allows you to list totals by level. The default is only the number of elements listed in the schedule per level with a grand total near the very bottom. When you get to the *Formatting* tab you will also learn how to make specific columns add up automatically. For example, what is the total for the *Cost* column?

Finally, for the *Sorting/Grouping* tab, you will look at the **Itemize Every Instance** option at the bottom of the *Sorting/Grouping* tab.

FIGURE 5-1.19 Schedule Properties: Sorting/Grouping tab revised

<Furniture Schedule - Do not itemize every instance>							
A	B	C	D	E	F	G	H
Type Mark	Model	Manufacturer	Description	Count	Cost	Recycled Content	Non Recycled Con
C1	LEAPV2	STEELCASE	OFFICE CHAIR	5	1325.00	0.1	0.9
D1	SC2401F	STEELCASE	OFFICE DESK	5	2550.00	0.35	0.65
Grand total: 10							

FIGURE 5-1.20 Furniture schedule NOT itemizing every instance

The two examples above, Figures 5-1.19 and 5-1.20, show sorting just by *Type Mark* and with *Itemize every instance* turned off. When *Itemize every instance* is turned off, the schedule only shows one row for each element in the model. That is, there is one row for the five chairs and one row for the five desks.

This example in Figure 5-1.20 has had the **Count** parameter added to the *Scheduled Fields* list on the *Fields* tab. Notice how the *Count* column automatically shows the correct number of items. This is a built-in parameter, not a *Calculated Value*. Finally, the Cost parameter still lists the per unit cost rather than a total cost for five units. This may well be the information you are looking for and nothing needs to change. However, in the *Formatting* tab discussion you will learn how to make the *Cost* column report the total for five units.

Schedule Properties: Formatting Tab

There are a number of items which can be controlled from the *Formatting* tab (Figure 5-1.21).

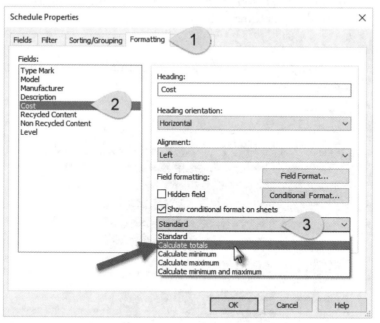

FIGURE 5-1.21 Schedule Properties: Formatting tab

	A	B	C	D	E	F	G	H
	Type Mark	Model	Manufacturer	Description	Count	Cost	Recycled Content	Non Recycled Con
	C1	LEAPV2	STEELCASE	OFFICE CHAIR	5	6625.00	0.1	0.9
	D1	SC2401F	STEELCASE	OFFICE DESK	5	12750.00	0.35	0.65
Grand total: 10						19375.00		

<Furniture Schedule - Do not itemize every instance>

FIGURE 5-1.22 Furniture schedule with Cost parameter set to "Calculate Totals"

Following is a brief description of the various parts of the *Formatting* tab (Figure 5-1.21):

Fields: This is a list of the *Fields* which have been added to the schedule on the *Fields* tab. All the settings on the right relate to the parameters (or fields) selected from this list.

Heading: This is the text that is used at the top of the column in the schedule. By default it matches the parameter name. However, it is possible to change the heading and not affect the parameter name. This is often done to shorten the heading, which saves room as the column does not need to be as wide. For example, **Room Number** might be changed to **RM. #**. This allows the column to be narrower because the numbers are often 3-4 digits (e.g., 1004).

Heading orientation: The header text may be horizontal, which is the default, or vertical.

> *FYI: Changing the heading to vertical is only apparent when the schedule is placed on a sheet.*

FIGURE 5-1.23 Type Mark heading set to be vertical

Alignment: This determines whether the information in a given column is left, right or center justified.

Field Format... (button): This option only applies to *Fields* with numbers. When the project-wide format settings (*Manage → Project Units*) do not work, they may be overridden for this one *Field*. Figure 5-1.24 shows the **Cost** parameter being changed and Figure 5-1.25 shows the result in the schedule.

Conditional Format... (button): The background for a cell can have its color changed if certain conditions exist. For example, if the number is below 20, make the background red; if the *Type Mark* is blank, make the background yellow, etc. The background color does not appear on the sheet, nor does it print.

FIGURE 5-1.24 Field Format options

Hidden field (check box): Figure 5-1.18 requires the *Level* field be in the schedule for grouping, but we do not want to also see a column listing the schedule as this would be redundant. Thus, we select the *Level* field and set it to be hidden.

> *FYI: You may also right-click in a column and see the hidden column options.*

Cost
$6,625
$12,750

FIGURE 5-1.25 Field Format results

Show conditional format on sheets (check box): This will make the filled cells, as defined via the *Conditional Format* button just discussed, actually print. If this is not checked, the filled cell will still appear in the schedule view, but it will not appear nor print on the sheet it is placed on.

Calculate totals (check box): This option does two things. If *Itemize every instance* is not checked, any fields set to **Calculate totals** will display the number times the count. Also, when a *Footer* with totals, or *Grand totals*, is selected on the *Sorting/Grouping* tab, a total value will appear below the column when **Calculate totals** is checked for that field or column (see Figure 5-1.22).

Multiple Values indication: This controls what appears in a cell when there is more than one value to show. This can happen when *Itemize Every Instance* is toggled off.

Schedule Properties: Appearance Tab

This last tab in the *Schedule Properties* dialog has several settings which are very straight forward. Following are a few things to note:

- *Build schedule*, Top-down, Bottom-up is only available for revision schedules.
- Some appearance settings do not show until placed on a sheet.
- Text options are based on text styles defined in the Revit project.

FIGURE 5-1.26 Schedule Properties: Appearance tab

Unfortunately, you cannot print a schedule unless it is on a sheet (see Exercise 18-1 for steps on how to place a view on a sheet). It is possible to export a schedule so it can be used, viewed or printed in a spreadsheet or database application. You must be in the schedule view, not the sheet, and then select **File Tab → Export → Reports → Schedules**.

In this chapter you will leverage much of the information just covered when developing the *Space Program*. Later in the book you will create a few more schedules: Floor Finish Schedule, Furniture Schedule, Room Finish Schedule, and a Sheet Index. All of these schedules are live tabular views of the model. In fact, it is possible to delete items in the model from the schedule.

There are a few things about phasing and *Ribbon* options related to schedules that have not been covered. Some of the *Ribbon* options will be covered later. It is recommended that you refer back to this information when needed.

Exercise 5-2:
Law Office Space Program

This section will walk you through the creation of a **Space Program**. This is a schedule based on *Room* elements, similar to a door schedule being based on *Door* elements. The schedule will list all the required rooms, what size they should be, what size they actually are, etc. All this will be done before modeling the first interior wall.

Room schedules are unique in that they can list something that has not been modeled yet. This allows you to develop the *Space Program* and then, later, assign each *Room* element to a space, i.e., an area enclosed by walls. Once all the *Rooms* have been used, the design has met the program requirements, assuming the square footages and other project requirements are met. You will learn how a scheduled *Room* is assigned to a space in the next chapter.

Some interior designers begin to develop the space program in a spreadsheet program such as Microsoft Excel. Others start directly in Revit to limit duplicated effort. There are many things Excel can do which Revit cannot, but for the purposes of this tutorial we will stick to Revit.

Creating a Schedule

The first thing you need to do is create a schedule. This process is based on the material covered in the previous section of this chapter.

1. **Open** your *Law Office* project. This file should have all the *Materials* in it from the previous chapter.

2. Select **View → Schedules → Schedule/Quantities**.

3. In the *New Schedule* dialog, do the following (Figure 5-2.2):

 a. *Category:* **Rooms**

 b. *Name:* **Space Program**

 c. *Selected:* **Schedule Building Components**

 d. *Phase:* **New Construction**

FIGURE 5-2.1
Schedule creation

4. Click **OK**.

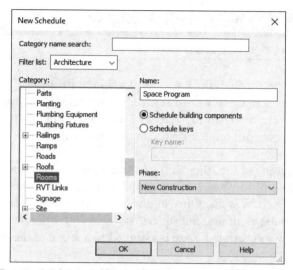

FIGURE 5-2.2 Schedule creation: new schedule dialog

You are now in the *Schedule Properties* dialog. Here you select the parameters you want on your schedules and specify how they should appear. You can also create new parameters if needed.

5. On the *Fields* tab, select **Name** on the left and then click the **Add Parameter(s)** button in the middle (i.e. green arrow pointing to the right).

TIP: You can also double-click on Name to move it to the right.

You now have one *Field* added to the schedule (Figure 5-2.3).

FIGURE 5-2.3 Schedule creation: adding fields

6. Add the following *Available fields* to the schedule:

 a. **Area**

 b. **Comments**

 c. **Level**

At this time you will complete the schedule just to see how it will look with the few steps taken thus far. This will also help the reader know how to get back into the *Schedule Properties* dialog should they need to stop working and return to it later. Following the previous steps would create another schedule rather than edit the existing one.

7. Click **OK** to create the schedule.

You now have a schedule named "Space Program" appearing in the *Project Browser*, under the heading *Schedules/Quantities*. You should see the schedule on-screen as shown in Figure 5-2.4. The columns start with a default width, which you will adjust later. Next you will add a custom parameter and a calculated parameter.

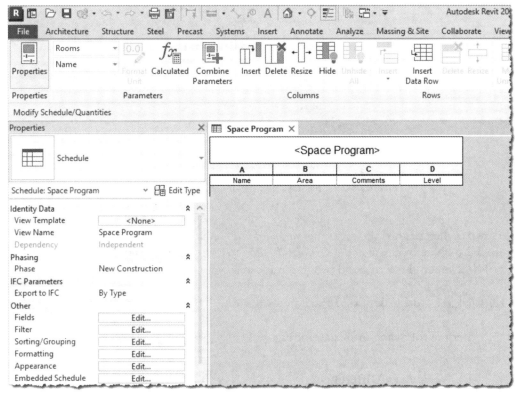

FIGURE 5-2.4 Schedule created

To get back into the *Schedule Properties* dialog you need to be "in," or viewing, the schedule view.

8. With the *Space Program* schedule view open and visible on the screen, select the **Edit** button next to *Fields*, in the *Properties Palette*.

 FYI: All five Edit buttons open the same dialog box, just to a different tab.

9. On the *Fields* tab, click the **New Parameter...** icon.

10. Create a new *Project Parameter*, entering the information shown in Figure 5-2.5.

11. Click **OK** to create the new parameter.

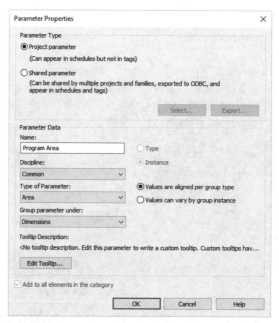

FIGURE 5-2.5 Creating a new parameter

The new parameter is now listed in the *Scheduled fields* column (Figure 5-2.6). Next you will add a *Calculated Value* to help keep track of how much the actual and programmed areas vary.

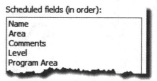

FIGURE 5-2.6
New parameter listed

12. Click the **Add Calculated Parameter** button.

13. Enter the following information (Figure 5-2.7):

 a. *Name:* **Area Difference**

 b. *Discipline:* **Common**

 c. *Type:* **Number**

 d. *Formula:*
 (Area/Program Area) - 1

FIGURE 5-2.7 Creating a calculated value

TIP: Clicking the "..." button to the right of the Formula *box allows you to pick from a list of available parameters. This is helpful because it must be typed exactly as it exists and is case-sensitive.*

14. Click **OK**.

15. Use the **Move Up** and **Move Down** buttons to make the *Scheduled Fields* appear in this order:

- Name
- Area
- Program Area
- Area Difference
- Comments
- Level

You will not see anything happen with these new parameters until rooms are added to the project. At that time you will be able to enter the desired square footage for each room.

16. Switch to the **Sorting/Grouping** tab.

First you will sort by level and then by room name.

17. Make the following changes to the *Schedule Properties* dialog (Figure 5-2.8).

FIGURE 5-2.8 Setting the sorting and grouping options

18. Switch to the **Formatting** tab.

If you were to finish the schedule now and add two rooms on Level 1 and one room on Level 2, the schedule would look like Figure 5-2.9A. We will use the options on the *Formatting* tab to adjust a few things. For example:

- Adjust the Area Difference column to round to two decimal places
- Adjust the Area Difference column title, making it shorter (thus allowing the column to be narrower)
- Add the total square footage for all the rooms on each level and for the entire building
- Hide the Level column as it is redundant information

colspan=6 align=center			<Space Program>			
A	B	C	D	E	F	
Name	Area	Program Area	Area Difference	Comments	Level	
Level 1						
Room	175 SF	185 SF	-0.054129		Level 1	
Room	175 SF	165 SF	0.060522		Level 1	
Level 2						
Room	175 SF	125 SF	0.399889		Level 2	
Grand total: 3						

FIGURE 5-2.9A Schedule prior to formatting edits

First you will make an adjustment to tell Revit you want the *Area* parameter to be totaled. Based on the options set in the *Sorting/Grouping* tab, this total will appear for each level and for the entire building.

19. Select **Area** in the *Fields* column (Figure 5-2.9B).

20. Select the **Calculate totals** option on the right (Figure 5-2.9B).

21. Select **Program Area** and again select the **Calculate totals** option.

22. Select **Area Difference** from the *Fields* list on the left (Figure 5-2.10).

23. Change the **Heading** section from "Area Difference" to just "**Diff**" This allows the column title to be different from the actual parameter name.

FIGURE 5-2.9B Schedule properties: Formatting tab

Next you will change the format of the *Area Difference* value so it shows fewer decimal places and reads as a percentage with a percent symbol. The default format comes from the project-wide unit settings; this may be accessed, and edited, via **Manage → Project Units**.

24. Click the **Field Format...** button.

25. Make the changes shown in Figure 5-2.10 below.

26. Click **OK**.

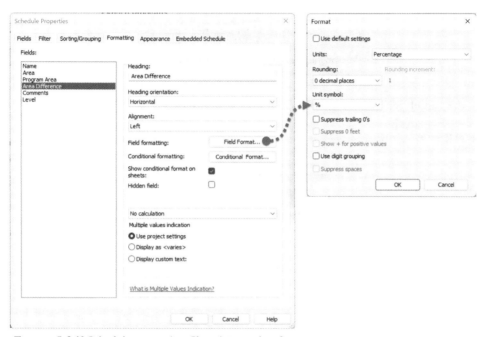

FIGURE 5-2.10 Schedule properties: Changing number format

If the percentage difference between the program area and actual area is more than a certain number we want the cell to change to a bright yellow to alert us to this problem. In the next few steps you will apply a *Conditional Format* to the *Area Difference* cells.

27. With *Area Difference* still selected, click the **Conditional Format...** button (Figure 5-2.10).

28. Make the following edits to the **Conditional Formatting** dialog (Figure 5-2.11):

 a. *Field:* **Area Difference**

 b. *Test:* **Not Between**

 c. *Value:* **-10% and 10%**

 d. *Background Color:* **Yellow** (click the color swatch to select color)

29. Click **OK**.

FIGURE 5-2.11 Schedule properties: adding conditional format

30. Click the **Level** field and select the *Hidden Field* option.

31. Click **OK** to close the *Schedule Properties* dialog and see the changes in the schedule view.

Your schedule would now look similar to Figure 5-2.12 shown to the right, but you have not added any rooms yet. You should see the *Area Difference* title changed to "Diff."

<Space Program>				
A	B	C	D	E
Name	Area	Program Area	Diff	Comments
Level 1				
Room	175 SF	200 SF	-13%	
Room	175 SF	165 SF	6%	
	350 SF	365 SF		
Level 2				
Room	175 SF	125 SF	40%	
	175 SF	125 SF		
Grand total: 3	525 SF	490 SF		

FIGURE 5-2.12 Schedule view with formatting applied

You will make one more format change before adding rooms. You will group the two area columns together and edit their heading titles. This will utilize a few tools on the *Ribbon* which have not been covered yet.

32. Open the *Schedule Properties* dialog again and select the **Formatting** tab.

33. Select *Area*, on the left, and change the *Heading* to **Actual**.

34. Select *Program Area*, on the left, and change the *Heading* to **Program**.

35. Click **OK**.

We now have one column labeled "Actual" and another "Program." Next you will add a combined heading over these two columns which reads "Area."

To create a **Grouped Header**, you must select the current headers in a specific way. Follow the instructions for the next step very closely.

36. Click and drag your mouse from the **Program** cell over to the **Actual** cell and _then_ let go of the mouse. You must click in the cells with these words (i.e., the header cells) not the cells below (i.e., the data cells). Of course, you should not have any data cells yet as you have not created or placed any rooms.

37. With the two header cells selected, click the **_Group_** icon from the _Ribbon_, located on the contextual tab _Modify Schedule/Quantities_ (Figure 5-2.14).

38. Click within the new grouped header cell and type **Area** (Figure 5-2.13).

You can add many combinations of grouped headers. You may remove a grouped header by clicking within the cell and selecting the **_Ungroup_** option on the _Ribbon_.

FIGURE 5-2.13 Grouped header added

You are now ready to start adding data rows to the schedule. Each row added will represent an individual room. So, if your design requires four restrooms, which need to be identical in size, you have to create four rows.

When adding rooms before placing them, be sure you have the **_Show_** option selected on the _Ribbon_. If this is not selected, you will not see any new rows because they are not placed in the model yet.

FIGURE 5-2.14 Schedule-related options on the _Ribbon_ when viewing a schedule

39. Click the **Insert → Data Row** tool in the _Rows_ panel (Figure 5-2.14).

You now have a new row, which represents a room in your building (Figure 5-2.15). Notice the default name is "Room" and the actual area says, "Not Placed" because the _Room_ has not been assigned to a space in the model yet.

A	B	C	D	E
		Area		
Name	Actual	Program	Diff	Comments
Not Placed				
Room	Not Placed			
	0 SF	0 SF		
Grand total: 1	0 SF	0 SF		

<Space Program>

FIGURE 5-2.15 New row added

40. Click within the room name cell; change the room name from "Room" to **Reception**.

41. Click within the *Program Area* cell and type **300**.

42. Click within the *Comments* cell and type **Front – first impression that serves as multipurpose space**.

43. Adjust the width of the *Name* and *Comments* column by dragging the vertical lines between the columns; adjust as shown in Figure 5-2.16.

> *TIP: Columns can also be resized using the Resize tool in the Columns panel on the Ribbon. Using this tool, you can type in an exact number for each column.*

A	B		C	D	E
		Area			
Name	Actual		Program	Diff	Comments
Not Placed					
Reception	Not Placed		300 SF		Front – first impression that serves as multipurpose space
	0 SF		300 SF		
Grand total: 1	0 SF		300 SF		

FIGURE 5-2.16 New row modified and columns resized

Now you will take the time to enter the rest of the rooms for the *Law Office* design project. The schedule shown in Figure 5-2.17 was created in Microsoft Excel and is formatted a little differently from what you are working towards in Revit.

44. Enter the remaining rooms as shown in Figure 5-2.17.

 a. Be sure to add one space for the quantity indicated; for example, add two *Small Conference Rooms* because the quantity column indicates two are required.

 b. Add the comments as they appear.

 c. Do not worry about the special headings (i.e., private offices, accessory spaces).

 d. Adjust the column widths as needed to make sure all the information is visible.

 e. Do **not** add any rooms with zero square feet listed as a total.

Our schedule is set to sort by level. However, the *Rooms* have not been assigned to spaces in the model yet, so Revit does not know which level they are on. In addition to sorting by level, it is also possible to sort by department. This would allow the schedule to be organized similar to the one shown in Figure 5-2.17.

Space/Group	QTY	SF	Subtotal	Comments
Conference Center (The front entry)				
Reception	1	300	300	Front first impression that serves as multipurpose space
Waiting / lobby	1	300	300	
Large Conf / Board Room	1	500	500	16 to 24 persons
Medium Conference Room	1	250	250	Seat 8 to 10 persons
Small Conference Room	2	100	200	Office size that could change use if needed. 2 to 4 persons
Impromptu meeting space	2	100	200	Informal interactions spaces that do not disturb others
Public Restroom	2	50	100	One Male one Female
Subtotal	0		1,850	
Office and Workstations (about 1/3 the space utilization)				
Private offices				
Partner / principal office	4	235	940	
Associate office	6	140	840	Partner to associate ratio 1:2
First year associate/paralegal	1	125	125	
Retired partners' office	1	235	235	
Visitor office	1	125	125	
Workstations				
Paralegal	5	81	405	Attorney to paralegals 9:1 to 4:1
Secretary	3	64	192	Attorney to secretary ratio 3-4:1 better 2.5:1 (the command center)
Assistant	2	64	128	
Intern	1	64	64	
Office manager	1	81	81	
Subtotal			3,135	
Accessory Spaces				
Mock courtroom training center	1	450	450	
Lunchroom staff (the café)	1	350	350	On building perimeter natural daylight with higher end finishes, adj to lib.
Library	1	300	300	
Video Conference room	1	0	0	Video conference integrated into all conference rooms
Full time graphic design office	1	140	140	Visualizes the case for arguments
File room	1	250	250	12 - 15 lateral file drawers per attorney
Subtotal			1,490	
Building support services spaces				
Private Restroom M/F pair	2	60	120	
Technology closet	1	100	100	
Janitor closet	1	75	75	
Workroom	1	300	300	
Recycling room	0	100	0	Area designated in mechanical room
Fitness center	1	200	200	
General Storage	1	200	200	
Archive files	0	0	0	Off property at independent storage facility
Subtotal			995	
Total Programmed SF			7,470	
35% circulation			2,615	**Includes mechanical, electrical & elevator**
Total SF			10,085	

FIGURE 5-2.17 Law Office space program

When finished adding the rooms to the schedule, it should look like Figure 5-2.18 below. Double check your total at the bottom to make sure things match up before moving on. It is not really possible to add an "extra" 35% area as shown in the Excel schedule, but you could do the math and add a dummy room with that amount of area – the name of the dummy room could be "35 percent circulation."

<Space Program>				
A	B	C	D	E
	Area			
Name	Actual	Program	Diff	Comments
Not Placed				
Assistant	Not Placed	64 SF		
Assistant	Not Placed	64 SF		
Associate office	Not Placed	140 SF		Partner to associate ratio 1:2
Associate office	Not Placed	140 SF		Partner to associate ratio 1:2
Associate office	Not Placed	140 SF		Partner to associate ratio 1:2
Associate office	Not Placed	140 SF		Partner to associate ratio 1:2
Associate office	Not Placed	140 SF		Partner to associate ratio 1:2
Associate office	Not Placed	140 SF		Partner to associate ratio 1:2
File room	Not Placed	250 SF		12 - 15 lateral file drawers per attorney
First year associate/paralegal	Not Placed	125 SF		
Fitness center	Not Placed	200 SF		
Full time graphic design office	Not Placed	140 SF		Visualizes the case for arguments
General Storage	Not Placed	200 SF		
Impromptu meeting space	Not Placed	100 SF		Informal interactions spaces that do not disturb others
Impromptu meeting space	Not Placed	100 SF		Informal interactions spaces that do not disturb others
Intern	Not Placed	64 SF		
Janitor closet	Not Placed	75 SF		
Large Conf / Board Room	Not Placed	500 SF		16 to 24 persons
library	Not Placed	300 SF		
lunchroom staff (the café)	Not Placed	350 SF		On building perimeter natural daylight with higher end finishes, adj to lib.
Medium Conference Room	Not Placed	250 SF		Seat 8 to10 persons
Mock courtroom training center	Not Placed	450 SF		
Office manger	Not Placed	81 SF		
Paralegal	Not Placed	81 SF		Attorney to paralegals 9:1 to 4:1
Paralegal	Not Placed	81 SF		Attorney to paralegals 9:1 to 4:1
Paralegal	Not Placed	81 SF		Attorney to paralegals 9:1 to 4:1
Paralegal	Not Placed	81 SF		Attorney to paralegals 9:1 to 4:1
Paralegal	Not Placed	81 SF		Attorney to paralegals 9:1 to 4:1
Partner / principal office	Not Placed	235 SF		
Partner / principal office	Not Placed	235 SF		
Partner / principal office	Not Placed	235 SF		
Partner / principal office	Not Placed	235 SF		
Private Restroom Female	Not Placed	60 SF		
Private Restroom Male	Not Placed	60 SF		
Public Restroom	Not Placed	50 SF		One Male one Female
Public Restroom	Not Placed	50 SF		One Male one Female
Reception	Not Placed	300 SF		Front – first impression that serves as multipurpose space
Retired partners office	Not Placed	235 SF		
Secretary	Not Placed	64 SF		Attorney to secretary ratio 3-4:1 better 2.5:1 (the command center)
Secretary	Not Placed	64 SF		Attorney to secretary ratio 3-4:1 better 2.5:1 (the command center)
Secretary	Not Placed	64 SF		Attorney to secretary ratio 3-4:1 better 2.5:1 (the command center)
Small Conference Room	Not Placed	100 SF		Office size that could change use if needed. 2 to 4 persons
Small Conference Room	Not Placed	100 SF		Office size that could change use if needed. 2 to 4 persons
Technology closet	Not Placed	100 SF		
Visitor office	Not Placed	125 SF		
Waiting / lobby	Not Placed	300 SF		
Workroom	Not Placed	300 SF		
	0 SF	7470 SF		
Grand total: 47	0 SF	7470 SF		

FIGURE 5-2.18 Law Office space program in a Revit schedule

Notice how the schedule layout can significantly change if you uncheck "Itemize every instance" and add the built-in *Count* parameter via the *Schedule Properties* dialog (Figure 5-2.19). Because the *Program Area* was set to *Calculate Totals*, that column actually lists the total for the quantity listed. You should give this a try but be sure to make a copy of the schedule first; right-click the title in the *Project Browser* and select *Duplicate View* → *Duplicate*.

	A	B	C	D	E	F
			Area			
	Name	Count	Actual	Program	Diff	Comments
Not Placed						
	Assistant	2	0 SF	128 SF		
	Associate office	6	0 SF	840 SF		Partner to associate ratio 1:2
	File room	1	0 SF	250 SF		12 - 15 lateral file drawers per attorney
	First year associate/paralegal	1	0 SF	125 SF		
	Fitness center	1	0 SF	200 SF		
	Full time graphic design office	1	0 SF	140 SF		Visualizes the case for arguments
	General Storage	1	0 SF	200 SF		
	Impromptu meeting space	2	0 SF	200 SF		Informal interactions spaces that do not disturb others
	Intern	1	0 SF	64 SF		
	Janitor closet	1	0 SF	75 SF		
	Large Conf / Board Room	1	0 SF	500 SF		16 to 24 persons
	library	1	0 SF	300 SF		
	lunchroom staff (the café)	1	0 SF	350 SF		On building perimeter natural daylight with higher end finishes, adj to lib
	Medium Conference Room	1	0 SF	250 SF		Seat 8 to10 persons
	Mock courtroom training center	1	0 SF	450 SF		
	Office manger	1	0 SF	81 SF		
	Paralegal	5	0 SF	405 SF		Attorney to paralegals 9:1 to 4:1
	Partner / principal office	4	0 SF	940 SF		
	Private Restroom Female	1	0 SF	60 SF		
	Private Restroom Male	1	0 SF	60 SF		
	Public Restroom	2	0 SF	100 SF		One Male one Female
	Reception	1	0 SF	300 SF		Front – first impression that serves as multipurpose space
	Retired partners office	1	0 SF	235 SF		
	Secretary	3	0 SF	192 SF		Attorney to secretary ratio 3-4:1 better 2.5:1 (the command center)
	Small Conference Room	2	0 SF	200 SF		Office size that could change use if needed. 2 to 4 persons
	Technology closet	1	0 SF	100 SF		
	Visitor office	1	0 SF	125 SF		
	Waiting / lobby	1	0 SF	300 SF		
	Workroom	1	0 SF	300 SF		
			0 SF	7470 SF		
Grand total: 47			0 SF	7470 SF		

FIGURE 5-2.19 Law Office space program revised

Sorting by Department

Before moving on, the last thing you will do is adjust the schedule to sort by *Department* rather than *Level*. This will show you how easy it is to duplicate a schedule and then adjust it. You will use the department field to enter the group header text shown in Figure 5-2.17: Conference Center, Accessory Spaces, etc.

45. In the *Project Browser*, right-click on the **Space Program** schedule and select *Duplicate View* → *Duplicate*.

46. Right-click on the new view, **Copy of Space Program**, and select *Rename*.

47. Enter **Space Program by Department** and click **OK**.

You now have a new room-based schedule, but it is looking at the same 47 rooms that exist in the model. You will change how this information is sorted in the new schedule view.

48. Click *Edit* next to *Fields* in the *Properties Palette*.

49. Add **Department** to the list and remove *Level*.

50. Adjust *Sorting/Grouping* to <u>sort by</u> **Department** and <u>then by</u> **Name**; remove *Level*.

51. Check *Header* and *Footer*, count and totals, for *Department*.

52. Click **OK**.

Now, as you begin to enter text into the *Department* field, the schedule will begin to sort by that value.

53. Locate the *Room* named **Reception** and enter the following into the *Department* column: **Conference Center (The front entry)**.

54. Press **Enter** on the keyboard.

Notice the room moved to the bottom of the list and has a heading which equals the department name. New groups will be added as unique department names are entered.

Visitor office	Not Placed	125 SF			
Waiting / lobby	Not Placed	300 SF			
Workroom	Not Placed	300 SF			
46	0 SF	7170 SF			
Conference Center (The front entry)					
Reception	Not Placed	300 SF		Front – first impression that serves as multipurpose space	Conference Center (The front entry)
1	0 SF	300 SF			
Grand total: 47	0 SF	7470 SF			

FIGURE 5-2.20 Law Office space program grouped by department

55. Locate **Waiting / lobby** and click within its *Department* field.

You do not have to type the same word(s) again. Rather, you may pick it from a drop-down list.

56. Click the down-arrow that appeared to the right, when you clicked in the *Department* field, and select **Conference Center (The front entry)**.

57. Per Figure 5-2.17 you should enter the remaining department labels. The list below summarizes the departments to be used:

 a. Conference Center (The front entry)

 b. Office and workstations

 c. Accessory Spaces

 d. Building support services spaces

58. Edit the *Department* field to be **hidden** via the *Formatting* tab in *Schedule Properties*.

When finished, your schedule should look similar to Figure 5-2.21 below. Notice the field "Department" is not really being used in its full literal meaning. We are repurposing this parameter a bit here. We could have created a new parameter for this, but that is not really necessary. Verify your totals with Figure 5-2.17.

<Space Program by Department>				
A	B	C	D	E
	Area			
Name	Actual	Program	Diff	Comments
Accessory Spaces				
File room	Not Placed	250 SF		12 - 15 lateral file drawers per attorney
Full time graphic design office	Not Placed	140 SF		Visualizes the case for arguments
library	Not Placed	300 SF		
lunchroom staff (the café)	Not Placed	350 SF		On building perimeter natural daylight with higher end finishes, adj to lib.
Mock courtroom training center	Not Placed	450 SF		
5	0 SF	1490 SF		
Building Support Services Spaces				
Fitness center	Not Placed	200 SF		
General Storage	Not Placed	200 SF		
Janitor closet	Not Placed	75 SF		
Private Restroom Female	Not Placed	60 SF		
Private Restroom Male	Not Placed	60 SF		
Technology closet	Not Placed	100 SF		
Workroom	Not Placed	300 SF		
7	0 SF	995 SF		
Conference Center (The front entry)				
Impromptu meeting space	Not Placed	100 SF		Informal interactions spaces that do not disturb others
Impromptu meeting space	Not Placed	100 SF		Informal interactions spaces that do not disturb others
Large Conf / Board Room	Not Placed	500 SF		16 to 24 persons
Medium Conference Room	Not Placed	250 SF		Seat 8 to10 persons
Public Restroom	Not Placed	50 SF		One Male one Female
Public Restroom	Not Placed	50 SF		One Male one Female
Reception	Not Placed	300 SF		Front – first impression that serves as multipurpose space
Small Conference Room	Not Placed	100 SF		Office size that could change use if needed. 2 to 4 persons
Small Conference Room	Not Placed	100 SF		Office size that could change use if needed. 2 to 4 persons
Waiting / lobby	Not Placed	300 SF		
10	0 SF	1850 SF		
Office and Workstation				
Assistant	Not Placed	64 SF		
Assistant	Not Placed	64 SF		
Associate office	Not Placed	140 SF		Partner to associate ratio 1:2
Associate office	Not Placed	140 SF		Partner to associate ratio 1:2
Associate office	Not Placed	140 SF		Partner to associate ratio 1:2
Associate office	Not Placed	140 SF		Partner to associate ratio 1:2
Associate office	Not Placed	140 SF		Partner to associate ratio 1:2
Associate office	Not Placed	140 SF		Partner to associate ratio 1:2
First year associate/paralegal	Not Placed	125 SF		
Intern	Not Placed	64 SF		
Office manger	Not Placed	81 SF		
Paralegal	Not Placed	81 SF		Attorney to paralegals 9:1 to 4:1
Paralegal	Not Placed	81 SF		Attorney to paralegals 9:1 to 4:1
Paralegal	Not Placed	81 SF		Attorney to paralegals 9:1 to 4:1
Paralegal	Not Placed	81 SF		Attorney to paralegals 9:1 to 4:1
Paralegal	Not Placed	81 SF		Attorney to paralegals 9:1 to 4:1
Partner / principal office	Not Placed	235 SF		
Partner / principal office	Not Placed	235 SF		
Partner / principal office	Not Placed	235 SF		
Partner / principal office	Not Placed	235 SF		
Retired partners office	Not Placed	235 SF		
Secretary	Not Placed	64 SF		Attorney to secretary ratio 3-4:1 better 2.5:1 (the command center)
Secretary	Not Placed	64 SF		Attorney to secretary ratio 3-4:1 better 2.5:1 (the command center)
Secretary	Not Placed	64 SF		Attorney to secretary ratio 3-4:1 better 2.5:1 (the command center)
Visitor office	Not Placed	125 SF		
25	0 SF	3135 SF		
Grand total: 47	0 SF	7470 SF		

FIGURE 5-2.21 Law Office space program grouped by department

Exercise 5-3:
Developing Bubble Diagrams

Bubble Diagram

The *Bubble Diagram* is probably one of the most used diagrams in the Architectural Profession when it comes to design layout of a building. It is a quick way to start thinking about where various rooms should go in a building. As you can imagine, a circle is quicker to draw than a square or rectangular shape – no corners!

The circles are not so much to scale as they are proportionally correct relative to each other. So, if a Living Room needed to be 800sf and a Dining Room 400sf, your circle for the living room would be twice the size of the dining room. All the circles should be labeled so you can refer back to them later and make sense of them.

Arrows can also be added, as shown in the sketch that follows (Figure 5-3.1), to indicate relationships of circulation. So an arrow connecting two spaces (i.e., circles) may indicate the two spaces need to be near each other, like the *Visitor Office* and the *Mock Courtroom*, or it may indicate a desire to walk from one room directly into another, by a door or opening in the wall between the two rooms. This is helpful when you are trying several options; you may be tempted to move a room that should be near another room that you were not planning to move, which would create a problem.

The example provided has been developed to be presented to a client. Color and text have been added.

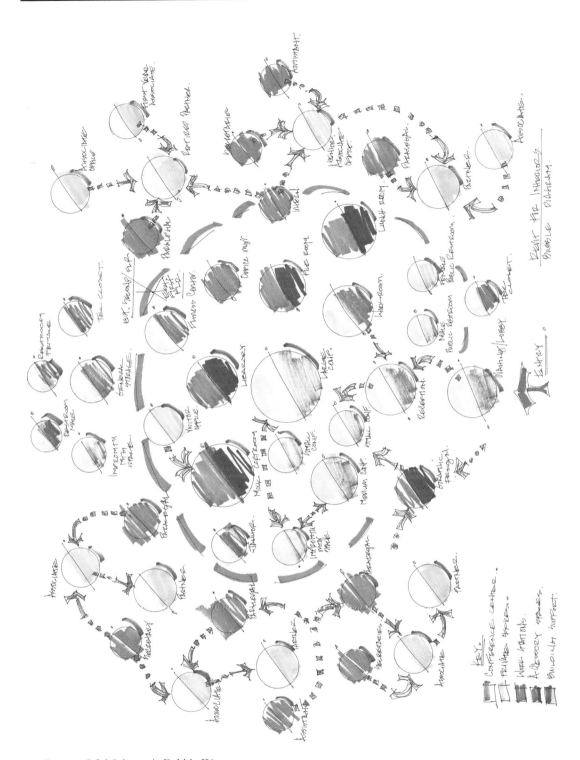

FIGURE 5-3.1 Schematic Bubble Diagram

Block Diagram

This is a great diagrammatic drawing to show the client and get their initial impression, because 1) it does not take too much time so if big changes are needed it will not be a problem, and 2) the client will not think the drawing is done and won't be afraid to request changes. A computer drawn plan loaded with dimensions would give the impression that the drawings are set in stone and cannot be changed, causing the client to possibly compromise on a change that otherwise might not have been a problem.

Some would prefer to show a drawing like this to a client rather than the more abstract sketch shown on the previous page – it really comes down to preference. On larger projects where there are more client meetings, you may show both: the *Bubble Diagram* at an earlier meeting and then this loosely sketched *Block Diagram* at a follow up meeting.

FIGURE 5-3.2 Level 1 Block Diagram

FIGURE 5-3.3 Level 2 Block Diagram

Related Revit Options

There are a number of ways to begin to develop block diagrams in Revit. It is possible to create a rectangular *Family* which would be copied around and adjusted in size for different spaces in the building. This can be printed in plan and 3D views. You can create simple *Rooms* in plan by just drawing *Room Separation* lines, but *Rooms* do not appear in 3D views.

Revit also has a ***Massing*** tool which can be used to block out overall building shapes. You can then set the *Mass* to automatically represent floors at each *Level*; you actually select the desired levels from a list. These floor areas can then be scheduled. Again, this is done by creating one or more *Masses*, selecting *Levels* and then creating a **Mass Floor Schedule**. You never actually create a floor element. *Floors, Walls and Curtain Walls* can be created later, based on the faces of the *Mass* elements.

Tracing Paper

Even in our highly digital world, sketching is still an important part of the process for many designers.

A favorite "tool" in the architectural designer's toolbox is *tracing paper*. Tracing paper is a quick and easy way to clean up a drawing or start another design option based on a previous option; this would be a **design iteration**. Tracing paper is very lightweight and easy to see through, which makes tracing the page below it easy. If you did not have tracing paper, you would need to use a light table, which some designers have but is usually inconveniently away from their desk as it is often shared by the office.

Tracing paper is available in two forms: by the sheet or on a roll. Most designers use tracing paper from a roll, often having a handful of roll widths on hand. The nice thing about a roll is that you can tear off a piece of paper of any length that you need. This is especially useful when working on two-point perspectives and your vanishing points are widely spaced. The final sketch-over can then be done on a smaller piece of sketching paper (or higher quality paper) where the actual vanishing points do not need to appear.

The image below shows the tracing paper concept. It is handy to use drafting tape to hold the various overlays into place to keep them from moving around.

Self-Exam:

The following questions can be used as a way to check your knowledge of this lesson. The answers can be found at the bottom of the next page.

1. Changing an *Instance Parameter* will affect all instances of the selected type. (T/F)

2. When creating a formula, you do not need to worry about uppercase or lowercase letters. (T/F)

3. You can create custom parameters in schedules. (T/F)

4. The actual area displays _____ for a room which has not yet been placed in the model.

5. When two or more column headers are selected, you can select *Group* on the *Ribbon* to create a header above them. (T/F)

Review Questions:

The following questions may be assigned by your instructor as a way to assess your knowledge of this section. Your instructor has the answers to the review questions.

1. When creating a schedule, you have to tell Revit which category to look for elements in. (T/F)

2. Calculated values only appear in schedule, not tags. (T/F)

3. When creating a formula, Revit lets you select from a list of Fields/ Parameters. (T/F)

4. The cost parameter cannot be used in formulas. (T/F)

5. A room finish schedule can tell you what the material assigned to a wall is. (T/F)

6. A schedule can be filtered to only show items which cost more than $100. (T/F)

7. To only show one of each type within the model, uncheck _____ _____ in the *Schedule Properties*.

8. Revit has a *Massing* tool which can be used to block out overall building shapes. (T/F)

9. _____ can be used to change the color of a cell.

10. Schedules cannot be sorted by *Level*. (T/F)

Notes:

Lesson 6
Floor Plans:

In this chapter you will begin to develop the floor plan layout using Autodesk® Revit®. This mainly involves adding *Walls* and placing *Rooms*. Before you begin creating these elements, the first two sections of this chapter will provide an introduction to *Walls* and *Rooms*. When finished with this chapter, all the *Rooms* created in the previous chapter will be accounted for.

Exercise 6-1:
Walls

In Chapter 3 you quickly learned the basics of adding walls in a model. Here is an overview of those basic steps:

- Select the *Wall* tool
- Select the **wall type** from the *Type Selector*
- Set the desired **wall height** on the *Options Bar*
- Set the **Location Line**, also on the *Options Bar*
- Pick points within the model

Now we will delve in a little deeper and learn how to better control the appearance of walls.

Top and Base Constraints

First, we will look at how the top and bottom positions of the wall are located. The bottom of the wall has to be connected to a level, any level. By default, a new wall is connected to the level associated to the view you are working in; that is, the Level 1 Floor Plan view is associated (i.e., connected) to the Level 1 datum.

Every wall instance in a model has a *Base Constraint* and a *Top Constraint*. When a wall is selected, these values may be seen in the *Properties Palette* (Figure 6-1.1). As the name implies, the top or base can be constrained, such that the wall will move, grow or shrink with a level datum change.

FIGURE 6-1.1 Properties Palette with wall selected

In addition to top and base constraints, you also have a top and base offset. The *Top Offset* and *Base Offset* are offsets which remain relative to a level (which has been specified in the *Top Constraint* or *Base Constraint* field). You will see examples of all of this now, which will make things crystal clear!

FIGURE 6-1.2 Walls with various top and bottom constraints and offsets

The image above (Figure 6-1.2) shows several instances of the same wall. Each has a variety of adjustments for the four parameters under consideration: *Top Constraint, Top Offset, Base Constraint* and *Base Offset.* The *Properties Palette* shown in Figure 6-1.1 relates to Wall Number **2** above. The base of the wall is constrained to Level 1 and the top to Level 2. Notice no offset has been entered for the top or base. Thus the top and base of the wall align with the levels in question.

Wall Numbers **3-7** all have a *Base* and/or *Top Offset* applied; the dimensions shown indicate the offset amount. Walls 3 and 4 have a negative offset entered for the *Top Offset.* Wall number 6 has a negative *Base Offset* entered.

The final variation is Wall Number **1**. The *Top Constraint* is set to **Unconnected** rather than a level. This setting allows the top of the wall to be fixed, relative to the *Base Constraint.*

One would use this for an existing wall or a wall that only extended up a few inches above a ceiling. When *Unconnected* is selected for a wall, the *Top Offset* becomes inactive (i.e., grayed out) and the *Unconnected Height* parameter becomes editable.

These wall constraints are part of what makes Revit a parametric building modeler. We are editing parameters (i.e., parametric) which directly or indirectly control geometry. For example, if Level 2 were changed from 8'-0" to 5'-0", all the walls would adjust except for Wall Number **1**. An example of this is seen in Figure 6-1.3 below.

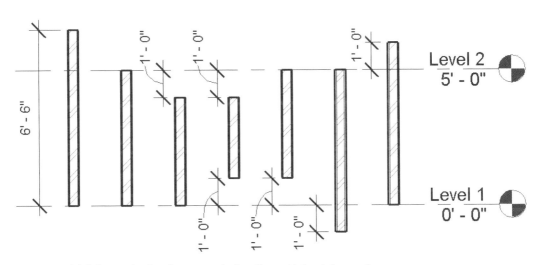

FIGURE 6-1.3 Constrained walls automatically adjust with level datum change

FYI: *Revit has two types of properties:* **Instance** *(shown in the* Properties Palette*) and* **Type** *(seen via the* Edit Type *button):*

Instance Properties: This is information that can vary by instance (i.e., individual walls, doors, windows, etc.); for example, the height to the bottom of a window, or the number of a door. These values can vary from instance to instance, hence the name.

Type Properties: This is information that is the same for every instance of a specific type. For example, the width and height of window type "W1" should be the same for all, and the width of wall type "S6" should be the same for all. Any modification to these values changes all previously, and yet to be, drawn objects of that type in the model.

Type Properties

A project can have multiple wall *Types*. Each type defines the construction of the wall and other unique characteristics such as cost and fire rating. There are three ways to see the *Type Properties* for a wall.

- Select a wall in the model and click **Edit Type** from the *Properties Palette*.

- With the *Wall* tool active, pick a wall type from the *Type Selector* and then click **Edit Type** from the *Properties Palette*.

- Right-click the wall type in the *Project Browser* (under Families → Walls → Basic Walls) and select **Type Properties**.

Any changes made to the *Type Properties* (Figure 6-1.4) will automatically be applied to every instance of that wall type in the model. For example, the seven walls shown in Figure 6-1.3 would change because they are all the same wall type.

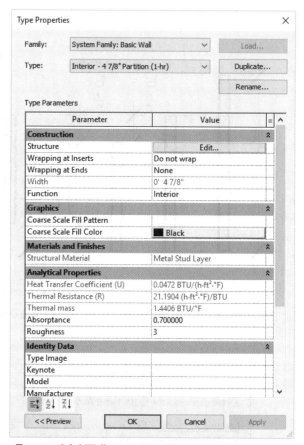

Let's take a look at a few type properties relative to walls:

Type:
First, make sure you know which wall type you are editing. At the top of the dialog box you see the *Type* is set to **Interior – 4 7/8″ Partition (1-hr)**.

Structure:
Clicking *Edit…* opens another dialog box which allows you to modify the wall construction. You will learn more about this shortly.

Wrapping:
Wrapping at *Inserts* (i.e., doors and windows) and *Ends* relates to how the

FIGURE 6-1.4 Wall type properties

finish materials terminate at their ends. Do they just stop or do they wrap around the corner? An example of this is shown in Figure 6-1.5. Wall "1" has the *Wrapping at Ends* set to **None**, the default setting, and wall "2" is set to **Interior**; in this example the **Exterior** option would yield the same results.

Function:

Interior, Exterior, Foundation, Retaining, Soffit, Core-Shaft are the options. This refined categorization can be used in schedules and view filters.

Coarse Scale Fill Pattern & Color:

The name of these two parameters is a little confusing. They relate to how the wall appears when a view's *Detail Level* is set to *Coarse*. When the *Detail Level* is set to *Coarse*, only the outline of the wall is shown. Additionally, using the *Coarse Scale Fill Pattern* and *Color*, you could make the wall have a solid fill and be black or a shade of gray, for example. This is great for preliminary floor plans. In Figure 6-1.6 you see the wall in *Coarse* view; for Wall Number **1**, notice the gypsum board is gone. For Wall Number **2**, the *Coarse Scale Fill Pattern* is set to solid and the color is set to black.

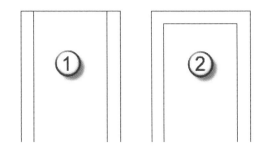

FIGURE 6-1.5 Wall end wrapping options (plan view)

FIGURE 6-1.6 Wall options at coarse level of detail

Assembly Code

This is a list of building categories based on the *Construction Specification Institute's* (CSI) *Uniformat Classification* system. Clicking in the *Value* cell causes a button to appear; clicking the button opens the dialog box shown in Figure 6-1.7. Because you have a wall selected, the list is narrowed down to just walls. Note the drop-down list at the top of the screen. These codes can be used in schedules and by some specification programs.

FIGURE 6-1.7 Selecting an assembly code

Each of the remaining *Type Properties* for the wall is fairly straightforward. Next, you will explore the structure of the wall.

Structure

Now we will go back and learn about the wall **Structure**. This mainly has to do with the layers of material that make up the wall's construction (i.e., the assembly), rather than just the structural part of the wall. First we will start with the fundamentals and then look at a few more detailed things like *Sweeps* and *Reveals*.

So, we have a wall selected and have clicked *Edit Type* (on the *Properties Palette*) and are now viewing the wall's *Type Properties*. To edit the wall's composition, you click the **Edit...** button next to the *Structure* parameter.

You should now be viewing the ***Edit Assembly*** dialog box (Figure 6-1.8). In this example, we continue to use the standard commercial interior wall type: **Interior – 4 7/8″ Partition (1-hr)**.

Here you can see the wall has three components, or *Layers*, which make up the Assembly:

- 5/8″ Gyp. Bd.
- 3 5/8″ Mtl. Studs
- 5/8″ Gyp. Bd.

This dialog is the same for both interior and exterior walls. Note that technically one side is identified as *Exterior* and the other *Interior*.

In plan view, the *Flip* arrow always appears on the exterior side of the wall when the wall is selected. For a symmetrical wall, such as this one, the orientation does not matter. However, when one side has a special finish or resilient channels to help with noise, the "special" side can be on what Revit thinks of as the *Exterior* side. Thus, when a wall is selected in plan view the *Flip* arrow can be used to quickly verify the orientation of the wall without needing to zoom way in (Figure 6-1.9).

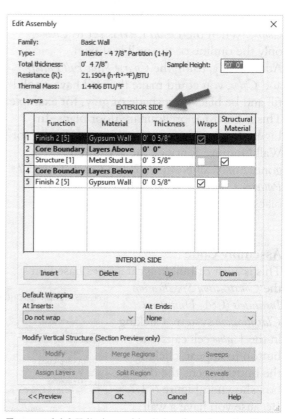

FIGURE 6-1.8 Edit Assembly dialog for selected wall

FIGURE 6-1.9 Wall flip-arrow on exterior side of selected wall

Each **Layer** in a wall has five settings: *Function, Material, Thickness, Wrap and Structural Material* and an important concept called *Core Boundary*. This is what they are for:

- **Function**

 Controls how wall *Layers* are joined together at intersections.

- **Material**

 This is the same list of *Materials* you studied in Chapter 3. The *Material* controls the appearance of the *Layer* in plans, sections and elevations, plus renderings. Also, when walls connect with each other, walls with the same *Material* and *Function* clean up; that is, the *Layer* appears to continue without interruption.

- **Thickness**

 How thick is the material? All *Layers* must have a thickness except the membrane layer, which might represent an air infiltration barrier or vapor barrier.

- **Wrap**

 Does this *Layer* wrap at inserts and ends as previously discussed? Notice *Layers* within the *Core Boundary* do not have a wrapping option.

- **Core Boundary**

 Notice that one or more *Layers* in a wall occur within the *Core Boundary*. This is considered the structural part of the wall. In the wall we have been looking at, the metal studs are the structural portion of the wall. If the studs were used as furring on a concrete block wall, they would not occur within the wall's *Core Boundary*. You will learn a little more about this in a moment.

- **Structural Material**

 Used by the structural engineer.

For a typical symmetrical **interior** wall you would use the same *Function*, **Finish 2 [5]**, and the same *Material* as shown in Figure 6-1.8. This causes walls to clean up everywhere as in Figure 6-1.10 below. If the *Function* and/or *Material* are different, then you will see a line appear between two intersecting *Layers* within a wall (Figure 6-1.11).

FIGURE 6-1.10 Wall cleanup concepts. When the *Material* <u>and</u> *Function* are the same, wall layers clean up at intersections.

FIGURE 6-1.11 Wall cleanup concepts. When the material or function are different, the wall layers do not clean up, as shown here.

The following bulleted list is directly from Revit *Help*. *Layers*, within a wall assembly, can be assigned the following functions:

- **Structure [1]:** *Layer* that supports the remainder of the wall, floor, or roof.

- **Substrate [2]:** *Material*, such as plywood or gypsum board, which acts as a foundation for another material.

- **Thermal/Air Layer [3]:** Provides insulation and prevents air penetration.

- **Membrane Layer:** A membrane that commonly prevents water vapor penetration. The membrane layer should have zero thickness.

- **Finish 1 [4]:** Finish 1 is typically the exterior layer.

- **Finish 2 [5]:** Finish 2 is typically for the interior layer.

Generally speaking, the *Core Boundary* represents the real-world structural portion of the wall. Walls can be both drawn and dimensioned based on the faces and centerline of the wall's core. The *Core Boundary* has a higher precedence when walls intersect.

| Stud wall | Stud furring | Masonry cavity wall | CMU wall |

FIGURE 6-1.12 The wall's Core Boundary is highlighted in each wall type

Notice a few things about the walls above before moving on. The two walls on the left have the metal stud *Layer* within the *Wall Core*, but the masonry cavity wall does not. In the first two, the studs are the structural element holding up the wall; in the latter the CMU holds up the wall and the metal studs are just used to fasten the gypsum board to it, called furring. Finally, the fourth wall contains one *Layer* which is the *Wall Core*. Every wall must have at least one *Layer* within the *Core Boundary*. An example of two *Layers* within the *Core Boundary* might be the studs and plywood sheathing which is used to prevent shear, making both *Layers* part of the structural makeup of the wall. In this case you would never want any intersecting wall to interrupt the sheathing.

One final aspect you need to know about *Layers*: they can be added, moved and deleted using the four buttons located just below the rows of *Layers*.

Sweeps and Reveals

The last step you will learn about editing a wall assembly is how to add a *Sweep* or a *Reveal*.

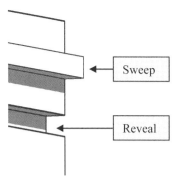

Both *Sweeps* and *Reveals* are initially defined by a 2D profile. This is a family with a simple 2D outline drawn in it. This profile is extruded horizontally along a wall at the defined height. It is possible to have more than one extrusion defined in the wall type.

Before you can begin defining *Sweeps/Reveals*, you have to do two things to make the tools active (Figure 6-1.13). First, you need to click the **Preview** button in the lower right so the preview is visible (Step 1). Next, you need to change the **View** from *Plan* to *Section* (Step 2). Now the **Modify Vertical Structure** tools are available (Step 3).

FIGURE 6-1.13 Making 'modify vertical structure' tools active

When you click on the *Sweeps* button, the ***Wall Sweeps*** dialog appears (Figure 6-1.14). It is possible to add multiple sweeps occurring on both sides of a wall. Each time you click the ***Add*** button a new row appears. Clicking in the *Profile* cell causes a down-arrow to appear.

This provides a drop-down list of profiles previously loaded into the project. The template you started from has a few. If the profile you need is not there, you can click the **Load Profile** button and load more from the *Profiles* sub-folder. If what you want is not there, simply create a new one: **File Tab → New → Family → Profile** (family template). If you are not sure what a setting does, just change it and click *Apply* to see the preview update.

> **FYI:** *Cuttable means the extrusion will be interrupted at door and window openings.*

FIGURE 6-1.14 Adding a wall sweep

The *Reveals* are set up the same way as *Sweeps*. The only difference is they cut a portion out of the wall rather than add to it. This "cut" appears in elevations and sections.

When adding *Sweeps* and *Reveals* to the wall type, you may want to **Duplicate** the wall type first. Otherwise, this reveal may appear in locations you did not intend it to!

That is all we will cover regarding wall types at this time. Next, you will learn about a few ways in which you can control the appearance of a wall once it has been created.

Wall Joins

When two walls intersect, they do not always clean up the way you want them to (Figure 6-1.15). Revit has a tool called *Wall Joins* to help with this problem. When the tool is active, you click the intersection of two walls. A box will be placed around the intersection.

> *TIP: Modify* tab, *Geometry* panel, *Wall Join.*

Notice the *Options Bar* has a number of related options (Figure 6-1.16). Here you can click **Butt** or **Miter**. When *Butt* is selected you can cycle through the options using the *Previous* and *Next* buttons.

FIGURE 6-1.15
Wall cleanup issue

> *FYI: This actually changes the 3D geometry, not just how it looks in plan-view.*

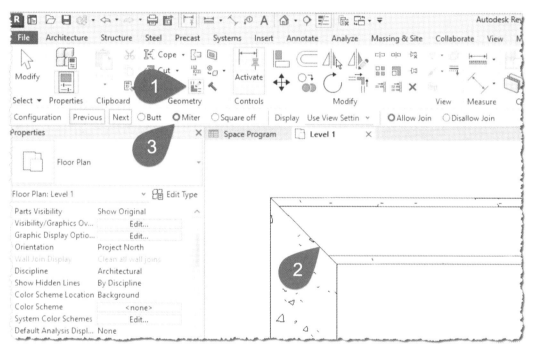

FIGURE 6-1.16 Wall cleanup being edited using the wall join tool

The *Display* option, on the *Options Bar*, has the ability to prevent wall cleanup all together at the given intersection. This is helpful between new and existing conditions and special situations.

Disallow Join

Revit has another feature one can use to control how walls join at intersections; it is called *Disallow Join*. It works much like the *Display* option just mentioned above, within the *Wall Joins* tool.

The *Disallow Join* feature is only available when a wall is selected. When a wall is selected, as in Figure 6-1.17 – Step #1, a small grip appears at each end of the wall. When you right-click on the grip, a pop-up menu appears as shown (Step #2). Selecting *Disallow Join* prevents the selected wall from cleaning up with other walls.

The final result would be as shown in Figure 6-1.18. However, immediately after selecting *Disallow Join*, the wall extends all the way to the grip. Therefore, you have to drag the grip back to the face of the other wall to get the result shown.

Another problem you might run into is if you have the *Wrapping* feature enabled at the wall ends. This will cause the wrap materials to appear at the end of the wall, which is probably not desired. The *Display* option in the *Wall Joins* tool works a little better in this case.

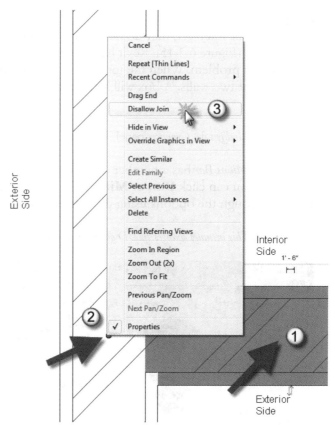

FIGURE 6-1.17 Wall cleanup concepts; disallow join feature

Cut Profile

The first thing to know about the *Cut Profile* tool is not to use it if at all possible! This tool allows you to modify how the *Layers* in a wall look in a plan view (Figure 6-1.19). However, these modifications are really just faked in the current view. Other views at the same level, or 3D views, do not show your modifications. Therefore, it is not something that should be used in a project as quantity and dimensions can be at variance within the model.

The only time the *Cut Profile* view might be useful is showing a really tricky existing condition and if the project has a tight budget or timeframe. But even then it is not ideal.

FIGURE 6-1.18 Disallow join applied to wall on right

Edit Profile

By default, the top and bottom of a wall are perfectly horizontal and the ends are vertical. As you are probably well aware, this is not the case for every wall. Some low walls have a sloped top, some walls have a tapered end, while still others have a hole in them that is not associated with a door or window.

This is where the *Edit Profile* tool comes into play. This tool allows you to redefine the perimeter of a wall when looking at it in section or elevation.

The steps are simple:

- Select the wall in *Section/Elevation/3D VIEW*.
- Select the **Edit Profile** tool from the *Ribbon*.
- In *Sketch* mode, redefine the perimeter or add an opening within the perimeter.
- Use **Trim** and **Split** for a clean perimeter and openings.
- Click the green check mark to finish.

FIGURE 6-1.19 Cut Profile used to modify gypsum layer within a normal wall type

When a wall is selected, you see a few modify tools on the *Ribbon* (Figure 6-1.20). The following four images capture the process outlined above. The arrows in Figure 6-1.22 highlight portions of the original line work which must be deleted before you can leave *Sketch* mode. You must have a single, end to end, perimeter. You may also have additional closed perimeters within the overall perimeter; these create holes in the wall.

FIGURE 6-1.20 Modify wall tools with wall selected

FIGURE 6-1.21
Original wall in elevation view

FIGURE 6-1.22
Edit Profile sketch mode (elevation view)

FIGURE 6-1.23A ABOVE
Edit Profile Sketch mode

FIGURE 6-1.23B RIGHT
Final wall as seen in a 3D view

When the original perimeter of a wall is trimmed, dashed lines appear, letting you know where the original perimeter is relative to your new sketch lines. As you can see in Figure 6-1.23b, the final wall has your edits in all views, including this 3D view. The holes even allow light to penetrate through them as seen in the shadows.

Anytime you want to change the edited profile, you simply select the wall again and click *Edit Profile*. You will return to *Sketch* mode where you can make additional changes. If you want to completely wipe the slate clean and remove all *Edit Profile* modifications, simply select the wall and click the **Reset Profile** button on the *Ribbon*, which is only active when the wall has an edited profile. Finally, if you select the *Edit Profile* command while in a plan view, you will be prompted to select from a list of elevation or section views, which Revit will automatically switch you to as the tool does not work in plan view.

Attach Top and Bottom

The top or bottom of a wall can be attached to a non-vertical surface, such as a roof or floor. This creates a parametric relationship and thus the wall will adjust accordingly if the roof is moved or the pitch changed.

The steps are simple:
- Select the wall in *Section/Elevation/3D VIEW*.
- Click the **Attach Top/Base** tool from the *Ribbon*.
- Select **Top** or **Base** from the *Options Bar*.
- Select the non-vertical surface to attach the wall to.

A common use of this tool is on a roof dormer. Both the top and bottom of the wall are attached to different roof elements. The following series of graphics illustrate this concept (Figure 6-1.24). Unfortunately, you cannot attach a wall to the underside of a stair, so you need to use *Edit Profile* for that.

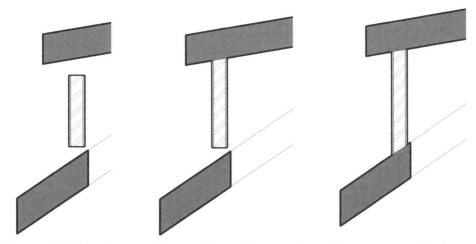

FIGURE 6-1.24 Attaching a wall's top to the roof above, and then its bottom to the roof below

While using the *Attach* tool, you need to specify *Top* or *Base* on the *Options Bar* as shown in Figure 6-1.25.

Modify | Walls Attach Wall: ◉ Top ○ Base

FIGURE 6-1.25 Options Bar while *Attach Wall* tool is active

The final result as seen in a perspective view (i.e., *Camera View*).

FIGURE 6-1.26 Three walls attached to the roof surfaces

Slanted Walls

There are two ways to make a wall sloped or slanted vertically. The first method is a newer feature in Revit.

Wall Properties Method:
Simply select that wall and adjust the **Cross-Section** parameter to **Slanted**, which reveals the **Angle From Vertical** parameter that appears. The wall will join with adjacent walls. Wall can also be **Tapered**, where selected layers in the wall change thickness vertically.

Mass Method:
First, you have to create a **_Mass_** element (*Massing and Site* tab on the *Ribbon*) with a slanted side which matches the height and length of the wall you want to create (Figure 6-1.27). Once the *Mass* is created, use the **_Wall by Face_** tool, select the wall type from the *Type Selector* and then pick the slanted face on the *Mass* (Figure 6-1.28). That is it; you have a slanted wall!

If the *Mass* is modified, you can select the wall and then click the **_Update to Face_** button on the *Ribbon*. The wall will automatically update to match the *Mass*. Deleting the *Mass* will not delete the wall, but it cannot be modified, so it is best to never delete the *Mass*; just hide it.

Update
to Face

FIGURE 6-1.27 Mass created with a sloped side **FIGURE 6-1.28** Walls applied by face to mass

In the example above, notice how the wall added "by face" is slanted AND the two side walls have an angled end to align with the slanted wall. The side wall modification is similar to what can be done with the *Edit Profile* feature, but this method tends to clean up better when a slanted wall is involved. Finally, you will notice the wall appears slanted even in the plan view, in which case a portion of the wall's surface is visible.

> **_FYI:_** Masses *should never be used to model furniture or things that will appear on the final drawing set.* Masses *are meant for conceptual use and are not visible by default in new views. Note that the* Mass *category will temporarily become visible without the box being checked in the* VV *dialog box.*

Exercise 6-2:
Rooms

Room and *Space* elements are one of the few things modeled in Revit that a person cannot put their hands on when the project is built. They are used to provide and store information about the various spaces within a building, such as area, volume, finishes, room name and number, etc. User created parameters, or information placeholders, can be created to keep track of project specific requirements such as "Government force protection required," "LEED daylighting required," "Medicare space tracking code," etc. Creating these custom parameters was covered in the previous chapter.

Revit "Architecture" uses *Rooms* and Revit "MEP" mainly uses *Spaces* but can also place *Rooms*. The two elements are identical in almost every way except that a *Space* can read values from a *Room* if they occupy the same area, even if the *Room* is in a linked model. Also, *Spaces* are primarily intended to hold engineering data such as heating and cooling requirements. We will only be talking about *Rooms* in this book. If the word "space" is used, we are talking about the physical area typically enclosed by walls.

When a *Room* is placed, its bottom sticks to the *Level* (datum) it was placed on and its sides search out for an enclosed perimeter. The top position can vary; this will be discussed more later on. The image below illustrates this concept. Note the *Room* element has been graphically displaced from the actual room to clarify this concept.

ROOMS BOUNDING WALLS ROOM ELEMENT

FIGURE 6-2.1 Room elements conform to room bounding elements

There are a number of practical concepts which need to be considered when using *Rooms* on a project. When not used properly, several problems can arise. Problems with phasing, schedules, room name and number tagging and cross-discipline coordination can impact the success of a project.

It is important to understand all of these concepts because your model will impact the other disciplines' models: architectural, structural and MEP.

The architectural model utilizes *Rooms* to manage room names and numbers, as well as areas and finishes. The Revit model should only have one *Room* element per space, per phase. *Room Tags* are view specific, so they need to be added to each view in which they are needed. Just be careful not to place another *Room* when you really only want a *Room Tag*. This happens to a lot of new users because *Room Tags* are automatically added to the view in which the *Room* element is created, so they think of the *Room* tool as a *Room Tag* tool. The **Tag All** tool can be used to quickly add a *Room Tag* to every *Room* in a view; *Rooms* already tagged will not be tagged again. However, it is possible to tag a room multiple times. This is necessary when match lines and dependent views are used to break up large plans.

Not everything in Revit is a **Room Bounding** element. Additionally, the elements which are *Room Bounding* can be set to not be. For example, in a housing project the closet area might need to be included in the bedroom area. In this case, the closet wall common to the bedroom can be selected and set to be "non-Room-Bounding" via the *Properties Palette*.

> *Room Bounding* elements.
> - Walls (curtain, standard, in-place, face-based)
> - Roofs (standard, in-place, face-based)
> - Floors (standard, in-place, face-based)
> - Ceilings (standard, in-place, face-based)
> - Columns (architectural, structural with material set to concrete)
> - Curtain systems
> - Room separation lines
> - Building pads

When placing a *Room*, in the architectural model, it is important to be on the correct *Level* and in the correct *Phase*. Once a *Room* is placed, the *Level* and *Phase* cannot be adjusted unless you employ a workaround, such as *Cut* and *Paste Aligned*. The current view dictates the *Level* and *Phase* the *Room* element is placed on. Therefore, it is necessary to have a view for each *Phase*: Existing, Demo and New Construction. In this book we are only working in the *New Construction Phase*. It is also required to have a view for each *Level* a person can walk on, even if the *Levels* are only six inches apart.

The image below (Figure 6-2.2) shows a common scenario for an existing building with one or more additions previously added to it. The **Level 1 - 1971** view might be the only view placed on a sheet, but a **Level 1 - 1953** view is required to place Walls, Doors, Rooms, etc. The floor plan view may look correct if everything was modeled in the **Level 1 - 1971** floor plan view but various problems would arise. For example, interior elevations and sections would not be accurate, especially if the walls and doors were drawn at the higher level; volumes would be off and the MEP designers would run into problems placing *Spaces*.

FIGURE 6-2.2 Room elements placed at wrong level

Looking at Figure 6-2.2 again, another problem might be placing a *Room* at the lower level, but all the walls have been modeled at the upper level. Things look correct in plan, but Revit cannot find a valid boundary at the lower level in which you are working. Revit only searches for a valid boundary at the level you are working on. A quick section cut will reveal this problem.

Change is always occurring during the design process (Figure 6-2.3). Revit has a few simple rules when it comes to model changes and *Rooms*. First, the point you pick when placing a *Room* has some significance. Where you click is where the center of the "X" is located. The "X" does not show up by default, except when using the *Room* tool. Both the "X" and colored fill can be manually turned on in a view's *Visibility/Graphics Overrides*; expand the *Room* category and check each sub-category (see page 6-41 for more on this topic).

Once placed in the model, to select a *Room* you must move your cursor around a room until the "X" is highlighted and then click. When deleting a *Room*, it is not actually deleted from the project! This is confusing to new users. The only way to delete a room is from a room schedule, which may not exist depending on the template the project was started from. This functionality is meant to maintain the information stored in a *Room* and help maintain the programmed spaces. It is possible to place unplaced *Rooms*.

When a *Room* is made smaller by an intersecting wall, the *Room* element will automatically get smaller. Which side of the new wall does the *Room* favor? The side with the center of the "X." See the following image. Also, when using the *Tag All* tool, the tag is placed at the center of the "X."

So it would be wise, then, to use some thought when picking the point within a space to place a *Room* element.

Whenever the boundary for a *Room* is broken, a *Room* "collapses" to a small rectangle. As soon as the boundary is "healed" the *Room* automatically floods the space.

FIGURE 6-2.3 Notice the Room reference point determines the location when the boundary conditions change

The height of the *Room* element has an impact on the overall BIM project. By default, Revit does not calculate volume and Autodesk's Revit "Performance Guide" recommends not doing this for performance reasons. However, this writer believes it is required to make *Spaces* work properly in the MEP model and avoid problems.

The image below shows how *Rooms* will extend to the level or height specified no matter what it passes through; this happens when "calculate volumes" is turned off.

FIGURE 6-2.4 Area and Volume Computations

FIGURE 6-2.5 Rooms extend through upper bounding elements when not calculating volumes

The next figure shows how the *Room's* top stops when it hits a room bounding element such as a ceiling or a floor. This is not important in the architectural model, unless you need to keep track of volumes, but will help keep *Spaces* working smoothly in the MEP model.

FIGURE 6-2.6 Rooms stop at upper bounding elements when calculating volumes is turned on

In the partial dialogs shown previously (Figures 6-2.5 and 6-2.6), one can see the ability to control how *Room* area is calculated. Note, when the wall is set to centerline, this also includes the centerline of the exterior walls. It is not possible to use the centerline of interior walls and the outside face of exterior walls.

The top of the *Room* conforms to soffits and sloped ceilings as can be seen in Figure 6-2.7 below. Note that the bottom of walls, i.e., the bulkhead, are not room bounding. This will throw the volume calculations off a little. See page 9-8 for more on bulkheads.

FIGURE 6-2.7 Rooms stop at upper bounding elements when calculating volumes is turned on

It is also possible to add *Room Tags* in a section view. *Rooms* can extend up past other levels in multi-story spaces, for example, atriums and stair shafts. This allows consistent tagging and scheduling.

When working on existing buildings, *Rooms* must be placed at least twice: once in the *Existing Phase* and once in the *New Phase*. A room element is needed for each phase. If you place all the existing *Rooms*, you can *Copy* and *Paste Aligned* to quickly do most of the redundant work.

Rooms can be further controlled by sketching **Room Separation Lines**. This special 2D line allows alcoves, waiting areas, open office areas, etc. to be carved out of a larger area within the building as in the following image (Figure 6-2.8). However, they must be on the same *Level* (datum) as the bottom of the *Room* element.

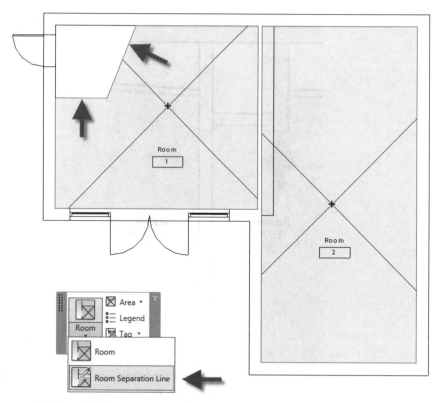

FIGURE 6-2.8 Room separation lines create boundary conditions where no built geometry exists

If a project has grouped unit plans, it is possible to have *Rooms* within the *Group*. Each instance of the *Group* can have a unique room name and number. If phasing is involved, each phase needs its own *Room* element within the *Group*.

Exercise 6-3:
Floor Plan Development

It is now time to develop the floor plans further based on sketches we have developed by hand. You will add walls, doors and rooms in this section. This will provide design validation feedback via your *Space Program* schedule; the actual area will appear as rooms are placed. In order to quickly verify space requirements, you will begin by using one common wall type everywhere. Later, specific walls can be changed to curtain wall.

Concept Sketches

The next two pages have hand drawn concept sketches which will be the basis of our Revit modeling. Notice a number of walls align with structural grid lines or are centered on a mullion of an exterior window. Remember, in this chapter we will only be adding walls, doors and rooms, so do not worry about the additional information provided.

> *FYI: The plans are rotated to maximize the image size in this book. Please do not try and rotate the building in Revit.*

Let's get started!

1. Open your *Law Office* project file.

2. Switch to the *Level 1 Floor Plan* view.

To minimize the unnecessary detail at this point in the project, you will change the view's *Detail Level* to *Coarse* so we just see the outline of the walls. This hides the internal linework representing the gypsum board panels, metal studs and more.

3. Set the *Detail Level* for the view to **Coarse**.

4. Select the ***Wall*** tool from the *Architecture* tab.

5. Pick *Basic Wall:* **Interior - 4 7/8" Partition (1-hr)** from the *Type Selector*.

6. Check the following settings:
 a. *Height* (aka, *Top Constraint*): **Level 2**
 b. *Top Offset*: **0'-0"**
 c. *Base Constraint*: **Level 1**
 d. *Base Offset*: **0'-0"**
 e. *Location Line*: **Wall Centerline**

FIGURE 6-3.1 Level 1 conceptual sketch (rotated view)

FIGURE 6-3.2 Level 2 conceptual sketch (rotated view)

You are now ready to add walls. Sometimes it is difficult to decide where to begin. This would be similar to modeling existing conditions, especially when all the dimensions are not provided. In this case, all required dimensions are provided. The easiest thing to do is start with known conditions, like the wall centered on a gridline and intersects an exterior wall. You can click at the intersection of the grids, or center of a column, and then draw a length of wall along the grid line. Do not worry about the exact length of the wall at first, as you can trim or extend the wall later.

7. Add the interior walls as shown below (Figure 6-3.5)

 a. The **Align** tool can be used to move a wall into alignment with a grid or mullion.

 i. For example: draw a wall near its final location (Figure 6-3.3).

 ii. Make sure the wall is orthogonal (i.e., perfectly vertical or horizontal).

 iii. Use the *Align* tool to quickly move the wall in alignment with a gridline or curtain wall mullion (Figure 6-3.4).

 iv. Walls should be centered on the window mullions.

 b. All dimensions shown are as follows:

 i. Centerline of an interior wall

 ii. Interior face of an exterior wall

 iii. On a gridline

 c. Dimensions can be added to help position the walls but should be deleted once the walls are properly positioned.

FIGURE 6-3.3 Wall quickly added

FIGURE 6-3.4 Align wall with mullion

Remember, walls without dimensions (in Figure 6-3.5 below) are either centered on a structural gridline or a mullion in an exterior window.

FIGURE 6-3.5 Level 1 interior walls to be added – West side (rotated view)

8. Add the remaining interior walls shown in Figure 6-3.6:

 a. Use the gridlines to keep track of where you are in the building.

 b. Notice a portion of wall needs to be deleted between the restrooms.

 c. Two walls get extended that were just added in the previous step.

FIGURE 6-3.6 Level 1 interior walls to be added - East side (rotated view to enlarge size on page)

9. Switch to **Level 2** and set the *Detail Level* to **Coarse**. In the *Properties Palette*, for the view, set the *Underlay (Base)* to **None** to hide the walls from the floor below.

10. Add the walls shown in Figure 6-3.7 below.

 a. Constrain tops of walls to roof level above.

FIGURE 6-3.7 Level 2 interior walls to be added West side (rotated view)

11. Add the *Level 2* walls shown in Figure 6-3.8 below.

 a. Note that two existing walls get modified (shown dashed).

 b. See the next step for making the three offices equal.

FIGURE 6-3.8 Level 2 interior walls to be added East side (rotated view)

When creating equally spaced walls, you can start out by adding the walls at random locations and then use the *Dimension* tool to reposition the walls. This is how that is done:

12. Add the two walls between Grids 3 and 4 without worrying about their exact location (Figure 6-3.9).

FIGURE 6-3.9 Level 2 interior walls added somewhat randomly

13. Use the ***Aligned*** dimension tool to add a string of dimensions; click all four walls and then click to position the string.

14. Click the ***EQ*** icon pointed out in Figure 6-3.10.

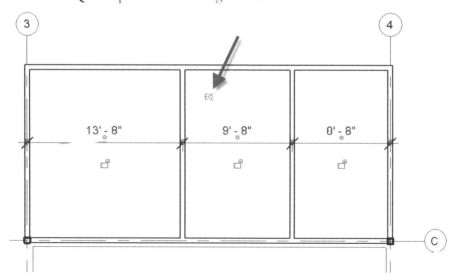

FIGURE 6-3.10 Dimension string added

The walls are now equally spaced (Figure 6-3.11). Notice the dimension text now says *EQ* rather than the length. This can be changed via the dimension's *Instance Properties*. Change the *Equality Display* from **Equality Text** to **Value**. Next you will delete the dimensions now that the walls are properly positioned. However, you will have to decide if the *Equality* constraint should remain even after the dimension is gone. In this case you will choose *Unconstrain*.

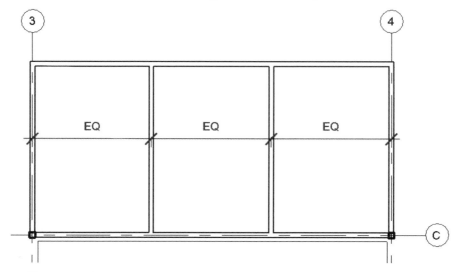

FIGURE 6-3.11 Dimension string equalized

15. Select the dimension string and **delete** it.

16. Choose **Unconstrain** from the prompt that appears (Figure 6-3.12).

FIGURE 6-3.12 Prompt to retain constraint

The dimension is now gone, and the walls are no longer constrained to be equally spaced. It is a best practice to limit the number of constraints to only essential conditions to prevent the project from bogging down and producing unnecessary warnings and error messages.

17. Erase the existing curtain wall (i.e. door with sidelights) on Gridline B; hover your cursor over the wall until the dashed line at the center appears. Also delete the room separation line just south of the curtain wall (be careful not to delete the floors).

18. Add the remaining *Level 2* wall as shown in Figure 6-3.13.

FIGURE 6-3.13 Level 2 interior wall to be added (rotated view)

Adding Doors

Now that you have sketched the walls, you will add some doors. A door symbol in plan view indicates which side the hinges are on and the direction the door opens.

New doors are typically shown open 90 degrees in floor plans, which helps to avoid conflicts such as the door hitting an adjacent base cabinet. Additionally, to make it clear graphically, existing doors are shown open only 45 degrees.

Door symbol Example – New Door drawn with 90 degree swing

Door symbol Example – Existing Door shown with 45 degree swing

Accessible Doors

Most building codes in the US require that a door's clear opening is a minimum of 32″ wide. If you look at a typical commercial door that is open 90 degrees, you will notice that the stops on the door frame, the door thickness and the throw of the hinges all take space from the actual width of the door in the closed position; therefore, 36″ wide doors are typically used to achieve the 32″ clear width required (see Figure 6-3.16).

FIGURE 6-3.14
Clear floor space at accessible doors

Building Codes can vary regarding clearances required at a door, but most are pretty similar. Typically, doors that have BOTH a closer and a latch must have sufficient space for a handicapped person to operate the door.

Pull Side of a door that is required to be accessible typically needs to have 18″ clear from the edge of the door, not the outside edge of the frame, to any obstruction. An obstruction could be a wall, a base cabinet or furniture.

Push Side of a door that is required to be accessible typically needs to have 12″ clear from the edge of the door to any obstruction.

Most building codes and the Americans with Disabilities Act (ADA) require slightly different dimensions depending on which direction a person is approaching the door.

In this book, all doors adjacent to a wall will have a 2″ dimension between the wall and the edge of the door frame (per Figure 6-3.15), and all other doors will be dimensioned. A note to this effect is usually found in a set of *Construction Documents*, which significantly reduces the number of doors that need to be dimensioned.

FIGURE 6-3.15 Typical hollow metal door jamb detail

FIGURE 6-3.16 Details to know about accessible doors

Door Stops are usually an integral part of the hollow metal door frame. Their main function is to "stop" the door when being shut. However, they also provide visual privacy and can have sound or weather strips applied. Door stops are NOT typically drawn in floor plans.

Hinge Throw is the distance the door is projected out into the opening as the door is opened, which varies relative to the hinge specified.

The **2″ Wall** dimension is the amount of wall many designers provide between the frame and any adjacent wall. This helps to ensure that the door will open the full 90 degrees regardless of the door hardware selected.

Now you will add the doors to your floor plans.

19. Using the *Door* tool, add the doors shown in Figures 6-3.17 and 6-3.18, *Levels 1* and *2* respectively. Be sure to note the following:

 a. Use the **Single-Flush** and **Double-Flush** door type, using 3'x7' and 6'x7' doors respectively. *FYI:* These doors are already loaded in the starter file.

 b. With the *Door* tool active, on the *Ribbon*, make sure the **Tag on Placement** toggle is **off**; you do not need to show door tags this early in the design process.

 c. The swing of the door is based on which side of the wall you click on. The hinge side can be adjusted by pressing the space bar before clicking.

 d. All doors near an adjacent wall should be 4″ away per Figure 6-3.16 above; note that the Revit doors do not show the door frame, so the 4″ takes into account the 2″ portion of wall and the 2″ hollow metal frame.

 TIP: Take a look back at Chapter 3 for instruction on positioning the door opening.

 e. Doors clearly not close to an adjacent wall should be placed approximately as shown; most of these doors will be adjusted later.

 f. The double doors are centered on the room they open into.

 TIP: Use the Dimension tool and then the Equalize feature, and then delete the dimensions.

FIGURE 6-3.17 Level 1 doors to be added (rotated view)

FIGURE 6-3.18 Level 2 doors to be added (rotated view)

Exercise 6-4:
Placing Room Elements in the Model

Now that most of the spaces are defined, it is time to start thinking about how they agree with the *Space Program*, both in terms of quantity and size. Did we forget a small conference room or an office? Are any of the spaces way too big or too small? Once the *Room* elements have been placed, you will take a look at the *Space Program* schedule and see how it has automatically updated to show actual square footage and more.

Placing Room Elements

Before you use the *Room* tool and begin clicking in spaces within your model, you need to understand how to select from the list of *Rooms* in your *Space Program*. If you do not do this step properly you will add additional new *Rooms* to the model. If you accidentally place a new *Room* element, the *Room* can be deleted from the floor plan view, but it will not be removed from the model. To remove the *Room* element from the model, it must be deleted from a room schedule view, such as your *Space Program* schedule.

You will begin by placing the *Room* elements in the enclosed spaces. Later in this section you will learn how to place *Rooms* which are not completely enclosed by walls.

1. Switch to your **Level 1** Floor Plan view.

2. Select the **Room** tool from the *Architecture* tab.

3. With the *Room* tool active, do the following (Figure 6-4.1):

 a. Make sure **Tag on Placement** is selected.

 b. Set the *Type Selector* to the room tag option that will show the room's area.

 c. Set the *Upper Limit*: **Level 2**, *Offset*: **0'-0"**

 FYI: When placing rooms on level 2, set *Upper Limit*: Roof, *Offset*: **0'-0"**

 d. From the *Room* drop-down list, select the **Large Conference Room**.

FYI: The room number will likely be different; ignore the room number for now. You will change the room numbers in chapter 16.

FIGURE 6-4.1 Using the *Room* tool to place programmed spaces

When you selected the "Large Conference Room" from the *Room* drop-down list, you are telling Revit to place a *Room* from the *Space Program* list rather than create a new *Room* element within the model. Setting the *Upper Limit* to the level above in conjunction with the "Calculate areas and volumes" enabled (see Page 6-21) will ensure all the rooms are tall enough to reach the ceiling or include elements such as a wall mounted TV bracket.

4. Click within the Large Conference Room (Figure 6-3.1) to place the *Room* element.

Your drawing should look similar to Figure 6-4.2 below, assuming the *Room* tool is still active. While the *Room* tool is active, the *Room* element is colored and the placement "X" is visible. As soon as the *Room* tool is finished, by clicking the *Modify* button or the *Esc* key, the color fill and placement "X" disappear. It is possible to turn these graphics on permanently: *type* **VV** → **Model Categories** *tab* → **Rooms** *category*, and then check **Interior Fill** and **Reference**. You do not need to make these changes now.

FIGURE 6-4.2 Room element placed in model

5. If your room area is off, double-check the wall location dimensions from the previous exercise.

6. Place the *Room* elements shown in Figures 6-4.3 and 6-4.4 per steps just covered.

FIGURE 6-4.3 Level 1 rooms added (rotated view)

FIGURE 6-4.4 Level 2 rooms added (rotated view)

Room Separation Lines

In Revit, the edges of a *Room* element are typically defined by a series of walls. Some spaces do not have a boundary defined by walls; for example, the lobby waiting area, or the impromptu meeting area which is basically in the hallway. Revit provides a way to deal with this: *Room Separation Lines*. A *Room Separation Line* is a 2D line which provides a boundary for *Rooms*. You will add several *Room Separation Lines* next and then place the remaining *Room* elements.

7. Switch to the Level 1 Floor Plan view.

8. Select the **Room Separator** tool from the *Room* drop-down list on the *Architecture* tab.

9. Place two *Room Separation Lines* as shown (dashed) in Figure 6-4.5.

 TIP: Click directly on the center of the intersection of the two walls at the corner.

FIGURE 6-4.5 Level 1 room separation lines added

You have now defined the perimeter of one of the *Impromptu Meeting* spaces. Now you can place a *Room* element here.

10. Using the **Room** tool, place one of the *Impromptu Meeting* spaces (Figure 6-4.6).

The new *Room* element is now bounded by two walls and two room separation lines. You will use this same technique to define an area for the reception area, the lobby and waiting area as well as a number of cubicles, some of which have no walls around them. Notice, as *Rooms* are placed, they are removed from the drop-down list on the *Options Bar.*

11. Add the four *Room Separation Lines* shown in Figure 6-4.7 on *Level 1*, and then place the **Rooms**.

 a. The dimension shown is 11'-5".

 b. The other lines align with adjacent model elements.

12. Add the *Level 2* **Room Separation Lines** and **Rooms** as shown in Figures 6-4.8 and 6-4.9.

FIGURE 6-4.6 Level 1 room added

FIGURE 6-4.7 Level 1 reception and waiting areas defined with room separation lines

FIGURE 6-4.8 Level 2 spaces defined with room separation lines (rotated view)

Now that you have all the *Rooms* placed in the model, there are none left to select from while the *Room* tool is active. You could only place new *Rooms* now.

Open your *Space Program* schedule: **Space Program by Department**. Notice the actual area has been added (Figure 6-4.10). Additionally, now that there is a value for *Actual Area*, the *Area Difference* column can be calculated and shaded yellow if the difference is too great or small.

Sometimes, when working within a fixed area you have to adjust the program. We are working with the idea that this is a new building and the perimeter could possibly change if needed, but we will not be doing so. In this case, you could also think of this as an existing building in which the perimeter could not be changed, unless an addition was in the scope of work. This project has not been set up with phasing, so this is not meant to be thought of as an existing building. If it were, you would have needed to "demo," rather than delete, the previously drawn walls earlier in this chapter.

In the next chapter you will continue to develop the floor plan, adding glass walls, which are also referred to as *Curtain Wall*.

FIGURE 6-4.9 Level 2 spaces defined with room separation lines

	Area			
Name	Actual	Program	Diff	Comments
Accessory Spaces				
File room	248 SF	250 SF	-1%	12 - 15 lateral file drawers per attorney
Full time graphic design office	95 SF	140 SF	-32%	Visualizes the case for arguments
library	251 SF	300 SF	-16%	
lunchroom staff (the café)	321 SF	350 SF	-8%	On building perimeter natural daylight with higher end finishes, adj to lib.
Mock courtroom training cente	389 SF	450 SF	-14%	
5	1304 SF	1490 SF		
Building Support Services Spaces				
Fitness center	215 SF	200 SF	8%	
General Storage	172 SF	200 SF	-14%	
Janitor closet	46 SF	75 SF	-38%	
Private Restroom Female	126 SF	60 SF	110%	
Private Restroom Male	126 SF	60 SF	109%	
Technology closet	70 SF	100 SF	-30%	
Workroom	342 SF	300 SF	14%	
7	1097 SF	995 SF		
Conference Center (The front entry)				
Impromptu meeting space	106 SF	100 SF	6%	Informal interactions spaces that do not disturb others
Impromptu meeting space	73 SF	100 SF	-27%	Informal interactions spaces that do not disturb others
Large Conf / Board Room	426 SF	500 SF	-15%	16 to 24 persons
Medium Conference Room	240 SF	250 SF	-4%	Seat 8 to10 persons
Public Restroom	149 SF	50 SF	199%	One Male one Female
Public Restroom	149 SF	50 SF	198%	One Male one Female
Reception	70 SF	300 SF	-77%	Front – first impression that serves as multipurpose space
Small Conference Room	123 SF	100 SF	23%	Office size that could change use if needed. 2 to 4 persons
Small Conference Room	102 SF	100 SF	2%	Office size that could change use if needed. 2 to 4 persons
Waiting / lobby	306 SF	300 SF	2%	
10	1743 SF	1850 SF		
Office and Workstation				
Assistant	60 SF	64 SF	-7%	
Assistant	60 SF	64 SF	-7%	
Associate office	108 SF	140 SF	-23%	Partner to associate ratio 1:2
Associate office	133 SF	140 SF	-5%	Partner to associate ratio 1:2
Associate office	109 SF	140 SF	-22%	Partner to associate ratio 1:2
Associate office	150 SF	140 SF	7%	Partner to associate ratio 1:2
Associate office	150 SF	140 SF	7%	Partner to associate ratio 1:2
Associate office	150 SF	140 SF	7%	Partner to associate ratio 1:2
First year associate/paralegal	91 SF	125 SF	-27%	
Intern	64 SF	64 SF	0%	
Office manager	80 SF	81 SF	-1%	
Paralegal	80 SF	81 SF	-1%	Attorney to paralegals 9:1 to 4:1
Paralegal	80 SF	81 SF	-1%	Attorney to paralegals 9:1 to 4:1
Paralegal	80 SF	81 SF	-1%	Attorney to paralegals 9:1 to 4:1
Paralegal	73 SF	81 SF	-10%	Attorney to paralegals 9:1 to 4:1
Paralegal	73 SF	81 SF	-10%	Attorney to paralegals 9:1 to 4:1
Partner / principal office	196 SF	235 SF	-17%	
Partner / principal office	242 SF	235 SF	3%	
Partner / principal office	247 SF	235 SF	5%	
Partner / principal office	227 SF	235 SF	-3%	
Retired partners office	252 SF	235 SF	7%	
Secretary	67 SF	64 SF	4%	Attorney to secretary ratio 3-4:1 better 2.5:1 (the command center)
Secretary	60 SF	64 SF	-7%	Attorney to secretary ratio 3-4:1 better 2.5:1 (the command center)
Secretary	60 SF	64 SF	-7%	Attorney to secretary ratio 3-4:1 better 2.5:1 (the command center)
Visitor office	118 SF	125 SF	-6%	
25	3007 SF	3135 SF		
Grand total: 47	7151 SF	7470 SF		

FIGURE 6-4.10 Space Program shows actual area now that rooms are placed in model

Self-Exam:

The following questions can be used as a way to check your knowledge of this lesson. The answers can be found at the bottom of this page.

1. A _____ can be used to add wall base in the wall type.

2. When placing a room, where you click has an effect on where tags are placed with *Tag All*. (T/F)

3. There is no way to create a sloped wall in Revit. (T/F)

4. Room elements automatically find the perimeter of a space. (T/F)

5. The bottom of a room aligns with level. (T/F)

Review Questions:

The following questions may be assigned by your instructor as a way to assess your knowledge of this section. Your instructor has the answers to the review questions.

1. Rooms are constrained by elements which have *Room Bounding* checked. (T/F)

2. The height of a room can vary. (T/F)

3. A *Room* is what the area of a space is derived from. (T/F)

4. The _____ tool allows you to modify the perimeter of a wall in elevation or section.

5. Rooms can measure area to the center or face of walls. (T/F)

6. The *Room Separation Line* tool is used to define the edge of a room where there are no walls. (T/F)

7. The *Wall Join* tool can apply one of two conditions: _____ or _____.

8. Doors with a 45° swing are new construction. (T/F)

9. When selecting a string of dimensions, click the "EQ" icon to make each segment equal. (T/F)

10. Use the _____ tool to make a wall extend up and conform to a roof or floor above.

Notes:

Lesson 7
Curtain Wall:

This chapter will continue the development of the floor plans using Autodesk® Revit®. The focus will be on a curtain wall (aka, *glazing*). Curtain wall is what you will use to model the large glass areas and walls.

Exercise 7-1:
Curtain Wall

In addition to what are often referred to as punched openings, or the windows, Revit also provides a 'wall' type called **Curtain Wall**. *Curtain Walls* are placed using the *Wall* tool, rather than the *Window* tool, but typically represent an expanse of glass, interior or exterior.

The commercial template, which this project was started from, provides three *Curtain Wall* types: Curtain Wall, Exterior Glazing and Storefront.

The first one, Curtain Wall, represents one extreme, that being a blank slate in which you totally design the mullion layout. When the wall is placed, you basically have a panel of glass the entire length and height of the wall (Figure 7-1.1).

The second wall type, Exterior Glazing, has the glass divided up, but does not provide any mullions. This would be like a butt-glazed system until mullions are added (Figure 7-1.2).

Thirdly, the Storefront wall type represents the other extreme where mullions are placed in both the horizontal and vertical directions at a specific spacing. The layout is locked down, i.e., pinned, and cannot be accidentally modified unless a mullion is manually un-pinned. For simplicity in this tutorial, this will be the system you use (Figure 7-1.3).

A *Curtain Wall* has four primary elements:

- Wall
- Panel
- Grid
- Mullion

Each of these elements must be individually selected in order to edit them. You often have to tap the **Tab** key to cycle through and select something specific; once the item is pre-highlighted, you click to select it.

FIGURE 7-1.1 Curtain wall

The blank slate *Curtain Wall* wall type represents the simplest condition: the wall and a single panel that coincides with the wall (see Figure 7-1.1).

To begin dividing up a curtain wall you need to add *Curtain Grids* (see Figure 7-1.2); Revit has a specific tool for this on the *Architecture* tab. When a *Curtain Grid* is added, the panel in which it was placed gets split.

The image to the right has:

8	Curtain Panels
4	Curtain Wall Grids
1	Wall

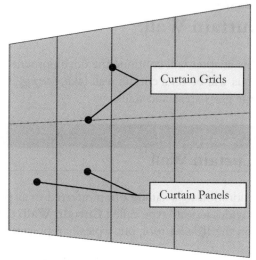

FIGURE 7-1.2 Curtain wall with grids added

When a **Curtain Grid** is selected, you may adjust its location with the *Move* tool or the temporary dimensions.

When a **Curtain Panel** is selected, you can swap it out with other Revit elements such as walls and special curtain wall doors.

Once *Curtain Grids* have been placed, you may use the **Mullion** tool to place *Mullions* on the *Curtain Grids*.

The image to the right shows *Mullions* at each of the four grids and along the perimeter. A portion of a *Mullion* was deleted at the bottom edge and a *Curtain Panel* was swapped out for a door (Figure 7-1.3).

The image to the right has:

7	Curtain Panels
4	Curtain Wall Grids
21	Curtain Wall Mullions
1	Door
1	Wall

FIGURE 7-1.3 Curtain wall with mullions added

Notice the large number of mullions in the list above. Mullions stop and start at each *Curtain Grid* intersection. This made it easy to select the mullion at the desired door opening location and simply delete it.

The next few steps will walk you through adding one of the curtain walls to the *Level 1* floor plan. This has already been done in your model, but the steps are identical to what you will do for the interior glazed areas. The following "lettered" steps **do not** need to be performed in your model.

A. Switch to the **Level 1** *floor plan* view.

B. Select the **Wall** tool on the *Ribbon*.

C. Select the wall style named **Storefront** from the *Type Selector* (Figure 7-1.4).

Before drawing the *Storefront* wall type you will review its properties.

Note the *Instance Properties* via the *Properties Palette*.

FIGURE 7-1.4 Curtain wall named Storefront

FIGURE 7-1.5 Storefront instance properties

Take a moment to notice a few things about the information presented (Figure 7-1.5). The *Base Constraint* is equal to the floor plan level you are currently on. Keep in mind **Level 1** refers to the *Level datum* and not the name of the *View* that you are in. The sill height is set by the *Base Offset* parameter and the *Head Height* is set by *Top Constraint*. Because these are *Instance Parameters* they can vary from location to location within the project.

D. Make the following adjustments:

- *Base Offset:* **2'-0"**

- *Unconnected Height:* **6'-0"**

FIGURE 7-1.6 Storefront Type properties

E. Click the ***Edit Type*** button to view the *Type Properties*.

Again, take a moment to notice a few things about the information presented (Figure 7-1.6).

Each wall type has a **Function** setting; this one is set to *Exterior*. Others may be set to *Interior, Foundation, Retaining, Soffit* or *Core-Shaft*. This setting has an effect on how the wall is drawn and may be used by energy analysis programs.

Automatically Embed allows one wall to cut a hole in another, similar to what a door or window does when placed in a wall. This is an important step!

Various rules on the mullions and their spacing are adjustable here.

F. Press **OK** to close out of the open dialog box.

G. In the Northwest corner of the building, snap to the middle of the wall, **4′-0″** down from the corner and then click (Figure 7-1.7).

H. Move your cursor straight downward, along the wall, and click when the temporary dimension reads **14′-0″** (see Figure 7-1.8).

The *Curtain Wall* is now placed in the West wall (Figure 7-1.9). The *Flip* control appears on the exterior side of the curtain wall when it is selected, similar to regular walls. If the *Flip* control is on the interior side of the wall (as in Figure 7-1.9), click the control to flip the curtain wall so the glass favors the exterior side of the wall.

FIGURE 7-1.7
Storefront – *first pick*

FIGURE 7-1.8 Storefront – *second pick*

FIGURE 7-1.9 Storefront – selected

FIGURE 7-1.10 Storefront shown in 3D camera view

Notice in the image to the left, the new curtain wall actually cuts an opening in the larger exterior masonry wall (Figure 7-1.10).

These steps used to place this exterior curtain wall in the model are nearly identical to what you will do in the interior walls. Often, the curtain wall only goes up to the top of the doors or the ceiling. Above that is a regular wall. This "regular" wall is what the ceiling will connect to. Therefore, the wall you have previously drawn will not be deleted, but added to.

Because a *Curtain Wall* is a wall, it is possible to use the various wall modify tools discussed in the previous chapter, such as **Edit Profile** and **Attach Top/Base**.

The image to the right shows how an interior door can be easily created using the *Edit Profile* feature (Figure 7-1.11b). The sidelight can have multiple mullions and be as wide as needed. This flexibility is not possible with the standard doors provided with Revit. The frame also shows in plan view (Figure 7-1.11a).

FIGURE 7-1.11A Curtain wall in plan-view

FIGURE 7-1.11B Curtain wall with edited profile

The **Attach Top/Base** feature can quickly conform the top of a curtain wall to a sloped roof element above as shown in Figure 7-1.12. Notice in both 3D images on this page that the original shape of the *Curtain Wall* appears when your cursor hovers over the edge.

FIGURE 7-1.12 Curtain wall with top attached to sloped roof element

Next we will take a look at how corner conditions are handled with *Curtain Walls*. If two "storefront" type curtain walls are drawn with a 90 degree corner, the result is two overlapping mullions as shown in Figure 7-1.13 and 7-1.14. This does not look good. Revit has a solution to this problem. There are special corner mullions which can interact with both walls, even though it is technically only attached to one of them.

FIGURE 7-1.13 Overlapping mullions at corner (plan-view) **FIGURE 7-1.14** Overlapping mullions

To change the way a corner is represented you need to select one of the two overlapping mullions and swap it out with a corner mullion. When selecting a mullion, if it is pinned, you must un-pin it before any changes or edits can be made (Figure 7-1.15).

FIGURE 7-1.15 Un-pin the mullion

Select one of the corner options from the *Type Selector* (Figure 7-1.16).

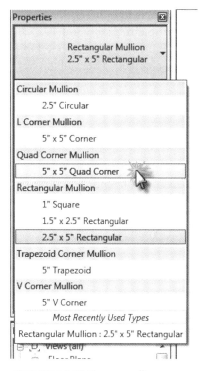

FIGURE 7-1.16 Corner mullion options

Once you select one of the corner mullion options, Revit will prompt you to delete the adjacent mullion – which you must accept or cancel the modification (Figure 7-1.17).

FIGURE **7-1.17** Prompt to delete adjacent, overlapping mullion

Now the corner condition should appear as shown in Figure 7-1.18 below. The dimensions of the corner mullion can be changed via *Edit Type*.

FIGURE **7-1.18** Corner mullion added

If you switch to a 3D view you will notice a problem (Figure 7-1.19). Only the mullion visible or selected in plan view has changed. The mullion above was not. Remember, each curtain panel has segmented mullions around it.

We could have changed the entire gridline of mullions if we would have selected them first. To do that you need to select the mullion, and then right-click; from the pop-up menu you need to select **Select Mullions → Gridline** as shown in Figure 7-1.20.

FIGURE **7-1.19** Not all mullions changed via plan-view edit

In Figure 7-1.21 below, you see how the **L Corner Mullion** looks when selected. Also, all of the corner mullions work with multiple angles – not just a 90 degree corner. An example of this is shown in Figure 7-1.22.

Finally, when a curtain wall edge gets too close to an adjacent wall, the curtain wall will automatically extend over to the adjacent wall. This can be undesirable when trying to specify an exact wall length. You can turn this feature off by selecting the *Curtain Wall* and then right-clicking on the end grip that appears. From the pop-up menu you can select **Disallow Join** (Figure 7-1.23). Now the curtain wall will not "jump" out of position. Repeat the process to **Allow Join**.

FYI: This does not affect how the wall is embedded within the host wall.

FIGURE 7-1.20 Mullion selection options

FIGURE 7-1.21 L Corner mullion option

FIGURE 7-1.22 Corner mullion at angle

FIGURE 7-1.23 Curtain wall selected; right-click on end grip options

Exercise 7-2:
Developing the Interior Fenestrations

Now you will add the *Curtain Wall* to your floor plans. This will represent large areas of glass. As mentioned earlier in this chapter, the curtain wall will use a feature called **Automatically Embed** so the curtain wall will cut a hole in the stud wall you have previously drawn. This way you do not have to delete any walls. Also, your curtain wall will only go from the floor to the ceiling, so the metal stud wall will extend from the ceiling to the structure above.

The curtain wall layout will be based on the hand sketches presented at the beginning of the previous chapter.

You will begin working with the *Level 1 Small Conference Room* identified in Figure 7-2.1.

1. Zoom in to the *Level 1 Small Conference Room* identified in Figure 7-2.1.

FIGURE 7-2.1 Begin in this area

The first thing you will do is create a new mullion type that is appropriate for an interior condition. Currently the project only has a 2 ½″ x 5″ mullion with a 1″ glazing element. You will make a 2″ wide mullion with a depth that matches the wall thickness. You will also create a ¼″ glazing option as the current 1″ option represents a double glazed insulated glass panel for exterior conditions.

2. In the *Project Browser*, expand **Families** and then **Curtain Wall Mullions** (Figure 7-2.2).

3. Right-click on the **2.5″ x 5″ Rectangular** *Family Type*.

4. Select **Type Properties** from the menu.

5. Click **Duplicate**.

6. Enter **2″ Interior Frame** for the name.

7. Adjust the two width dimensions to **1″** (Figure 7-2.3).

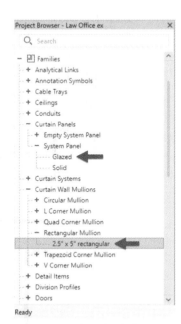

FIGURE 7-2.2 Right-clicking on Mullion type in the *Project Browser*

8. Edit the *Thickness* to match the typical interior wall thickness: **4 7/8"**.

This is not a typical aluminum frame depth, but graphically it looks better to align with the wall thickness, especially in this early phase of the project.

9. Click **OK** to create the new mullion type.

The *Project Browser* now lists a new mullion type. This will be used in your curtain wall system in a few minutes.

Next you will create a new curtain wall panel type. This will represent ¼" glass that will be centered on the mullions.

10. In the *Project Browser*, expand **Families** and then **Curtain Panels** and then **System Panel** (Figure 7-2.2).

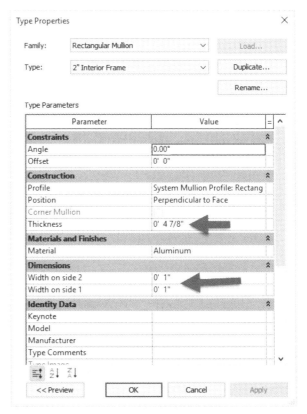

FIGURE 7-2.3 Type properties for new mullion

11. Right-click on the **Glazed** type and select **Type Properties**.

12. Click **Duplicate** and then enter **Glazed – Interior**.

13. Make the following *Type Properties* adjustments (Figure 7-2.4):

 a. *Offset*: **0"**

 b. *Thickness*: **1/4"**

You now have two sub-components ready for the curtain wall type you are about to edit. You will see in a moment how it was convenient to have these two items ready before editing the curtain wall system.

FIGURE 7-2.4 Type properties for new curtain panel

Here you will start the *Wall* tool and get ready to draw curtain walls.

14. Select the *Wall* tool.

15. Set the *Type Selector* to Curtain Wall: **Storefront**.

Because the exterior fenestrations use the "wall" type *Storefront*, you will duplicate it so you can adjust the mullion spacing without changing the exterior windows.

16. Click *Edit Type*.

17. Click *Duplicate*.

18. Enter **Interior Window Wall** for the name.

19. Make the following changes to the *Type Properties* (Figure 7-2.5):

 a. Set the *Function* to **Interior**.

 b. Make sure **Automatically Embed** is checked.

 c. Curtain Panel: *System Panel:* **Glazed - Interior**

 d. Adjust the vertical grid pattern as follows:

FIGURE 7-2.5 Type properties for new curtain wall type

 i. Layout: **Maximum Spacing** *and* Spacing: **4′-0″**

 ii. Adjust for Mullion Size: **Checked**

 e. Adjust the horizontal pattern as follows:

 i. Layout: **Fixed Distance** *and* Spacing: **7′-0″**

 ii. Adjust for Mullion Size: **Checked**

 f. Set all mullions to (6 total): *Rectangular Mullion:* **2″ Interior Frame**

 g. Set the Assembly Code to **C1010100**

20. Click **OK** to accept the changes.

You are now ready to sketch the first *Curtain Wall* in the *Small Conference Room*. Rather than trying to pick the exact start and end points near the adjacent walls, you will randomly pick your two points in a bit from the adjacent walls. Then you can adjust the wall (i.e., the embedded *Curtain Wall*) by dragging its end grips to the proper position or using *Align*.

21. Set the wall height, aka *Top Constraint*, to **Unconnected, 9′-0″** via the *Properties Palette*.

22. Draw the **Interior Window Wall** as shown in Figure 7-2.6; do not worry about the exact pick points for this step.

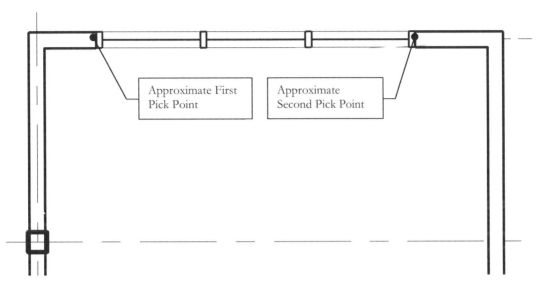

FIGURE 7-2.6 North wall of small conference room – curtain wall added

23. Click the **Modify** button to finish the *Wall* tool.

Next up: Use the *Align* tool to reposition the edge of the curtain wall with the adjacent wall.

24. Adjust BOTH sides of the curtain wall; select the **Align** tool from the *Modify* tab.

 a. Pick the edge of the adjacent wall.

 FYI: *The first pick is always the thing that should not move.*

 b. Pick the edge of the curtain wall; a short dashed line should appear before clicking.

FIGURE 7-2.7 North wall of small conference room – curtain wall adjusted

Notice that because the vertical spacing is set to *Maximum Spacing*, each glass size is the same for this curtain wall (Figure 7-2.7). The spacing will be as large as possible, without going over the 4′-0″ maximum. You can only see the horizontal mullion on the floor, but if you cut a section through the curtain wall, you will see another horizontal mullion at 7′-0″ and then a third one at the top at 9′-0″. The 7′-0″ height was set to accommodate a typical door height.

Next you will add glazing to the South wall of the same conference room. However, you will need to delete the door first. This is because the door in a curtain wall system is a little different from a regular door. It will still appear in the door schedule, but it is technically a *Curtain Panel*. On a real project, it would still be worthwhile to add the original door to help the preliminary plans read better, even though you plan on deleting it later.

25. **Delete** door into the *Small Conference Room*.

26. Add a curtain wall to the South wall, just as you did for the North wall.

 FYI: You can even copy the curtain wall from the North wall to the South wall (Figure 7-2.8).

FIGURE 7-2.8 South wall of small conference room

Now you will add a door. This involves adding another vertical mullion, then selecting the *Curtain Panel* and then picking a door option from the *Type Selector*. A *Curtain Panel* may be swapped out with any *Curtain Panel* loaded in the project or any wall type. The first thing you need to do is delete the frame on the floor so the doorway does not have a tripping hazard!

27. Draw a section as shown in Figure 7-2.8 above.

 a. Pick from left to right.

 b. Adjust the far clipping line as shown (drag grip).

28. Switch to the section view; double-click on the section bubble when the section mark is not selected.

Do not worry about adjusting the view's crop region as you will just delete this section when finished editing this curtain wall. Before adding a new vertical mullion, you need to add a *Curtain Grid* which will define the location of the *Curtain Mullion*.

29. Click **Architecture → Curtain Grid** from the *Ribbon*.

30. Select **One Segment** on the *Ribbon* (Figure 7-2.9).

This option will cause the mullion to stop at the 7'-0" horizontal mullion above.

FIGURE 7-2.9 Curtain grid placement options

31. Move your cursor near the bottom and anywhere within the area the door panel will go; **click** to place the *Curtain Grid*.

32. Click on the temporary dimension text, on the door side, and change it to **3'-3"** as shown in Figure 7-2.10.

33. Click **Modify** to finish *the Curtain Grid* tool.

The 3'-3" dimension is for the 3'-0", the 2" frame on the right, and half of the 2" frame on the left. You now have a 3'-0" x 7'-0" opening defined – well, except for the tripping hazard on the floor. You will fix that next.

34. Click to select the mullion on the floor; this can be done in the section or elevation view or in plan view.

FIGURE 7-2.10 Adding a curtain grid / mullion

35. Select the frame on the floor as shown in Figure 7-2.11.

36. Click the **Pin** icon to make the element editable, or 'delete-able' in this case.

37. With the mullion on the floor still selected, click the **Delete** key on the keyboard to remove the mullion from the model.

FIGURE 7-2.11 Deleting mullion on floor

Next you will adjust the way the new vertical mullion intersects the horizontal mullion above the door. By default the vertical mullion continues uninterrupted from top to bottom. This is ideal in that the vertical frame is usually the primary "structural" element of the window wall. Think of it like floor framing tipped up 90 degrees. The horizontal mullions span from vertical mullion to vertical mullion, so they are not continuous like the verticals need to be.

In the case of our new vertical mullion, the vertical should stop below the horizontal mullion above the door. You will do that next, as Revit allows you to edit all the mullion "join" conditions.

38. Select the new vertical mullion.

39. Click the *Toggle Mullion Join* icon that appeared at the top end (Figure 7-2.12).

The mullion now appears as desired (Figure 7-2.13).

The mullion on the floor is now gone, but the *Curtain Panel* still represents a piece of glass. You will change this *Curtain Panel* to represent a door. This can be done in your section view or the plan view; you will make the change in the plan view so you can see the door swing and adjust it if needed.

FIGURE 7-2.12 Toggle mullion join icon when mullion is selected

40. Switch back to the **Level 1 plan view**.

41. Select the *Curtain Panel*; use the **Tab** key to cycle through the options (i.e., entire curtain wall, mullion, curtain panel).

FIGURE 7-2.13 Toggle mullion result

42. Click the **Pin** icon to un-pin it as shown in Figure 7-2.14.

43. With the *Curtain Panel* still selected, click **Curtain Wall Sgl Glass** from the *Type Selector*.

FIGURE 7-2.14 Un-pin the curtain panel

The *Curtain Panel* door family is now showing. You will likely need to click the flip-arrows to adjust the hand/swing of the door. The door should open up against the adjacent stud wall (Figure 6-3.16).

44. Click the flip-arrows to adjust the door swing as needed (Figure 7-2.15).

45. Delete the section mark/view, by deleting the section mark from the plan view or from the *Project Browser*.

FIGURE 7-2.15 Door added and adjusted

Notice the 3D view of the curtain wall with door from the hallway (Figure 7-2.16), and the ceilings still need to be added. Also, the curtain wall door Revit provides is a little atypical in that it is a solid piece of glass. Most doors like this would have an aluminum frame around the perimeter.

FIGURE 7-2.16 3D view of small conference room from hallway

Because we do not like the type of curtain wall door Revit provides, i.e., a solid slab of glass, you will use one provided with this book. See the inside front cover for instructions.

46. Switch to the *Insert* tab on the *Ribbon*.

47. Select **Load Family**.

48. Browse to the location you saved the provided files; see inside front cover of this book for download instructions and for your unique access code.

49. Select **efco_d200_narrow_stile_doors_-_d200c002_10393.rfa**.

 EFCO is a manufacturer and the file name represents their specific model number and product construction. Follow this link to learn more about them: http://www.efcocorp.com/.

50. Click **OK** to load the family into your project.

51. Select the curtain wall door in the *Small Conference* via the plan view.

52. Pick the new door from the *Type Selector*; it is **D200C002**.

The door has now been replaced, but there is a slight problem with the way this *Family* was created. The 3D geometry should have been set to NOT show up in the plan views. Right now, you see the 3D line work showing the door open 45 degrees and you also see the 3D door in the closed position (Figure 7-2.17). You will quickly fix this issue in the *Family Editor*.

FIGURE 7-2.17 New door loaded, but has a problem

Edit
Family
Mode

53. Select the door in the model and then click the *Edit Family* button on the *Ribbon*.

You are now in the *Family Editor*. Your project is still open, but the door *Family* is now also open. It is possible to have multiple projects and families open at the same time. The main constraint is the capacity of your computer: processor speed and amount of RAM. You will learn more about creating families from the introduction to content creation in chapter 20.

54. Drag a window to select all the 3D geometry in the 3D view that opened.

Edit
Extrusion

55. Click the **Visibility Settings** icon on the *Ribbon*; this is only visible when one or more 3D elements have been selected.

56. Clear the two **plan** boxes as shown in Figure 7-2.18.

57. Click **OK**.

Your *Family* is ready to be loaded back into the project. It is possible to save the changes to a file and then use the *Load Family* feature again. However, all you really need to do is click the *Load into Project* button.

> *FYI: You should save the revised file to disk if you will want to use it in the future on other projects.*

FIGURE 7-2.18 Editing the visibility settings in the *Family Editor*

58. Click the **Load into Project** button on the *Ribbon*.

59. Click the **Overwrite the existing version** button when prompted.

> *TIP: You typically do not want to click the second option, i.e., overwrite parameters, as this could change project specific values.*

FIGURE 7-2.19 Door revised so 3D elements do not show in plan

The door now looks the way we want it to (Figure 7-2.19). Well, except for the door swing. As mentioned in the previous chapter, a 45 degree door swing typically represents an existing door. This can be corrected in the *Family Editor* as well, but we will skip that for now.

> *FYI: The door is still open in the* Family Editor. *You can* **Ctrl+Tab** *to it and close it now or when you close Revit. You don't need to save the file revision.*

Shown to the right is the updated 3D view from the hallway, showing the new aluminum curtain wall door.

A quick note about how this view was set up. To get the *Camera* far enough away, it was placed in the *Large Conference Room*. The wall common to the *Large Conference Room* and the hallway was hidden in this view (select it and then right-click → **Hide in View** → **Element**). To select the wall, you may have to select it in plan view and then use *Ctrl+Tab* back to the camera view.

The *Section Box* feature was also turned on and adjusted. Notice how the earth, aka, *Toposurface*, appears in section when cut by the *Section Box*.

FIGURE 7-2.20 Updated 3D view of small conference room, showing an aluminum door

Next you will add a curtain wall to the *Cafe*. This will have a few slight variations on what you have done previously.

60. **Zoom** in to the *Staff Lunch Room* on Level 1.

61. **Delete** the double doors into the *Lunch Room*.

62. Place a ***Curtain Wall*** in the West wall as shown in Figure 7-2.21; snap to the face of the adjacent wall to make the overall length of the curtain wall match the width of the lunchroom.

In this case you will be adding a double door centered on the curtain wall. However, you do not want a mullion in the center, so you will delete the center mullion, just at the door opening and not at the transom. Next you will adjust the two remaining mullions inward to make the overall double door opening exactly 6'-0", for two 3'-0" wide doors.

63. Cut a **temporary section** as shown in Figure 7-2.21, and then switch to that new view.

You can adjust the *Crop Region* if you want, but the view is temporary and will be deleted.

64. **Select** the center mullion as shown in Figure 7-2.22 and un-pin it.

65. **Delete** the selected mullion.

66. **Delete** the two mullions on the floor.

FIGURE 7-2.21 Curtain wall added to lunchroom

FIGURE 7-2.22 Curtain wall in elevation/projection – one mullion selected and un-pinned

The unneeded mullions are gone, but the *Curtain Grid* still divides the *Curtain Panels* in two. You need one large *Curtain Panel* to properly place a double door.

Next you will select the *Curtain Grid* and remove a portion of it from the *Curtain Wall*. This can be added back if needed later using the same method described next.

67. Select the vertical *Curtain Grid* as shown in Figure 7-2.23; use the **Tab** key as required (Step 1).

68. Un-pin the **Curtain Grid** (Step 2).

69. On the *Ribbon*, with the *Curtain Grid* selected, click the ***Add/Remove Segments*** button (Step 3).

70. Click the portion of ***Curtain Grid*** in the door opening to remove it (Step 4).

71. Adjust the remaining vertical mullion join above the door as shown.

FIGURE 7-2.23 Curtain *Wall* in elevation/projection – *Curtain Grid* selected

You should now have one large *Curtain Panel*. The next step is to adjust the overall width to exactly 6'-0" for the two 3'-0" doors. You will be using another provided family for the double door options. If the center mullion was to remain, you would be able to use the previously downloaded door. Center mullions are usually only used when a high acoustic rating is needed or for insulation requirements (usually only in exterior conditions). A double door without a center mullion offers a wider uninterrupted opening for moving large items in and out.

72. **Select** the *Curtain Grid* on the right side of the door opening and **un-pin** it (Figure 7-2.24).

73. **Adjust** the dimension to **3'-1"**.

This dimension is 3'-0" door and 1" for half of the 2" frame centered on the *Curtain Grid*.

74. **Adjust** the left *Curtain Grid* to be **3'-1"** as well.

75. Using the steps previously described, load the provided double door EFCO family: **efco_d200_narrow_stile_doors_-_d200c001_10393.rfa**.

76. Add the door to the *Curtain Wall* as shown in Figure 7-2.25.

 a. Use the **6" Bottom Rail** option in the *Type Selector*.

 b. Note that the *Family* does not need to be fixed in the *Family Editor* like the single door did.

 c. See steps 53-56 for more information.

FIGURE 7-2.24 Curtain wall in elevation/projection – adjusting curtain grid position

Your floor plan should now look like Figure 7-2.28. In this case you started with the default mullion layout defined in the *Curtain Wall* "wall type." However, you ended up changing just about everything except the perimeter framing.

It is possible to start this entire process from the "blank slate" curtain wall type mentioned earlier in this chapter. Either method would take close to the same amount of time. But this one is more consistent with the wall type and other locations to be developed; this is the route which was chosen.

You are now finally ready to add the rest of the curtain walls to your floor plans. The steps will be the same as those just covered.

FIGURE 7-2.25
Final layout

77. Add the remaining *Curtain Wall* on Level 1 per Figure 7-2.26.

 a. Use wall type *Curtain Wall:* **Interior Window Walls**.

 b. Door in *Large Conference Room* and windows in *Fitness Center* are centered on the *Impromptu Meeting* space, as denoted by the center line.

 c. Adjust all openings to make single doors exactly **3'-0"** wide and double doors **6'-0"** wide; add or adjust mullions as required.

 d. All dimensions are to the edge of the *Curtain Wall* or the center of the wall.

 e. *Curtain Walls* with no dimensions extend to adjacent wall(s).

 f. Set the end of each curtain wall to ***Disallow Join*** as needed to prevent unwanted joining (see pages 6-12 and 7-9).

TIP: *Sometimes it is helpful to draw Detail Lines (Annotate tab) to help determine the final location for Curtain Wall mullions. Be sure to delete the Detail Lines when they are no longer needed.*

 g. You may use the *Detail Line* to add the centerline as shown. Change the *Line Style* to **<Centerline>** from the *Ribbon* while the *Detail Line* tool is active.

78. Add the *Curtain Walls* on Level 2 per Figures 7-2.27 – 29.

 a. Follow all the steps and tips previously offered.

 b. Remember to use ***Disallow Join*** to prevent a curtain wall from moving when close to an adjacent wall.

3D view showing curtain layout for Level 2

FIGURE 7-2.26 Curtain Wall layout for Level 1 – rotated view

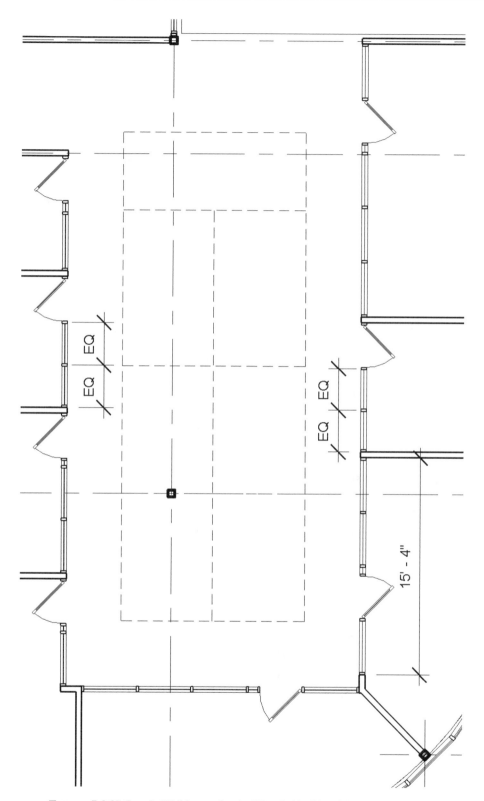

FIGURE 7-2.27 Curtain Wall layout for the West half of Level 2 – rotated view

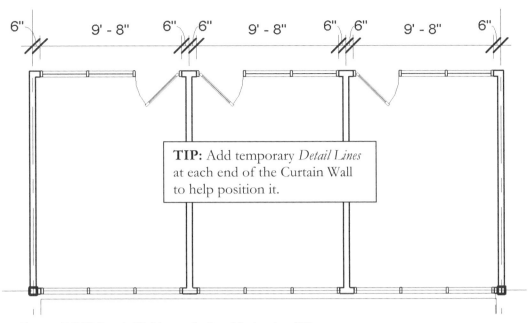

TIP: Add temporary *Detail Lines* at each end of the Curtain Wall to help position it.

FIGURE 7-2.28 Curtain Wall layout for Level 2 *Associates Offices*

FIGURE 7-2.29 Curtain Wall layout for *Retired Partner Office*

Self-Exam:

The following questions can be used as a way to check your knowledge of this lesson. The answers can be found at the bottom of this page.

1. Use the Wall tool to draw curtain wall. (T/F)

2. Revit's curtain wall door has a thick frame. (T/F)

3. When modeling curtain walls, it is helpful to cut a section on one side. (T/F)

4. Sometimes you need to unpin a mullion before it may be deleted from a curtain wall. (T/F)

5. A corner Mullion can interact with two curtain walls. (T/F)

Review Questions:

The following questions may be assigned by your instructor as a way to assess your knowledge of this section. Your instructor has the answers to the review questions.

1. You cannot use Edit Profile on curtain wall. (T/F)

2. To prevent a curtain wall from connecting to an adjacent wall, use

 _____.

3. Name the four primary curtain wall elements: _____ _____

 _____ _____

4. A curtain panel can only be glass or a door. (T/F)

5. What brand curtain wall door did you download: _____

6. When you click Load into Project, from the Family Editor, the family is closed. (T/F)

7. When adding Curtain Grid, use the One Segment option to limit the grid to one area. (T/F)

8. Sometimes it is helpful to draw detail lines to help determine the final mullion location. (T/F)

9. You changed the _____ in the curtain wall door family so the 3D door panel did not appear in plan.

10. Select _____ to make a curtain wall cut a hole in a regular wall.

Lesson 8
Stairs and Railings:

This chapter will cover stairs and railings in Autodesk® Revit®.

Exercise 8-1:
Introduction to Stairs

Stair Types

When the *Stair by Component* command is selected, the **Type Selector** lists three options (Figure 8-1.1); this example is based on the Commercial-default template. At this point it is helpful to compare these three options with their corresponding entries in the **Project Browser** under Families - marked with #1 in Figure 8-1.2. The fourth item marked with #2, "stair," contains everything related to the older *Stair by Sketch* command. The remaining items are the "components" for this newer stair tool. These components are selected within the properties of the three main options highlighted here.

FIGURE 8-1.1 Type Selector; Stair by Component tool active

FIGURE 8-1.2 Family\Stairs in Project Browser

Stair Component Types

To see how a Stair by Component type, listed above, makes use of the **component types** take a look at the Type Properties for the "Steel Pan Stair" in Figure 8-1.3. The selected component types, which have been highlighted, can each be seen in the Project Browser as shown in Figure 8-1.4, also highlighted.

Unlike the Stair by Sketch, which contains all required properties in one Type Properties dialog (covered in the previous section), the Stair by Component has Type Properties for the entire stair (Steel Pan Stair in this example) and then Type Properties for the various components: Run Type, Landing Type, Supports and Cut Mark Type.

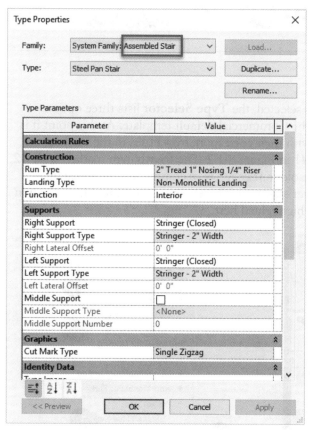

FIGURE 8-1.3 Steel Pan Stair type properties

FIGURE 8-1.4 Component families used by Steel Pan Stair

> **Supports defined:** Both a **Carriage** and a **Stringer** are able to support the stair. In Revit, a carriage is below the riser and treads while a stringer runs alongside them. A *Stair by Component* stair can have three supports, Left, Right and Middle, as seen in Figure 8-1.4. All three supports can be carriage-style but only the Left/Right can be stringers. **FYI:** There is some legacy terminology in Revit that is a bit confusing; a stringer is called "Closed" and a carriage is called "Open."

Examples of parameters associated with component types can be seen in Figure 8-1.5; here we see the *Run Type* named **2" Tread 1" Nosing ¼" Riser** and the *Right Support Type* named **Stringer – 2" Width**.

FIGURE 8-1.5 Type properties for two of the stair component types associated with 'Steel Pan Stair'

Notice in the image above that things like material and tread profile can be selected. This entire **Run Type** can be duplicated and modified when required—for example, a project has a simple utilitarian stair and a more formal open stair. Also notice that a profile can be selected for the supports: **Section Profile**. Thus, we can define a channel-shape which is fairly common in commercial construction. Remember, with the Stair by Sketch, the support can only be rectangular in shape.

All these settings will make more sense once you start using this tool. However, at this point you should have the basic understanding that there are properties for the entire stair system and then separate properties (i.e. types) for the individual components of the stair.

Creating Stairs using Stair by Component

While creating a stair with Stair by Component, the 3D "components" are created along the way. Using this tool, the individual sketch lines are not seen or editable (e.g., boundary and riser) like they are with the Stair by Sketch tool. This makes it easier to see what the stair will look like before clicking Finish Edit Mode. Also, because there are not sketch lines, it is possible to create a stair with multiple switch backs.

Here are the basic steps for the Stair tool:

- Start at the lowest level
- Select **Stair** tool
- Verify stair style in **Type Selector**
- Verify Base and Top Level settings in **Properties**
- Adjust settings on **Options Bar**
 - Location Line
 - Offset (horizontal)
 - Stair Width
 - Landing creation
 - See Figure 8-1.6 below
- **Pick points** to define one or more runs
- *Optional:* Add **Landing** to top or bottom if needed
 - select *Landing* on Ribbon while in Stair by Component tool
- *Optional:* Select and delete individual **Support** elements
 - For example, if there is a door at an intermediate landing
 - Use the *Support* option to replace if needed

The image below shows the **Options Bar** while the Stair by Component tool is active (Figure 8-1.6). Be sure to select the best option for Location Line. If in a stair shaft, one of the exterior support options is likely the more convenient option. Note that these settings can vary per run within the same stair instance.

Location Line: Run: Center ∨ Offset: 0' 0" Actual Run Width: 3' 0" ☑ Automatic Landing

FIGURE 8-1.6 Options bar for Stair by Component tool

Picking the points on-screen is similar to the Stair by Sketch tool. Two clicks could create a continuous single-run from floor to floor, or, as seen in the image below, multiple clicks creates multiple runs. All along, Revit indicates how many risers remain before reaching the next level.

FIGURE 8-1.7 Creating stair with Stair by Component tool

Once the components are created, and while still in the Stair by Component tool, individual components can be selected.

Stair Landings

In the first image (Figure 8-1.8) the intermediate landing is selected. Notice there are six grips. There is one grip on each side and one grip at each stair run.

The next image (Figure 8-1.9) shows how the landing can be modified when adjusting the grips by clicking and dragging. In this example, the three circled grips were moved to change the size and shape of the intermediate landing.

FIGURE 8-1.8 Stair by Component edit mode with landing selected

Landings are only automatically added between runs when **Automatic Landing** is checked on the Options Bar. If additional landing construction is required at the top or bottom of the stair, select the **Landing** option (see Ribbon image below) and then sketch a closed loop which is touching the adjacent stair run.

FIGURE 8-1.9 Landing adjusted using grips

The result is a landing with added supports on all three sides as seen in the upper part of Figure 8-1.10. Selecting the end support allows you to delete it. Notice how the side supports adjust from mitered to flush cut in the lower image once the support has been deleted.

When a landing is selected, its **Relative Height** and thickness are listed in the Properties Palette. Adjusting the elevation will automatically adjust the runs, plus the number entered will be changed to the closest riser position. However, the runs are not adjustable if any of them have been converted to a sketch (more on this in a moment).

FIGURE 8-1.10 Added landing and support deleted

Non-rectilinear Landings:

The intermediate landings automatically created by Revit can be modified. While in Stair Edit mode, select the landing component as shown in the first image in Figure 8-1.11. Click the **Convert** command on the Ribbon.

At this point, a message appears indicating the conversion is irreversible. Click **Close** to complete the conversion. The landing portion of the stair is now a Sketch-based stair element.

Original Landing

With the landing still selected, click the **Edit Sketch** command on the Ribbon. The lines which define the perimeter of the landing, in plan-view, are now editable. Use **Draw** and **Modify** tools, such as Line, Circle, Arc, Split, Trim, Offset, etc. to adjust the footprint of the landing. An example of a modified landing can be seen in the second image to the right.

When finished editing the landing sketch, click the **green checkmark** to exit sketch mode. If the sketch is not a clean outline with any overlapping or crossing lines Revit will create the landing. Unlike other sketches, like floors, there cannot be multiple closed loops (e.g. in a floor, a loop within a loop would create a hole in the floor). If a sketch error is shown, Revit will highlight the problem area. Simply click Continue and fix the problem.

Convert to Sketch and Modify

Click the green checkmark one more time to finish the entire stair instance; Revit will recreate any hosted railings as well. This can be seen in the last image to the right.

If the landing needs to be modified again, select the stair and click Edit Stair and then select the landing component and click Edit Sketch.

Custom Landing Result

FIGURE 8-1.11 Custom Stair Landing

Landing Properties:

A landing can have its main properties defined by the Run, which is the default setting, or independently.

Selecting a landing and then looking at its Type Properties we see **Same as Run** is selected (Figure 8-1.12). This tells Revit to inherit several properties from the stair run that the landing is connected to within the *Stair by Component* instance.

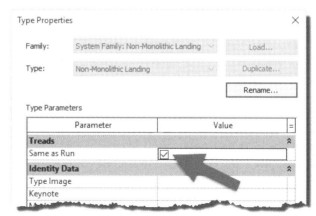

FIGURE 8-1.12 Landing type properties; 'Same as Run' checked

TIP: A component within the stair can be selected without entering edit mode. Simply hover your cursor over the component and then tap the **Tab** key and click to select once the component highlights.

When **Same as Run** is unchecked, several properties appear as shown in Figure 8-1.13.

Landing Material:

In addition to controlling the various tread properties, notice that a **material** parameter is also now available when **Same as Run** is unchecked. If the landing requires a

FIGURE 8-1.13 Landing type properties; 'Same as Run' unchecked

different floor finish, uncheck the **Same as Run** option and specify the material using the **Tread Material** parameter.

An example of the nosing and material matching the run can be seen in Figure 8-1.14.

FIGURE 8-1.14 Landing nosing and material example

Stair Supports

When a support (aka Stringer) is selected, there are three ways in which the end can be cut: Horizontal, Perpendicular and Vertical (Figure 8-1.15). The default is Vertical. This setting is only adjustable at the ends of a top or bottom run. Notice in the image that the setting can vary for each side of the stair.

These options for supports are more limiting than those for the Stair by Sketch tool. Looking back at that section, we see it is possible to control whether the stringer continues to slope or is flat, and its vertical dimension. With the Stair by Component tool we can only extend the stringer if there is a landing to host it.

Horizontal　　　　　*Perpendicular*　　　　　*Vertical*

FIGURE 8-1.15 Three options for a support <u>end cut</u>: Horizontal, Perpendicular and Vertical, respectively

Revit generally does a good job at how it supports transition. Sometimes there are issues as seen in the example below, on the left. This is not really Revit's fault. The issue is we did not give Revit enough room to make an appropriate transition. Once the landing has been adjusted, we see the buildable result on the right. As we will see later, this is true for railings as well.

Support Transition Issues　　　　　*Support Transition Correction*

FIGURE 8-1.16 Dealing with supports which do not transition correctly

The example in the previous image (Figure 8-1.16) shows two runs coming off of the same landing, up to the next level. The steps to accomplish this will be covered in the next section on stair runs.

Stair Runs

When an individual stair run is selected there are a few adjustment options per the grips as seen in Figure 8-1.17.

FIGURE 8-1.17 Stair run selected

Dragging the side grips modifies the width of the stair – just for the selected run. Each run can be a different width.

Dragging the top end (or bottom end) grip will depress the number of steps in the selected run. The opposite run will have the same number of steps added (or removed); see Fig. 8-1.18.

A selected run can also be **Nudged** using the Arrow keys. When nudging the stair away from the landing, Revit will automatically extend the landing to meet up with the new stair position.

FIGURE 8-1.18 Stair run grip adjusted

Additional runs can be modeled within the same stair instance as shown in the previous section. This may be for a grand stair or access to an adjacent roof level. Here is how this is done.

- While in edit mode, select the landing (Fig. 8-1.19)
- Copy the **Relative Height** to the clipboard; highlight and press **CTRL + C**
- Select the **Run** command on the *Ribbon* (Step #1, Fig. 8-1.20)

FIGURE 8-1.19 Height listed for selected landing

- Switch the *Properties Palette* drop-down to **Runs** (Step #2, Fig. 8-1.20)
- Paste the landing height into **Relative Base Height**; **CTRL + V** (Step #3, Fig. 8-1.20)
- Draw the stair run (Figure 8-1.21)

FIGURE 8-1.20 Adding an extra stair run from an intermediate landing

The result is a new run joined to the landing as seen in the two images that follow. Notice the run arrow is not able to connect multiple stair runs. Thus, make sure to model the desired primary path first. Also, if the secondary stair run needs to be repositioned, it will become un-joined with the landing. However, the landing can be adjusted and once it touches the run it will automatically join with it.

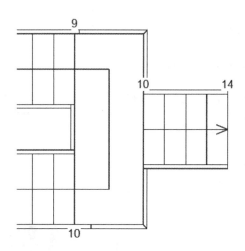

FIGURE 8-1.21 Extra run added to stair

FIGURE 8-1.22 3D view of modified stair

Stair Limitations

There are a few limitations with *Stair by Component* compared to *Stair by Sketch* when it comes to modeling a stair run or landing. For example, the edge of the stair cannot be angled as shown in the previous section. The bottom tread is not able to be extended out and rounded either.

The way to deal with these limitations is to convert the stair component to a sketch-based run or landing. As covered a few pages back for a custom landing, with a stair run selected, while in edit mode, click the **Convert** option on the Ribbon. As the following message indicates, this is irreversible. The main drawbacks are not being able to drag grips to resize the components, but runs and landings still join properly. Thus, the drawbacks are minimal when compared to the need to model the stair correctly. Once a component is converted, select it and click **Edit Sketch** on the Ribbon to modify a run or landing. At this point you are two "edit modes" deep—meaning the green checkmark, for finish, must be clicked twice to fully complete the stair modification.

Stair Annotation

There are a few annotation tools related to Stair by Component to be aware of.

The **Stair Path** tool, found on the Annotation tab, adds the UP or DN text and an arrow. Keep in mind that every tool on this tab is 2D and view specific. When a stair is initially created, the Stair Path annotation is automatically added to every plan-view. When subsequent plan views are created (or copied without detailing), stair arrow annotations are not automatically added to those.

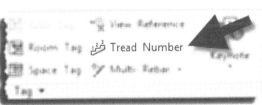

The Stair Path element can be selected and then repositioned or even deleted in each view. If the annotation is deleted, or a new plan-view is created, use the Stair Path tool to add it to a stair instance.

The visibility of the text and arrows can be controlled separately and per view. Figure 8-1.23 shows the **Visibility/Graphic Overrides** dialog with the Stair Paths category expanded on the Annotation Categories tab.

The text and arrow can be moved separately. The text, "UP" or "DN," can be moved in any direction. The arrow can only be moved perpendicular to the run of the stair.

The **Tread Number** tool, also found on the Annotation tab, enumerates each tread for a selected run. This tool only works with *Stair by Component* elements; it will not work with a stair created with *Stair by Sketch*. However, if a run is converted to a sketch within a *Stair by Component* stair this command will still work.

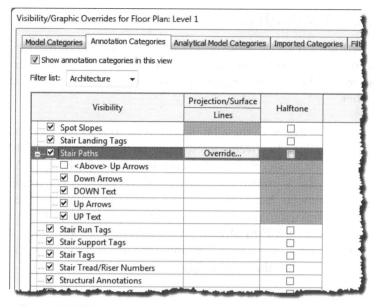

FIGURE 8-1.23 Stair path category

This tool is simple to use. To add tread numbers to a stair, simply start the command and then select a stair run within a plan or section view. Each separate run needs to be selected, even within one Stair by Component instance. The numbers only appear in the current view.

An example of the results can be seen in Figure 8-1.24 for both section and plan views.

Once placed, select it to access several formatting options via the Properties Palette. The Tread Number element is selected in Figure 8-1.24; notice the dashed line running through the numbers. When not selected, there is no visible line.

Plan View

Section View

FIGURE 8-1.24 Tread Number tool results

Warnings

If a stair has a problem, for example a stair instance does not reach a floor level, the **Show Related Warning** button will appear whenever that stair is selected.

Clicking this button will open a dialog describing the problem as shown in the image below (Figure 8-1.25). This can happen if a stair is created and there are "remaining" risers listed prior to clicking finish. Also, if a stair type is switched from non-monolithic to monolithic this can appear.

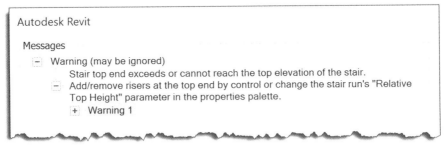

FIGURE 8-1.25 Warning message for selected stair

To see a full list of warnings for the entire model, select **Manage → Warnings** (Figure 8-1.26). This should be done occasionally to ensure there are no major issues with the model. Keeping this list free of major errors can improve the performance of the model as Revit is not constantly checking this situation.

FIGURE 8-1.26 Warnings list for entire project

Exercise 8-2:
Introduction to Railings

This "exercise" will provide a basic introduction to the railing tool in Revit. There are no required steps for the law office project in this section. However, it is highly recommended that this information be reviewed prior to completing the steps required later in this chapter.

Railing Types

When the *Railing* command is selected, the railing Types in the current project are listed in the **Type Selector**. Compare this list to the same items listed in the **Project Browser** under Families\Railings\Railings. The items listed in the Project Browser can be Duplicated or Deleted.

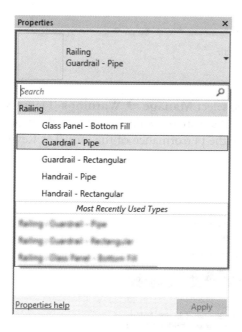

FIGURE 8-2.1 Type Selector; Railing tool active

FIGURE 8-2.2 Family\Railing in Project Browser

Railing Component Types

Similar to the way Stairs work, Railings have several high-level properties (i.e. Type Properties). These can be seen in Figure 8-2.3. Notice the option to select a Top Rail and Handrail type. Also, there is an option for two handrails on the same railing. This allows a Handrail to be placed on both sides of a railing.

The options available in the Top Rail and Handrail component drop-down lists is based on the Types created in the current project. These options can be seen in the Project Browser as shown in Figure 8-2.4. These Types can be Duplicated or Deleted via the right-click menu.

FIGURE 8-2.3 Guardrail – Rectangular railing properties

FIGURE 8-2.4 Component families used by railing

Conceptually, take note that a Top Rail and Handrail are more complex than just selecting a profile, as this will help you understand Revit's internal structure for these element types. The properties associated with the Top Rail and Handrail selected in the previous image can be seen in Figure 8-2.5.

In Revit, a Profile is a simple 2D closed outline—a series of connected lines and/or arcs. An example can be seen in Figure 8-2.7. Similar to other families, profiles can be parameterized. Thus, this simple circle is able to represent multiple diameters. The Project Browser shown in Figure 8-2.6 shows two sizes (i.e. types) for this profile. Just like doors, these types can be duplicated and new sizes created without the need to create a new profile family. However, creating a new profile is easy: **New → Family → Profile.rft** template.

FIGURE 8-2.5 Type properties for two of the railing component types associated with "Guardrail – Rectangular"

FIGURE 8-2.6 Profiles families
used by handrail & top rail types

FIGURE 8-2.7 Profiles Circular handrail profile
as seen in the family editor

In Figure 8-2.5, notice both Handrails and Top Rails have an option called **Terminations**. Handrails also have an option for **Supports**. In this example they are all set to "None." Figure 8-2.8 shows that a termination (aka escutcheon) occurs at the end of a Handrail (or Top Rail) and a Support is what holds the Handrail in place. Each of these items are separate families loaded into the current project.

FIGURE 8-2.8 Termination and Supports example

Now that the various components and nested families in a railing have been covered, we need to back up to the beginning and look at how the rest of the railing is defined. This is the pattern, either horizontal, vertical or both, that fills that area from the stair/floor up to the Top Rail. You

are about to see two of the most complex dialogs in Revit! We will compare the settings between the two guardrail types Pipe and Rectangular to help understand how the various options and settings work.

Rail Structure (Non-Continuous)

When a railing has horizontal/parallel rails, in addition to the Top Rail and Handrail, they are defined in the **Edit Rails (Non-Continuous)** dialog shown in Figure 8-2.10. To access this dialog, select a railing, click Edit Type, and then click Edit next to Rail Structure (Non-Continuous).

Looking at Figure 8-2.9 we notice the "Pipe" railing only has vertical elements. Thus, the Edit Rails dialog is empty. Contrast this with the "Rectangular" railing type which is mainly composed of horizontal/parallel rails.

Guardrail - Rectangular Guardrail - Pipe

FIGURE 8-2.9 Graphically comparing two railing types found in the Commercial-Default project template

FIGURE 8-2.10 Defining horizontal elements within a railing

The numbers added to the "Rectangular" railing example, in Figure 8-2.9, correspond to the numbered rows in Figure 8-2.10. Notice each row has five settings as defined below.

Name

Each row can be named to help keep things organized. Each row must have a unique name. These named rows appear in the Edit Baluster Placement dialog covered in the next section.

Height

The Height of the rail is from the floor/landing or the front edge of the tread as shown in Figure 8-2.11. Compare the dimensions shown in this image to those listed in the Edit Rails dialog (Figure 8-2.10). Note that the height is to the top of the rail in each case. Dimensions from the tread edge and to the top of the Top Rail and Handrail relate to how building codes define stair, guardrail and handrail requirements.

Offset

This allows each rail to be offset horizontally from the main railing if needed. In previous versions of Revit this is how a handrail was defined. There are many creative ways this option can be used. Keep in mind there are two other "offset" options for a railing: the instance parameter *Tread/Stringer Offset* and the type parameter *Baluster Offset*.

Profile

This defines the shape continuously extruded along the handrail. Speaking of continuously, it is unclear why the dialog says "Non-Continuous"! Load and/or create additional Profiles in the current project for more options. Note that this could include something that looks like a stair stringer to maintain a consistent look when the railing at a stair transitions to a railing at an open floor edge (Figure 8-2.12).

Material

This defines the material applied to the shape created by the extruded profile.

FIGURE 8-2.11 Railing rail dimensions

FIGURE 8-2.12
Railing with stringer

In the Edit Rails dialog, new rows can be inserted, and current rows can be duplicated or deleted. The order of the rows does not matter as the Height parameter dictates its position, but having them in the same relative order is helpful—thus the Up and Down buttons. Finally, expanding the preview option allows you to see the results without closing the dialog or clicking apply. However, clicking Apply will update the model without the need to close the dialog. This shows the changes in your model rather than a generic sample preview.

Baluster Placement

Where the *Rail Structure* dialog defines all the horizontal/parallel elements (except for top and handrail) the **Edit Baluster Placement** dialog defines all the underlined vertical elements within a railing system. This dialog, as seen in Figure 8-2.13, has two main sections: **Main Pattern** and **Posts** as pointed out.

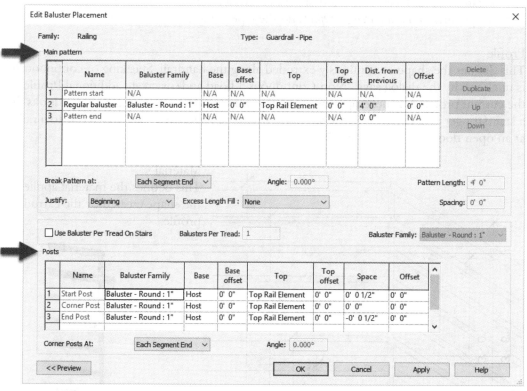

FIGURE 8-2.13 Edit Baluster Placement dialog; Guardrail – Pipe (4'-0" main pattern spacing)

Main Pattern
In the example above, for Guardrail – Pipe, there is not a "main pattern" per se. This is because the "main pattern" is defined by horizontal/parallel bars. Thus, in this example the Main Pattern is the "structural" supports for the railing system which are spaced 4'-0" (note the highlighted value in Figure 8-2.13). **FYI:** The 1" diameter pipe at 4'-0" on center may or may not be adequate to develop the code required lateral resistance at the top edge of the railing.

Comparing the two guardrail examples (Figure 8-2.9) we see the "Pipe" type has the "main pattern" spaced at **4'-0"** (Figure 8-2.13), whereas the "Rectangular" option spacing is only **4"** (Figure 8-2.14).

FIGURE 8-2.14 Edit Baluster Placement dialog; Guardrail – Rectangular (4" main pattern spacing)

Main Pattern settings per row:

- **Name**
 User defined name for each baluster in the main pattern. This is mainly to help keep things organized.

- **Baluster Family**
 This drop-down lists all the baluster families (post or panel) loaded in the current project. Note that *None* is also an option.

- **Base**
 Select where the bottom of the baluster is placed. Built-in options are *Host* and *Top Rail Element*. Also, all the named rails (from the Edit Rails dialog) appear here as well. **FYI:** The host can be a stair, floor, ramp, roof or toposurface.

- **Base Offset**
 Use this setting to move the baluster vertically relative to the selected *Base*. The value can be positive (for up) and negative (for down).

- **Top**
 Similar to the *Base* settings, the Top setting specifies the position of the top of the baluster. The height of the baluster can vary depending on the context, e.g. hosted on a stair (sloped) versus hosted on a floor (flat).

- **Top Offset**
 Use this setting to move the baluster vertically relative to the selected *Top*. The value can be positive (for up) and negative (for down).

- **Dist. From previous**
 Determines the spacing between the balusters.

- **Offset**
 Horizontally offset the baluster relative to the sketched path. The value can be positive and negative.

Both of these examples are simple and easy to understand. The example shown below depicts a more complex baluster pattern. Note that the Base and Top options selected relate to named rows in the Edit Rails dialog. This entire "main pattern" is repeated along the railing length. Note the **Justify** and **Excess Length Fill** settings as well.

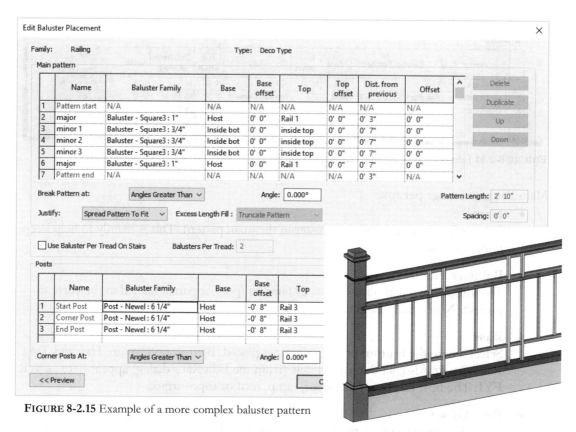

FIGURE 8-2.15 Example of a more complex baluster pattern

Another variation on **baluster placement** is to place them **per tread** as shown in the example below Figure 8-2.16. In this residential-type application there are two balusters per tread. Notice the more elaborate newel post family selected as a starting post.

FIGURE 8-2.16 Baluster per tread

A **baluster family** is unique in that its top and bottom must be able to slope, at stairs, or be flat at a floor or landing. An example of the Baluster – Square family can be seen in the image below—shown in the family editor (Figure 8-2.17). Notice the **Top Cut Angle** and **Bottom Cut Angle** parameters. All of the required parameters are provided in the **Baluster – Post** family template file. If a new baluster family needs to be created, either copy (the RFA file) or start from the provided template.

FIGURE 8-2.17 Baluster example in the family editor

For more information on Stairs and Railings in Revit:
- Visit the Autodesk Help website
- Visit www.revitforum.org, www.augi.com or https://forums.autodesk.com
- Check out **Tim Waldock's** in-depth posts on Revit's stairs and railings on his blog RevitCat: http://revitcat.blogspot.com/

Posts

The Posts section is more straightforward.

FIGURE 8-2.18 Posts section of Edit Baluster Placement dialog

Posts settings per row:

- **Name**

 There are only three built-in options named First, Corner and End.

- **Baluster Family**

 This drop-down lists all the baluster families (post or panel) loaded in the current project. Note that *None* and *Default* are also options.

- **Base**

 Select where the bottom of the post is placed. Built-in options are *Host* and *Top Rail Element*. Also, all the named rails (from the Edit Rails dialog) appear here as well.
 FYI: The host can be a stair, floor, ramp, roof or toposurface.

- **Base Offset**

 Use this setting to move the post vertically relative to the selected *Base*. The value can be positive (for up) and negative (for down).

- **Top**

 Similar to the *Base* settings, the Top setting specifies the position of the top of the post. The height of the post can vary depending on the context; e.g. hosted on a stair (sloped) versus hosted on a floor (flat).

- **Top Offset**

 Use this setting to move the post vertically relative to the selected *Top*. The value can be positive (for up) and negative (for down).

- **Space**

 Reposition the post along the length of the railing, if needed. One example might be to align the edge of the post with the vertical cut edge of the stringer below.

- **Offset**

 Horizontally offset the post relative to the sketched path. The value can be positive and negative.

Railing Tools

There are two main ways a railing can be added to a Revit project:

- In conjunction with the **Stair** or **Ramp** tools
- Separately, using the **Railing** commands
 a. Sketch Path
 b. Place on Host

Railing Placed with Stair Tool

When a new project is created using one of the default Revit templates, and a stair (or Ramp) is created, a railing is automatically placed on both sides. The image below (Figure 8-2.19) shows what the default results look like in a 3D view. These two railings are hosted by the stair and will update if the stair is modified. There are a number of things we can change to convey the design intent when needed.

FIGURE 8-2.19 Default railing created with a new stair

> If all railings have been deleted, use the **Railing → Place on Host** tool to quickly recreate the railings hosted by the stair. This tool only works if all railings have been deleted from the stair.

There are two settings related to railings we can adjust prior to creating the Stair (or Ramp). While creating the stair, click the **Railing** button on the Ribbon. This opens the Railing dialog shown in Figure 8-2.2. Here the railings type can be changed and its position relative to the stringer and treads. Note that both of these settings can be changed later.

Railing

Given the railing is hosted by the stair there are only a few things that can be done to modify the path.

The railing system can be continued at the top or bottom (or both) of the stair as shown in the image below (Figure 8-2.21).

FIGURE 8-2.20 Railing options within stair tool

FIGURE 8-2.21 Extending railing at top of stair – along open floor edge

To do this (after the stair and railings are created):
- Start at the top or bottom level of the stair (or 3D view)
- Select the railing
- Click **Edit Path** on the *Ribbon*
- Select one of the **Draw** options (i.e. Line, Arc) from the *Ribbon*
- **Sketch** lines from the end of the stair hosted railing
- **Select** the newly sketched line (one at a time)
- Set the **Slope** option to **Flat** on the *Options Bar*
- See Figure 8-2.22

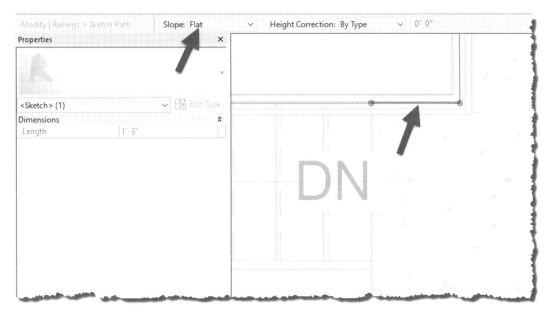

FIGURE 8-2.22 Extending railing at top of stair – along open floor edge

If the railing changes at the floor edge, for example the handrail stops or a stringer needs to be added to the railings as in Figure 8-2.12, then a separate railing element needs to be added—not part of the stair hosted railing. The trick in this case is adjusting the sketch lines so there is no odd overlap or gaps between the two railings.

When two sketch segments are selected, the Options Bar changes to show the **Rail Join** options. This can be changed if the transition between rails does not look as desired. This is similar to the **Edit Joins** tool on the Ribbon while in sketch mode.

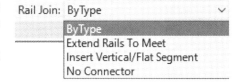

Railing → Sketch Path Tool

When a railing needs to be created apart from a stair or ramp, use the **Railing → Sketch Path** tool. When this command is selected, you simply sketch a path for the railing system to follow. If the railing has a Handrail, the side it is placed on is determined by which direction the sketch is created. While in sketch mode, click the **Preview** option to see the railing as it is to be sketched (similar to how Stair by Component works).

A sketched railing can be hosted to a floor—even a sloped floor. While in sketch mode, select the **Pick New Host** option on the Ribbon and then select the floor. The default host for a sketched railing is the level associated with the current plan-view.

The next two topics, **Handrails** and **Top Rails**, are relatively new features associated with a Railing type in Revit. Prior to Revit 2013, these elements were defined within the Edit Rails dialog. Now they are their own sub-element and can be selected separately from the main railing in the model (tapping the Tab key may be required). It is helpful to know this in case you are working in older project files—even if they have been upgraded. These old format railings do not allow the handrail or top rail to be selected.

Handrail

A Handrail, in Revit, is a component used within the Railing type. It cannot be used on its own. Keep in mind that any changes made to a Handrail type will change it wherever it is used. It is easy to edit a Handrail type while working in the context of a specific railing type, not realizing the same Handrail type is used in another railing type. In some cases, the Handrail type needs to be duplicated so changes do not affect other railing types.

Here are the main options related to a handrail type:
- Profile
- Material
- Extension (top and bottom)
- Terminations
- Supports (including spacing)

A few of these options have been adjusted to produce the results shown in the image below (Figure 8-2.23). The steps required to accomplish this will be covered next.

FIGURE 8-2.23 Handrail with several parameter adjustments

Railing Type Settings:

First, we notice the Railing type is where the Handrail is specified (Figure 8-2.24). In this case, two railing types are used, one for **Handrail 1** and one for **Handrail 2**. Creating multiple Handrail types is required if you want to have a different profile or position them at a different height as shown in this example.

For a center railing on a wide stair, Handrail 1 and 2 might be the same type but with different positions (one Left and one Right). This would place a handrail, at the same height, on both sides of the railing.

In this case, for two railing heights, the **Position** is the same.

This double-handrail example is not uncommon in elementary-grade schools. The normal higher rail is for adults and the lower for students.

Creating a duplicate handrail is accomplished by **right-clicking** on a handrail type in the Project Browser and selecting **Duplicate** (Figure 8-2.25). Once a new type is created, the various properties covered on the next few pages can be adjusted as required.

FIGURE 8-2.24 Handrail type and position

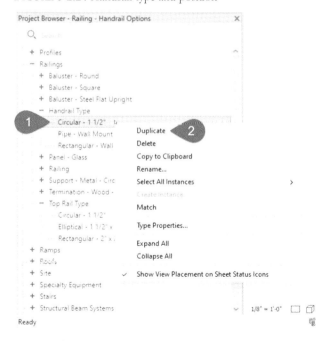

FIGURE 8-2.25 Duplicating a handrail type

Handrail 1 Properties:

The upper handrail, which is the default handrail as seen in Figure 8-2.9, has been modified per the following (see Figure 8-2.26).

- **Material:** Set to <u>Wood</u>

- **Extension Style:** Set to <u>Wall</u> (other options are None, Floor, Post). Default setting is None. This setting makes the handrail automatically return to the wall.

- **Length** (Bottom Extension): <u>1'-0"</u>. This will extend the length of the handrail from its default end, which aligns with the first riser. This is typically required to make the handrail comply with building codes.

- **Plus Tread Depth** (Bottom Extension): <u>Checked</u>. This makes the handrail continue along the slope the length of one tread. Checking this option makes the flat portion, controlled by the previous "Length" parameter, the correct height off the floor.

FIGURE 8-2.26 Upper handrail settings (i.e., Handrail 1)

- **Bottom Termination:** <u>Termination – Wood – Rectangular</u>. Places a family at the end of the handrail.

- **Family** (Supports): <u>Support – Metal – Circular</u>. Select the desired family for the handrail support.

- **Layout:** <u>Align with Posts</u>. Determines how the supports are placed/spaced. Other options are Fixed Distance, Fixed Number, Maximum Spacing and Minimum Spacing. Some of these options activate subsequent parameters such as Spacing and Justification.

The families listed for Terminations and Supports are created with a special subcategory designation Railing\Terminations and Railing\Supports.

Handrail 2 Properties:
The **lower handrail** is a duplicate of the previous handrail. Just the variations will be highlighted here (see Figure 8-2.27).

- **Height:** 2'-4"

- **Extension Style:** Set to Floor

The extension might more realistically just be set to wall like the upper handrail. However, this example helps to depict the options.

FIGURE 8-2.27 Lower handrail settings (i.e., Handrail 2)

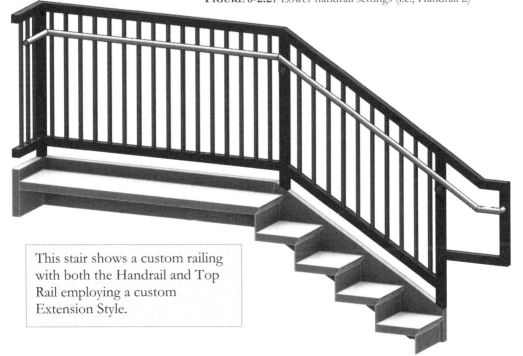

This stair shows a custom railing with both the Handrail and Top Rail employing a custom Extension Style.

Top Rail

The Top Rail is defined via a Railing's *Type Properties* as shown in Figure 8-2.28. Additional Top Rail types can be created by duplicating types listed in the *Project Browser*; **Families → Railings – Top Rail Type**. This is also where the height is defined.

The Top Rail has settings similar to the Handrail, but no supports, and can be adjusted separately from the Handrail. In the image below the Top Rail's **Extension Style** is set to **Post**. It can also be set to Wall, Floor or None just like the Handrail (Fig. 8-2.29).

FIGURE 8-2.28 Top Rail defined in railing's type properties

FIGURE 8-2.29 Top Rail extension style set to 'Post'

Remember that changes made to the Top Rail properties will affect all Railing types which use it. Notice the Top Rail type selected in Figure 8-2.28 is different than the Top Rail type specified in Figure 8-2.24 (also compare Top Rail pictures in the Handrail section).

Edit Rail (Top Rail or Handrail)

When a Handrail or Top Rail is selected, an option to edit the path of the railing appears on the Ribbon. This option only allows editing the path in a vertical plane aligned with the railing.

The Handrail or Top Rail can be selected by hovering the cursor over the Top Rail, being careful not to move the mouse, and then tapping the Tab key and clicking to select once highlighted. Once selected, the rail will be a transparent blue.

FYI: Stairs upgraded from older Revit projects or template files (anything prior to Revit 2013) do not have the Top Rail and Handrail options. Those elements were defined in the Edit Rail structure dialog. Thus, they are not selectable elements within the railing. In this case, the railing would need to swap it out for a new supported type.

When the Extension Style options, Wall, Floor, Post and None, do not cover what you want, use the **Edit Rail** option. Clicking this from an elevation, section or 3D view allows a custom path to be defined.

The image to the right, Figure 8-2.30, shows the Top Rail path being modified. Notice the geometry appears as each line is sketched.

The custom result can be seen in Figure 8-2.31.

Unfortunately, at this time the railing cannot change planes/direction—to wrap around a column for example.

Selecting the Top Rail and clicking **Reset Rail** on the Ribbon will undo any changes.

FIGURE 8-2.30 Editing the Top Rail path

FIGURE 8-2.32
Handrail only with supports at 6'-0" O.C. max.

FIGURE 8-2.31 Custom Top Rail defined

Handrail Only

A railing or guardrail is not required when a stair, or ramp, is against a wall (Figure 8-2.32). In this case we just need a handrail without a railing. Unfortunately, this is not directly possible as there is no Handrail tool, just the Railing tool. Here is how this can be accomplished…

The first step is to **Duplicate** a railing—Guardrail – Rectangular is a good option as it requires fewer changes. In the new railing's Type Properties, set the **Top Rail Use Top Rail** to **unchecked** (Fig. 8-2.33). Next, click Edit for **Rail Structure** and Delete all rows. Finally, click Edit for **Baluster Placement** and set everything, Main pattern and posts, to **None** (Fig. 8-2.34).

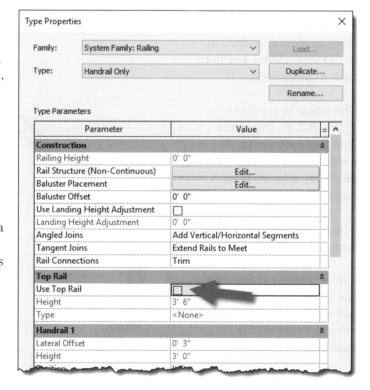

FIGURE 8-2.33 Omitting the Top Rail from a railing

FIGURE 8-2.34 Omitting Balusters from a railing

Handrail Supports

Within the Handrail <u>Type</u> supports can be selected (Figure 8-2.35). The **Family** drop-down is based on special families currently loaded in the project file. The **Layout** option selected determines which of the remaining parameters are editable. In this example, Maximum Spacing causes the Spacing parameter to be active, but Justification and Number are inactive (i.e. greyed out).

The automatically placed supports do not always land in the ideal position. Fortunately, they can be repositioned and even deleted individually. To modify a support it

FIGURE 8-2.35 Handrail support settings

must be selected. To select a Support, hover the cursor over a support (don't move it) and then tap (not press) the Tab key until the Support highlights—and then click to select it. Select the pin icon to unpin that instance (Figure 8-2.36). It can now be moved or deleted as needed.

FIGURE 8-2.36 Support selected and un-pinned

Railing Hosting Opportunities

Revit has the ability to host railings to floors with modified points and, as shown below, the top of walls—even when they have been modified with Edit Profile. Railings can also be hosted by toposurfaces and roofs. Thus, a railing could be used to represent a fence on the site or lightning protection rods (using a highly customized railing type) on the roof.

Sample Railing File

Make sure to examine the stair and railing sample file available with the provided Revit content; look in the **System Families** folder (click the *Imperial Library* shortcut on the left, in the Open dialog). While in this Revit project file, select a stair or railing and view its properties to see how it works. It is also possible to Copy/Paste selected examples into another open Revit project file. Download Autodesk content: https://autode.sk/30EJYkX

FIGURE 8-2.37
Stair and railing sample file
provided with Revit installation

Exercise 8-3:
Modeling the Stairs

Open Stair:

An open stair is often the primary stair used for vertical circulation in a building. This type of stair offers visual way-finding and can make an aesthetic statement for the building when well-located and designed. In our project, the open stair is located near the reception desk in the lobby so access to the upper level can be visually monitored.

Below is an example of a well-designed multi-story open stair. The railings are glass with a wood cap, the treads and risers are high quality terrazzo and the walls are lined with wood panels. The stair is located next to windows which help the users know where they are in the building relative to the outside, helping with way-finding. The photo was taken through a large piece of glass that makes the stair visible at each floor within the building. In the event of a fire, the fire-shutters come down and close off this view!

OPEN STAIR EXAMPLE High quality stair with interesting design

To round off the overall concept of your design, you will add an open stair in the lobby like the one shown below. A stair like this was not always possible to create in Revit. This can now be done with the introduction of the *Stair by Component* tool.

FIGURE 8-3.1 Open stair in lobby

The first thing you will do is lay out the stair concept using *Detail Lines*; these are two-dimensional, view specific lines. This step is not always required, as it is possible to simply start with the *Stair* tool, create the model and tweak the sketch later as needed. However, as you start learning to use the *Stair* tool it can be helpful to understand what you want before jumping in. It is also helpful for describing the layout desired in this book.

Looking back at page 8-3 we see that there are 23 risers required. In our design, we want an open stair in the lobby to extend up a certain distance and then split into two directions. The stair will look aesthetically interesting while providing proper egress from the second level as required by code. The first run will be longer, 14 risers, so the landing is high enough for people to walk under. The two runs going from the landing to *Level 2* will each be 9 risers. The width of the stair will change to make the design a little more interesting and this exercise a bit more challenging.

Let's get started…

1. In the Level 1 Floor Plan view, use the **Detail Line** tool (*Annotate* tab) to sketch the lines shown in Figure 8-3.2.

 a. Set the *Line Style* to **Hidden**.

FIGURE 8-3.2 Stair layout with detail lines

You may want to add the dimensions to verify your drawing, but they can be deleted once the lines are done and correct.

2. Be sure the two lines at the bottom of the 5'-2" wide runs, pointed out by the heavy lines and arrows in Figure 8-3.3, are separate lines, and not a continuation of the 1'-6" line adjacent to them. You will need to be able to pick the midpoint of these lines when sketching the run from the landing to *Level 2*.

FIGURE 8-3.3 Detail lines should not be continuous here

You are now ready to use the *Stair* tool.

3. Select **Architecture → Stair → Stair** from the *Ribbon*.

4. Be sure the *Type Selector* is set to *Assembled Stair:* **7" max riser 11" tread**.

5. Click the **Railing** button on the *Ribbon*.

FIGURE 8-3.4 Railing selection

6. Select **Guardrail – Pipe** and pick the **Stringer** option for *Position* (Figure 8-3.4).

7. Be sure the *Base Level* and *Top Level* are **Level 1** and **Level 2** respectively, with both offset values set to 0'-0".

8. On the *Options Bar* set the following:
 a. *Location Line:* **Run: Center**
 b. *Offset:* **0'-0"**
 c. *Actual Run Width:* **4'-0"**

9. Pick the midpoint of the line as shown in Figure 8-3.5.

FYI: You may not see the Snap symbol, but the tooltip should appear when snapped to the midpoint of the line.

10. Snap to the midpoint of the line at the other end of the first run of stairs.

Notice how Revit is reporting the number of risers that will be created and how many are remaining based on where your cursor is and where you pick.

FIGURE 8-3.5 Adding the open stair in the lobby – first run of stairs

Next you will draw the second run of stairs which heads towards the West; an intermediate landing will automatically be created!

11. On the *Options Bar*, change the *Actual Run Width* to **5'-2"**.

12. Draw the second run as shown in Figure 8-3.6.

FIGURE 8-3.6 Adding the open stair in the lobby – second run of stairs (heading West)

Notice the fourth pick is beyond the end of the last run; when picking the last point, you can pick beyond the end of the run to be sure you include all possible steps. Revit will not add more than are needed, based on the specific floor to floor height and maximum riser and minimum tread dimensions.

Your drawing should currently look like Figure 8-3.7. Notice the landing needs to be adjusted. You will do that next.

> *FYI: However, you could finish it now without error, and then select the stair and pick* Edit Stairs *on the* Ribbon *to get back into the* Stair Sketch *mode.*

FIGURE 8-3.7 Width of two stair runs adjusted

13. Select the landing, not one of the edge support stringers.

14. Drag the East side of the landing over to the detail line via the grip pointed out in Figure 8-3.8.

 a. The grip should snap to the detail line. Undo and try again if you have problems. You can also try zooming in more.

The landing is now the correct size based on the detail lines. Next you will mirror the second run. When this is done, the stringers, i.e. supports, around the landing will automatically clean-up.

15. Select the second stair run, which is 5'-2" wide.

16. Use the **Mirror – Draw Axis** tool to mirror the second run:

 a. Select the second run.

 b. Select the **Mirror – Draw Axis** tool from the *Ribbon*.

 c. Draw a vertical axis line.

 i. Snap to the midpoint of the north side of the landing.

 ii. Pick the second point straight down, "snapped" to the vertical.

 d. See Figure 8-3.9.

FIGURE 8-3.8 Adjusting landing size

First Pick Point

Approx. Second Pick Point

Select this stair first

FIGURE 8-3.9 Mirroring second stair run

TIP: If landing and second stair do not connect after mirror, make sure you are snapping to the correct point, so the mirrored stair "lands" in precisely the correct location.

The new stair run, heading East from the landing, should connect to the landing perfectly. The stringer should now be adjusted as shown in the image above. If not, you may not have picked the middle of the landing for the mirror line.

17. Click the **green check mark** to complete the stair model.

The image below shows your stair as it stands right now.

FIGURE 8-3.10 Stair model

The next thing you will do is modify the second floor to fill in the gaps at the top of each stair. Another way of doing this would be to create a new floor type called "Stair Landing" and add two new floor elements at the top of the stairs. However, for simplicity you will just quickly modify the existing Level 2 floor element.

18. Switch to the Level 2 floor plan and then pick the edge of the Level 2 floor to select it.

Edit
Boundary
Mode

19. Click **Edit Boundary** from the *Ribbon*.

20. Use *Trim* and the *Boundary Line* draw option to modify the floor edge as shown in Figure 8-3.11 below. The heavy dashed lines represent the original floor edge. The finished floor can be seen in Figure 8-3.12.

21. Click the **Green Checkmark** to complete the floor.

22. Select **Yes** to first prompt, *Attach Walls,* and **Unjoin** for the second prompt.

FIGURE 8-3.11 Editing Level 2 floor boundary

Next you will add railings along the Level 2 floor edge. First you will quickly add a basic railing to see how easy this is, and then you will modify it to more closely match the stair.

23. Select the **Architecture → Circulation → Railing → Sketch a Path** tool.

24. Select *Guardrail - Pipe* from the *Type Selector.*

25. Draw the three lines, one for each segment of railing shown in Figure 8-3.12, on the West side.

 a. The railing should align with the stair railing when done.

26. Click the **Green Checkmark** to finish the railing.

 a. If the handrail is on the wrong side, click the flip-arrow icon and then select **Edit Path** to reposition the sketch lines back to the correct position. The side the handrail is placed on is based on the direction the first line was drawn. Use a camera view to verify the railing is correctly positioned.

27. Repeat these steps for the East side as well.

When finished your model should look similar to Figure 8-3.12 below. The stair obviously lacks the continuation of the stair stringer. You will add this feature to the railing next.

FIGURE 8-3.12 Stair railings extended at Level 2 floor edge

The first step in adding a stringer to the railing is to create a profile of the stringer in the *Family Editor*. As you will see, this is very simple. You open a profile template and draw four 2D lines.

28. Select **File Tab** → **New** → **Family**.

29. Select *English-Imperial* **Profile.rft** from the list of template files.

30. Click **Open**.

31. Pick **Create** (tab) → **Line** from the *Ribbon*.

32. Draw the four lines shown in Figure 8-3.13 to the right.

 a. Note the intersection of the two dashed lines represents the origin of the family. This should be centered on the top edge as shown.

FIGURE 8-3.13 Stringer Profile

33. Save the *Family* as **Railing Stringer.rfa**. Do not close the file yet.

34. Click the **Load into Project** button on the *Ribbon*.

The profile family is now loaded into your project and can be seen from the *Project Browser*. Next you will use this profile in your railing.

35. In your law office project, select the West railing just drawn at the level 2 floor edge.

You need to duplicate the railing so the railing on the stairs does not change.

36. Click **Edit Type** from the *Properties Palette*.

37. Click the **Duplicate** button.

38. Enter the name: **Guardrail – Pipe with Stringer**.

39. Click the **Edit** button for *Rail Structure (Non-Continuous)*.

40. **Insert** a new row and edit it to match Figure 8-3.14.

41. Close the open dialog boxes.

FIGURE 8-3.14 Stringer Profile

Notice how the existing railing is defined and that a *Material* may be added to each row. Your railing should now look like Figure 8-3.15.

FIGURE 8-3.15 Stringer added to railing

42. Select the East railing and change it to the new railing, **Guardrail – Pipe with Stringer**, via the *Type Selector*.

Finally, you will learn how to adjust the *Top Rail* extension at the bottom of the stair.

43. Switch to the Level 1 floor plan.

44. Draw a section parallel to the stair as shown in Figure 8-3.16.

45. Adjust the Far Clip line as shown so you only see one railing and not the one beyond.

46. Switch to the new section view.

47. Select the **Top Rail** as shown in Figure 8-3.17; press *Tab* if needed to select just the *Top Rail* and not the entire railing system.

 FYI: The top rail cannot be selected for upgraded railings prior to Revit 2015.

48. Click the **Edit Rail** button on the *Ribbon*.

FIGURE 8-3.16 Section added

Now you will edit the path of the top rail profile.

49. Select **Edit Path** from the *Ribbon* (Figure 8-3.18).

50. Select the straight **Line** option on the contextual tab.

51. Sketch the two lines shown in Figure 8-3.19.

 a. Start at the bottom edge of the existing railing, drawing the line down 1'-0" and aligned with the existing rail slope.

 b. Click *OK* to any warning about the railing not being continuous.

 c. The second line can simply snap to the top edge of the stringer.

FIGURE 8-3.17 Top rail selected

FIGURE 8-3.18 Edit Top Rail options

52. Click the **Green Checkmark** twice to complete the *Top Rail* modification.

53. Reposition the section in the Level 1 plan-view to the other railing and make the same adjustment.

Your railings should look like Figure 8-3.20. Similar edits can also be made to the handrails. Next you will look at another way to add rail extensions to the *Top Rail*.

FIGURE 8-3.19 Edit path for Top Rail **FIGURE 8-3.20** Modified top rail

54. In the *Project Browser*, go to **Families →
 Railings → Top Rail Type**.

55. Right-click on **Circular – 1 ½"** and
 select **Duplicate**.

56. Right-click and *Rename* the new *Type* to
 Circular – 1 ½" with Extensions.

57. In a 3D or Camera view, select the back
 Railing (the entire element), not the *Top
 Rail* or *Handrail,* shown in Figure 8-
 3.22. *TIP:* If pinned, click to un-pin the
 element.

58. Click **Edit Type** and then **Duplicate**.

59. Enter the name: **Guardrail – Pipe
 with Extensions**.

60. Change the *Top Rail Type* to **Circular –
 1 ½" with Extensions** (Figure 8-3.21).

61. Click **OK**.

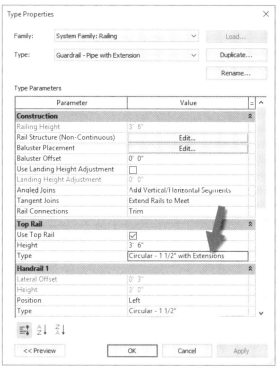

FIGURE 8-3.21 Modified top rail

62. Select the **Top Rail** as shown in Figure 8-3.22.

63. Click **Edit Type**.

64. Change the *Top* and *Bottom Extensions* to **Post** and the *Length* to **1'-0"**.

65. Click **OK**.

FIGURE 8-3.22 Editing top rail extension

Your railing should now look like the one shown in the image above. You had to duplicate the top rail and railing type so the other top rails and railings did not automatically get a one foot extension on them.

66. Select one of the handrails.

67. Click **Edit Type**.

68. Set the *Supports Family* to **Support – Metal – Circular**.

69. Click **OK** or **Delete Type**, also click *OK* to any errors you might get.

FIGURE 8-3.23
Handrail support

All your handrails should now have the supports as shown in Figure 8-3.23. Per steps covered in chapter 2, try switching to a 3D view and turn on the **Section Box**. Adjust the *Section Box* so you have a good view of your stair.

There are a number of other things which could be covered regarding stairs, but this is all we have time for now!

The Utilitarian Stair:

The back stair is required for egress and is often referred to as a utilitarian stair due to the simple and durable finishes.

Utilitarian stairs are not readily visible and are primarily meant for emergency egress; their existence is primarily dictated by code. They certainly may be used as a means of vertical circulation within the building.

The finishes and light fixtures are simple, often painted CMU walls and surface mounted lights.

In stairs, building codes often require a gate at the ground level and at the upper level, if the stair continues to the roof, as shown in the photo below. The reason for this is to help guide people in an emergency. Each floor in the utilitarian stair shaft looks the same, so it would be easy for a person, especially in a panicked state, to pass the ground level exit and continue down one or two more

Utilitarian Stair Example
Durable, average cost stair with simple design

floors to the basement, extending their time in a dangerous situation. If the stair is used for normal, day-to-day, vertical circulation, some building officials will allow the gate to be placed on a magnetic hold-open which allows the gate to close in the event of a fire.

Most commercial codes also require that a sphere 4″ or larger not be able to pass through the guardrail. This has a large impact on design but creates a safe environment for the users; looking at the photo to the right, it is not difficult to see how older codes allowed for unsafe designs!

Self-Exam:

The following questions can be used as a way to check your knowledge of this lesson. The answers can be found at the bottom of this page.

1. Stair landings are added automatically. (T/F)

2. A stair sample file can be downloaded from SDC Publications. (T/F)

3. You must draw detail lines before creating stairs. (T/F)

4. A stair's railing cannot be deleted. (T/F)

5. Stair types are listed in the *Project Browser* under *Families*. (T/F)

Review Questions:

The following questions may be assigned by your instructor as a way to assess your knowledge of this section. Your instructor has the answers to the review questions.

1. Revit is not able to make a T-shaped stair like the one you created in this exercise. (T/F)

2. The riser is typically calculated to be as large as the building codes will allow. (T/F)

3. Stairs have multiple sub-categories which can be used to control stair visibility. (T/F)

4. To change a stair, select it and then click _____ on the *Ribbon*.

5. A railing extending past the end of a stair run must always slope. (T/F)

6. To extend the stair's stringer you draw _____ lines.

7. What size is the Revit file on your hard drive? _____

8. After starting the *Stair* tool, the first thing you need to do is set the _____ level and the _____ level.

9. Railings can be modified to be shorter than the length of the stair. (T/F)

10. The riser height is based on the floor-to-floor height. (T/F)

Lesson 9
Ceilings:

With your floor plan developed to the point where you and the client are happy with the layout, the program is generally being met, the doors, glazing, and stairs are located, you can begin to develop the ceiling finishes. This chapter will cover all the basics related to modeling ceilings, plus offer a few tips and tricks to make your design ideas a reality using Autodesk® Revit®. You will be creating the USG specialty ceilings shown in the two images below.

Exercise 9-1:
Introduction to Revit Ceilings

Creating ceilings in Revit is very much like creating floors (floors are covered in Chapter 12). They are both horizontal planes that are composed of one or more *Layers*. Each *Layer* represents a building material, e.g., gypsum board, metal studs, acoustic ceiling tiles, etc. This informational exercise will cover the various parameters, options, and workflows used to design and document ceilings.

The basic steps are as follows:
- Select the *Ceiling* tool from the *Ribbon*.
- Pick the desired **Ceiling Type** from the *Type Selector*.
- Specify the **Ceiling Height** via the *Properties Palette*.
- Click within any room enclosed by walls (*Automatic Ceiling* is the default)
 - Or, toggle to *Sketch Ceiling* mode and manually define the perimeter.
- Select the *Modify* button to clear the *Ceiling* command.

The Revit templates have four predefined ceiling types, organized in two categories:
- *Compound Ceiling:*
 - 2′ x 2′ ACT System
 - 2′ x 4′ ACT System
 - GWB on Mtl. Stud
- *Basic Ceiling:*
 - Generic

Of course, additional types can be made. Just select the type closest to the one you want, in terms of construction, and click *Duplicate* in the *Type Properties* dialog. This can be initiated by right-clicking on the *Type* in the *Project Browser*, or by selecting a ceiling in the model and selecting *Edit Type* from the *Properties Palette*.

The **Basic Ceiling** should never be used, as it is a 2D plane with no thickness and creates several limitations. Unfortunately, it is not possible to delete it from the project, as one type must exist. But you could rename "Generic" to "Do Not Use."

The **Compound Ceiling** is placed in a plan view, usually a ceiling plan view. However, it is possible to place them in a floor plan view, though they will not show up if they are above the cut plane. If so, Revit provides a message indicating the element was created but does not appear in the current view. Once placed, the ceiling will appear in elevations, sections, 3D views and schedules. Even though the ceiling can only be created in plan views, once created it can be manipulated in other views. For example, a ceiling can be selected in a section, and its height changed via the *Properties Palette* or the temporary dimension on-screen. In a *Camera* view you can select the ceiling and change the type via the *Type Selector*. Many of the ceiling's parameters can also be edited in a schedule view.

Depending on how the ceiling was created, it will maintain a parametric relationship with the bounding walls, meaning the ceiling perimeter will move when a wall is moved. This

associativity is accomplished when *Automatic Sketch* mode is used. If a wall is deleted which defines the perimeter of a ceiling, the associativity will be lost.

The ACT ceiling types (ACT = Acoustic Ceiling Tile) have a *Material* specified which uses a **Model Pattern** to represent the ceiling grid. Refer back to the *Materials* chapter for more on *Model Patterns*. When the ceiling is initially placed in plan view, the grid is centered on the room. If the grid is not square, the longer tile will align with the longer direction of the room in which the tile is being placed. An example of this is shown in Figure 9-1.1 below.

FIGURE 9-1.1 Ceiling grid pattern centered and rotated upon placement

Adjusting Grid Layout

When you select the ceiling in plan view you will most likely be selecting the grid pattern (i.e., *Model Pattern*) rather than the ceiling element itself. This is by design, making it easier for the designer to adjust the grid layout. The rule-of-thumb with ACT layout is the tile should not be less than half a tile. If the initial layout creates this situation, you simply select a grid line and use the *Move* command to reposition the tile. **TIP:** *Move it half the tile size.* There are a few other reasons the grid is moved: sometimes the grid is positioned to align with a wall or the corner of two walls (Figure 9-1.4); this might be based on prominent sight lines. Another situation is centering the grid pattern on a corridor, when the ceiling spills out into a larger room (Figure 9-1.2, 9-1.3 and 9-1.4).

FIGURE 9-1.2 Ceiling grid centered on extents of space

FIGURE 9-1.3 Ceiling grid manually adjusted for corridor on left

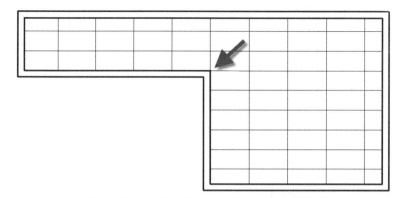

FIGURE 9-1.4 Ceiling grid manually adjusted to align with wall corner

Placing a ceiling is similar to placing rooms in that Revit automatically searches for the perimeter of the space in which you click. The main difference is that the perimeter is defined at the offset height entered for the ceiling, rather than being defined at the floor. Most of the time, the perimeter is identical for both the floor and the ceiling in a given room. However, sometimes a soffit or other room bounding element can make them different. Therefore, it is important to be sure the perimeter is properly defined at the ceiling elevation.

Sloped Ceilings

While in *Sketch* mode you can make a ceiling sloped by clicking the **Slope Arrow** tool on the *Ribbon* (Figure 9-1.5) and picking two points within the sketch area; this places an arrow in the sketch (Figure 9-1.6). The arrow element can be selected and has properties which can be used to control the slope of the ceiling (Figure 9-1.7).

FIGURE 9-1.5 Ceiling *Sketch* mode options

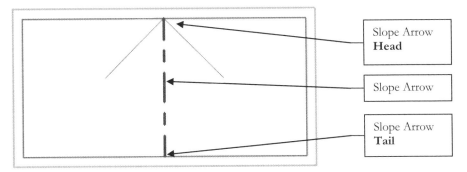

FIGURE 9-1.6 Slope arrow added to ceiling sketch

By default, you adjust the elevation of the "head" and the "tail" relative to the height specified for the selected ceiling. The head is the arrowhead side of the slope arrow, and the tail is the other end (Figure 9-1.6). You can change this to define the "slope" similar to the roof; e.g. 4″/12″. When "slope" is selected, you can also adjust the slope outside of *Sketch* mode when the ceiling is selected, via the *Properties Palette* (Figure 9-1.7).

For example, if the ceiling height is 8′-0″ for the example above and the properties are set as shown to the right, the result will be as shown in section in Figure 9-1.8. The head remains at 8′-0″ because the *Height Offset at Head* is 0′-0″ – i.e., no change. However, the tail side of the ceiling is at 9′-0″ because the *Height Offset at Tail* is set to 1′-0″.

FIGURE 9-1.7 Properties for selected slope arrow

The grid dimension for a sloped ceiling will be adjusted to the selected slope. That is, the grid will measure less than 2'-0" or 4'-0" in plan view; see image to right. This is because the grid is truly foreshortened on a slope, and not perpendicular to the plan view as it otherwise normally would be. The dimension in the other direction is unchanged.

Hosted light fixtures and mechanical items have the ability to conform to the sloped ceiling. There can be problems related to consistent graphics due to how the content was created and the fact that you may actually begin to see the side of some content.

Foreshortened grid dimension

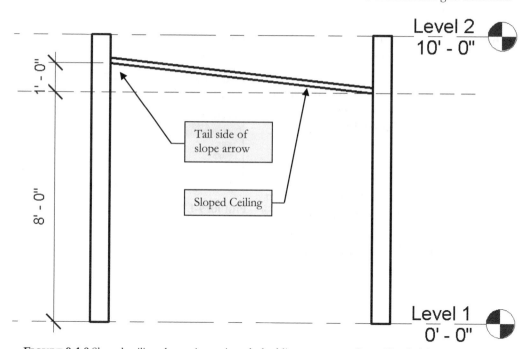

FIGURE 9-1.8 Sloped ceiling shown in section; dashed line represents flat ceiling height

One last comment about *Slope Arrows*: you can only have one per ceiling element and it does not have to be drawn from edge to edge (any slope arrow drawn will act as if it was drawn edge to edge). It does not need to be perpendicular to the sketch lines. It can be at an angle, thus creating a ceiling that slopes in two directions relative to the walls in a rectangular room. If a ceiling needs to slope in two directions, you have to create two ceiling elements and make them each slope as needed. Also, when cutting a section perpendicular to that shown above, be sure to pay attention to your *Far Clipping* and *Far Clip Offset* to better understand why additional ceiling lines may appear; these would be lighter lines seen beyond the cut plane (see Page 2-6). *Far Clipping* is best set to **Clip without line**.

Additional Ceiling Types

Creating additional ceiling types is very similar to the process for walls (and floors, which are covered later). You pick the *Ceiling* tool or select a ceiling in the model, and then click **Edit Type → Duplicate**. The ceiling may or may not have the structure as part of the ceiling type. For example, if you are adding furring and ½" gypsum board to the bottom of the floor joists above, you would not include the joists in the ceiling. The ceiling would just be the furring and the gypsum board. The joists are either included in the floor above or actually modeled individually, with the latter being the ideal option for proper coordination. On occasion, the ceiling may even be part of the floor assembly. However, that would mean the ceiling is the same everywhere the floor exists, which is often not the case.

An example of when you would include the structure in the ceiling type would be when the ceiling is self-supporting or suspended from the structure above by wires. In this case, the structure would be metal studs running horizontally, from wall to wall, or a suspension system similar to that used for an ACT system. Including the overall thickness of the ceiling assembly helps to ensure coordination with structural beams and mechanical ductwork.

The suspended acoustic ceiling tile (ACT) system has main runners and cross tees which are usually about 1½" thick (Figure 9-1.9). Most ceiling tiles are slightly less than one inch thick; a few can be as much as 1½" thick. The most common thicknesses are ⅞" and ¾".

FIGURE 9-1.9 ACT suspension system detail from USG website. Your Revit ceiling assembly should take into account the overall thickness of the system to ensure proper coordination.

Sometimes the ceiling tile exists within the overall thickness of the support tees, as shown on the left and right in Figure 9-1.11 on the next page. If this is the case, your ceiling would be thicker than needed if set up as shown in Figure 9-1.10.

Unlike the floor systems and metal decking, it is not possible to tell Revit the two products, tees and ceiling tile, overlap as shown in the center example in Figure 9-1.11.

FIGURE 9-1.10 Edit Assembly dialog for ceilings

Finally, small things that are very flexible in terms of exact location in the field are rarely modeled. So, the suspension wires are not modeled. They are accounted for in the project manual, which includes the specifications.

FIGURE 9-1.11 Various tee and panel configurations impact the overall ceiling type thickness

Soffits and Bulkheads

When laying out ceilings, there are many factors in determining the height of a ceiling. Larger rooms need to have a higher ceiling as do "louder" rooms, such as a band room, to absorb the sound. Also, things such as structure and ductwork limit the height.

In some cases, the ceiling height may change at a location where there would not otherwise be a wall. In this situation you may add a **bulkhead** (Figure 9-1.12). This is simply a wall suspended from above. It is seen in the ceiling plans and sections but not the floor plans. A bulkhead is a regular wall in Revit; it just has the **Base Offset** set to 8′-0″ in the example below. In this example, for an 8'-0" bulkhead, when placing a ceiling, the ceiling height must be 8′-0″ or higher for Revit to detect the bulkhead.

TIP: The lower ceiling is often an inch higher than the bulkhead to account for construction tolerances.

FIGURE 9-1.12 Section showing bulkhead and soffit conditions

Another common ceiling condition is a **soffit**, shown on the right in Figure 9-1.12. This is horizontal and vertical metal stud framing that then receives gypsum board. These often enclose a duct, as in the example above, or a deep beam. A soffit allows the main ceiling to be higher, while enclosing a lower element in part of the space. In Revit, a soffit is modeled with a ceiling element for the horizontal and a wall for the vertical locations. Both the wall and ceiling elements only have gypsum board on one side.

To get the ceiling and wall to join and clean up as shown above you have to create a special wall and ceiling type that has the gypsum in the "core boundary" rather than the studs. This will allow the ceiling and soffit walls to be joined properly. The ceiling element must also be adjusted to overlap the soffit wall. You will do this later in this chapter when modeling the ceilings in the *Law Office* project.

Deleting a Ceiling

The hardest part about deleting a ceiling is selecting it first. As previously discussed (see Page 9-3), the ceiling grid is selected by default. You cannot delete a ceiling by selecting the grid. In a plan view, you have to place your cursor near the perimeter of the ceiling and "tap" the **Tab** key on the keyboard until it is highlighted and then click to select it. You can, however, easily select a ceiling in a section view.

Resources

Be sure to take a look at *The Gypsum Construction Handbook* for practical information on designing walls and ceilings. Great information is provided on metal stud framing size and height limitations, as well as how ceilings and soffits are constructed (Figure 9-1.13). This manual is a must-have-reference in a design firm. Learn more here:

https://www.usg.com/content/usgcom/en/resource-center/gypsum-construction-handbook.html

The company that produces this manual is United States Gypsum (USG) who manufactures many building products such as gypsum board and acoustic ceiling tiles (ACT). Later in this chapter you will download some of their provided Revit content and use it in your model. There are certainly other manufacturers with similar information and products, but we will only be covering this one due to space limitations in this book.

USG offers a few other valuable resources as well:

Access several case studies and tools to aid development of walls, floors, ceilings and roofs: http://www.usgdesignstudio.com

Learn more about sustainable USG ceiling options and high-recycled content products: http://www.sustainableceilings.com

FIGURE 9-1.13
Sample page from USG's Gypsum Construction Handbook

Exercise 9-2:
Modeling Ceilings

In this section you will model the ceilings throughout the law office project. Here you will do some basic ceiling design and add bulkheads and soffits. You will also learn how to develop a few 3D ceiling elements. Finally, notes, tags and spot elevations will be placed to keep track of design decisions; this is not for construction documentation as we are still in the early design phase.

The first thing we will do is modify the ceilings, already in the model, located in the toilet rooms. If you recall, we modified the walls which define the perimeter of the toilet rooms. Therefore, the ceilings are wrong. As previously mentioned in this chapter, if a wall simply moves, the ceiling will automatically update. But if new walls are added, or walls are deleted, the ceiling cannot update. It is not like a *Room* element which is continually searching for the perimeter of a space, for better or worse.

Keep in mind that you should typically try not to delete things and recreate them. It is better to modify and swap *Types* for elements, as this maintains any project specific information that may have been added by you or someone else on the design team; e.g., adding "provide acoustic insulation above ceiling" to the instance parameter ***Comments***. Additionally, any mechanical or electrical equipment hosted by your architectural element becomes un-hosted and will not update unless re-hosted manually to your "new" elements.

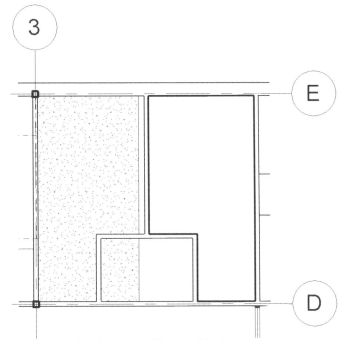

FIGURE 9-2.1 Toilet room ceiling modification

1. Open your *Law Office* model to the *Level 1* **ceiling plan** and zoom in on the toilet room area (Figure 9-2.1).

2. Select one of the ceilings, by the exposed edge, and click ***Edit Boundary*** from the *Ribbon*.

 TIP: *You can also toggle on* ***Select Elements by Face*** *to select the ceiling within its perimeter.*

3. **Erase** the sketch line not aligned with any wall.

To create a ceiling that will update with wall movements, you will want to use the ***Pick Walls*** tool rather than just using the manual ***Line*** option.

4. Click the **Pick Walls** option on the *Ribbon* (Figure 9-2.2).

5. Click the walls which don't already have a sketch line.

 a. Click *Unjoin Elements* if prompted.

 b. You may need to use the *Flip* arrow if the line is placed on the wrong side of the wall.

FIGURE 9-2.2 Edit ceiling boundary – *Pick Walls* option

6. Use the **Trim** option to clean up the corners and make a continuous perimeter.

While in *Sketch* mode, we notice the ceiling height is 8'-0" in the *Properties Palette*. It is decided that these ceilings should be 9'-0", so we can change this now.

> *FYI: You do not have to be in* Sketch *mode to edit the ceiling height.*

7. Adjust the ceiling height to **9'-0"** via the *Height Offset From Level* parameter.

8. Click the **green check mark** to complete the ceiling edit.

9. Repeat this for the other three restrooms, two of which are on *Level 2*.

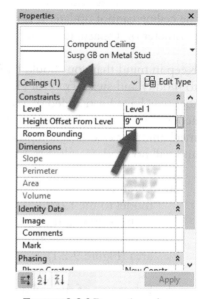

FIGURE 9-2.3 Properties palette with Ceiling tool active

Next you will add a gypsum board ceiling in the small room added within the toilet room area. This involves the most basic steps: select your ceiling type, specify the height and pick within the space.

10. Switch back to the **Level 1 Ceiling Plan** view; select the **Ceiling** tool from the *Architecture* tab.

11. Select **Susp GB on Metal Stud** from the *Type Selector* (Figure 9-2.3).

12. Set the *Height Offset From Level* to **9'-0"**.

13. Pick within the *Janitor's Closet* to place the ceiling.

FIGURE 9-2.4 New ceiling placed

The new gypsum ceiling has been placed at 9'-0". Because this ceiling has a *Surface Pattern* specified in the *Material*, we can easily see the ceiling (Figure 9-2.4).

The next two images are the interior designer's sketch we will use as the basis for our ceiling design. Keep in mind that sketches are preliminary and not always dimensionally accurate, so our final design may change as we develop it. However, these types of hand drawings are often the quickest way to develop your initial ideas.

FIGURE 9-2.5 Level 1 reflected ceiling plan – preliminary hand sketch developed by interior designer

TIP: A copy of the colored version of these sketches has been provided on the publisher's website. Looking at it will help to identify where one ceiling type stops and another begins.

FIGURE 9-2.6 Level 2 reflected ceiling plan – preliminary hand sketch developed by interior designer

14. Delete the previously placed 2′x2′ ACT Ceiling on Level 2 in the Northeast corner of the building.

 TIP: Tap the Tab *key if needed to highlight the edge and select it.*

Looking at the concept sketches, we notice several things. There is a gypsum board soffit along many of the corridor walls and we are using a few specialty ceiling systems. Some of the ceiling systems cannot easily be cut in the field like a basic ACT can, so we want to make sure our gypsum board soffits or basic ACT area take up any odd dimensions. To this end, you will lay out several *Detail Lines* in your ceiling plan to get an idea of what goes where. This is not totally required, but like the stair exercise, it gives you extra practice using the tools in Revit and will make it easier to model some of the ceiling elements later.

15. Select Annotate\ **Detail Line** tool.

16. Set the *Line Style* to **<Hidden>** via the *Ribbon* (Figure 9-2.7).

17. Add the *Detail Lines* to your Level 1 and Level 2 reflected ceiling plans per Figures 9-2.8 through 9-2.11.

FIGURE 9-2.7 Detail Line tool active – Line Style options on *Ribbon*

 a. To save time, this is a long chapter, insert the linework from the provided files; via Insert → Insert from File → **Insert 2D Elements from File** and browse for **Level 1 Ceiling Linework.rvt**. Align via grid lines A & E.

 b. You should add the dimensions to help position the lines. Remember, when an element is selected (like your line) you can use the dimensions to drive its location.

 c. The double dashed lines represent a bulkhead. Show the lines **6″** apart from each other. Just draw the two hidden lines for now.

 d. All dimensions are to the face of the wall.

 e. When dimensions do not tie into a wall, the line work should be centered on the room.

 TIP: Draw temporary reference lines, in the center of the room, from which to offset.

Later, when you are done modeling the ceilings and soffits, you will delete the *Detail Lines* and dimensions. Seeing as these elements are view specific, you could have *Duplicated* the two ceiling plans to create temporary working views. When done modeling the ceilings and soffits, you simply delete the views and all the 2D view-specific lines and dimensions go away. The 3D model elements would still exist, of course.

In our case, we will have to select everything on the screen and use the *Filter* tool, or use **VV** to find the 3D ceiling elements and then select the remaining 2D line work.
The dimensions with stars next to them, in the following images, are at locations where the ceiling element has a specific size. For example, the special wavy 3D ceiling tiles cannot be cut so we must provide for the 2′x2′ spaces and the 1′x5′ panels needed to have the 1′-0″

width accounted for. This will make the design look much cleaner and create less waste. It is not really practical to do this for every space as it would make the room layout too restrictive, therefore making some rooms smaller or larger than needed. However, whenever possible, a good designer is thinking about using standard dimensions for ceilings, masonry and more. This uses less labor, creates less waste and looks better.

FIGURE 9-2.8 Partial Level 1 reflected ceiling plan showing preliminary ceiling layout – rotated view

FIGURE 9-2.9
Partial Level 1 reflected ceiling
plan showing preliminary ceiling
layout

FIGURE 9-2.10 Partial Level 2 reflected ceiling plan showing preliminary ceiling layout – rotated view

FIGURE 9-2.11 Partial Level 2 reflected ceiling plan showing preliminary ceiling layout

The next step will be to lay out the soffits along the corridor walls. This will involve adding two walls, with the *Base Offset* adjusted up, and a ceiling between them. However, the first thing we will do is create a special new soffit wall type that only has gypsum board on one side of the metal studs.

18. Select the *Wall* tool.

19. Set the *Type Selector* to the **Interior – 4⅞″ Partition (1-hr)** option.

20. Click **Edit Type** and then **Duplicate**.

21. Enter the name: **Interior - Soffit Wall**

22. Click **Edit** next to *Structure*.

23. **Delete** the ⅝″ gypsum board on the interior side of the wall (Figure 9-2.12).

FIGURE 9-2.12 New wall type for soffit locations

By having the gypsum board on the "exterior" side of the wall, the *Flip* arrow will help you quickly determine if the wall is properly orientated while in plan view. Remember, the *Flip*

arrow appears on the "exterior" side of the wall. Of course, this wall does not really have an exterior side, but Revit assumes one side is "exterior" and the other "interior." Looking at the graphic to the lower right, we immediately know the gypsum board is on the same side as the *Flip* arrow. If that is the wrong side, you simply click the *Flip* arrow to mirror the wall.

Next you will add a sweep to the bottom of the wall. This is to deal with the fact that Revit cannot add a finish, such as gypsum board, to the bottom or top of the wall. The first thing you need to do is create a profile.

24. Click **OK** to close the open dialogs.

25. Using the steps covered in the previous chapter (see Page 8-47) create a **5/8" x 3 5/8"** profile as shown in Figure 9-2.13.

26. Save the profile as **Gyp Bd Profile** and load it into your project via the **Load into Project** tool.

27. Back in your Law Office project, select the wall tool and then, with <u>Interior –</u>

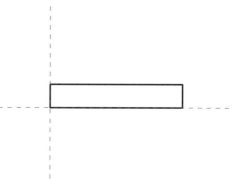

FIGURE 9-2.13 New profile created

Soffit Wall selected in the *Type Selector*, click **Edit Type**.

28. Click the **Edit** button, next to *Structure*.

29. If you do not see the graphic preview on the left, click the **<< Preview** button at the bottom.

30. Switch the *View* drop-down list, below the graphic preview, to **Section: Modify Type** (Figure 9-2.14).

31. Click the **Sweeps** button.

32. In the *Wall Sweeps* dialog, click the **Add** button.

33. Make the following adjustments to the new row (Figure 9-2.14):
 a. *Profile:* **Gyp Bd Profile**
 b. *Material:* **Gypsum Ceiling Board**
 c. *Distance:* **0'-0"**
 d. *From:* **Base**
 e. *Orientation:* **Perpendicular to Face**
 f. *Side:* **Exterior**
 g. *Offset:* **-4 ¼"** (make sure you add the minus sign)
 h. *Flip:* **Not checked**
 i. *Setback:* **0'-0"**
 j. *Cuts Wall:* **Checked**
 k. *Cuttable:* **Not checked**

FIGURE 9-2.14 Adding sweep to wall type

34. Click **OK** to close the open dialogs.

35. Change the *Function* of the wall type to **Soffit** (Figure 9-2.15).

You are now ready to draw the soffits! Notice the previous steps are assuming you are starting to instinctively know when to click **OK** in a dialog box in order to proceed.

Looking at the concept sketches, notice all the gypsum soffits are at 8'-6". Thus the soffit walls you are about to draw will go from 8'-6" (*Base Offset*) up to Level 2 (*Top Constraint*).

FIGURE 9-2.15 Setting function for new wall type

36. With the *Wall* tool active and the new soffit wall type selected, set the **Base Offset** to **8'-6"** in the *Properties Palette*.

37. Make sure the **Top Constraint** is set to **Up to Level: Level 2**.

38. Set the **Location Line** to **Finish Face: Exterior**.

At this point you could start drawing soffits by picking points in the drawing window. However, seeing as you have line work positioned exactly where the soffits go, you can use the *Pick Lines* option to quickly place a soffit exactly at each line location when you select it.

39. Select the ***Pick Lines*** option on the *Ribbon* (see image to the right).

40. Click the *Detail Line* pointed out in Figure 9-2.16.

 a. Watch for the dashed line to appear on the correct side of the *Detail Line* before clicking the mouse button. The dashed line indicates the side the wall will be placed on. This relates to the fact that the *Location Line* is set to one side of the wall rather than at the center.

 b. Again, notice the *Flip* arrow is on the side of the gypsum finish. The gypsum soffit is lower than the adjacent ceilings, so the side walls need to close off the ceiling space above.

 c. It is fine that the soffit wall runs through the main walls. As you can see they clean up because the base elevation varies.

FIGURE 9-2.16 First soffit wall placed along a detail line

41. Place the remaining soffit walls as shown in Figure 9-2.17 below.

 a. The thickness of the soffit wall needs to be within the overall footprint of the soffit. See Figure 9-2.26 for a section view of the end result we are looking for.

 b. The walls have been filled in with a solid color using the **VV→Filter** feature. You do NOT have to do this. This was done to make them stand out.

 c. Do not add a soffit wall at the double line locations. These areas need a bulkhead wall with gypsum board on both sides.

 FYI: The curved bulkhead shown will be added in a later step.

FIGURE 9-2.17 Soffit walls placed along a detail line

42. Add the remaining soffit walls on Level 1 and Level 2.

 a. Skip the areas below the two stairs for now.

Next you will close off the bottom of the soffit by placing a ceiling element. The first step in this process will be creating a new ceiling type. You need a gypsum board ceiling system with 3⅝" steel studs to frame the bottom of the soffit, similar to Figure 9-2.18.

43. Activate the *Ceiling* tool; select the **Susp GB on Metal Stud** ceiling type.

44. Click **Edit Type** → **Duplicate**.

45. New ceiling type name: **Gypsum Soffit**.

Braced soffit

FIGURE 9-2.18 Soffit detail from USG's Gypsum Construction Handbook

46. Edit the wall **Structure** as shown in Figure 9-2.19.

 a. Be sure the finish layer's *Material* is *Gypsum Ceiling Board*.

FIGURE 9-2.19 New soffit ceiling type

47. With the *Ceiling* tool active, and the *Gypsum Soffit* type selected, click *Sketch Ceiling* from the *Ribbon*.

a. **8'-6"** from Level 1.

48. **Sketch** the ceiling as shown in Figure 9-2.20 below, sketching along the backside of the soffit wall.

> *FYI:* *If you tried to create one large sketch for the entire soffit system it would be hard to manage later in the project, so you will just do small sections at a time. Once Joined, the soffit will look the same as if all drawn as a single element.*

49. Click to finish the sketch.

FIGURE 9-2.20 Sketch ceiling

Next we will zoom into a portion of the soffit we just created and notice a few issues.

Notice in the image to the right that the soffit walls are showing and that the curtain wall frame across the top is no longer visible (Figure 9-2.21). Both of these issues are a problem and need to be dealt with. The soffit walls just need to be *Joined* with the soffit ceiling element. The curtain wall issue relates to a design error we created in the model.

FIGURE 9-2.21 Ceiling soffit element added

50. Select **Modify → Join** from the *Ribbon*.

51. Select the soffit ceiling element and then one of the visible soffit walls.

52. Repeat the previous step until all soffit walls are no longer visible (Figure 9-2.22).

FIGURE 9-2.22 Ceiling joined to soffit walls

Now you will cut a section and see what is going on with the curtain wall. It is always a good idea to cut a temporary section anytime you are working in a specific area, especially when troubleshooting.

53. Add a temporary section as shown in Figure 9-2.22, and then open the new section view.

Looking at Figure 9-2.23 we immediately see the issue. The curtain wall is higher than the soffit. When the curtain walls were initially drawn we did not know how high they needed to be. Then later, when the ceiling was developed we consulted with the structural and mechanical engineers to get an idea of how high we could make the ceilings and still account for the beams, joists and ductwork. Now we will adjust the curtain wall elements to be 8'-6" high to match the soffits. You could just select the curtain wall in this section view and adjust its height. But we know there are several others that need to be fixed, so let's do them all now so we do not forget any!

54. Switch back to your Level 1 Floor Plan view.

55. Holding the **Ctrl** key, select all the interior curtain walls you added.

FIGURE 9-2.23 Section a new soffit

56. With the curtain walls selected, change the height from 9'-0" to **8'-6"** in the *Properties Palette*.

57. Switch to the **Level 2 Floor Plan** view and adjust the height of the curtain walls to **8'-6"** there as well.

58. Switch back to your **temporary section view**.

The curtain wall should now be properly aligned with the soffit ceiling element. Next you will adjust the stud wall in the center to stop at the bottom of the soffit. The 3⅝" soffit framing will adequately support the lateral, horizontal, loads of the wall below it. You should also notice the soffit walls extend through the floor above. We will fix this.

59. Use the *Join* command to join the soffit walls and the floor element above.

60. Select the stud wall between the two soffit walls, not the curtain wall.

61. Change the following in the *Properties Palette*:

 a. *Top Constraint*: **Unconnected**

 b. *Unconnected:* **Height: 8'-6"**

Once this is applied you may not notice a change. This is because the wall is "attached" to the floor above. This happened when prompted to attach walls after editing the Level 2 floor element. You will break this association next.

62. With the center stud wall still selected, click **Detach Top/Base** from the *Ribbon*.

Detach
Top/Base

63. Now select the element you want to detach the wall from: **select the Level 2 floor**.

Your section should now look like Figure 9-2.24. The curtain walls stop below the soffit, the soffit walls stop below the floor and the center wall also stops below the soffit.

> *FYI: We would not want to delete the center stud wall because this wall shows up in plan-view between the curtain wall locations.*

64. Add the remaining soffit ceiling elements on Level 1 and Level 2. Either move your temporary section mark around in plan view or create a new one as needed to verify the vertical conditions.

 a. Use the *Join* tool to join any soffit ceiling elements together to remove the line between them.

 b. In the two story lobby area, just extend the ceiling out to the edge of the Level 2 floor edge for now.

FIGURE 9-2.24 Soffit section condition

The image below shows a partial Level 1 plan view with the soffits added.

FIGURE 9-2.25 Ceiling soffit element added throughout the model and joined with soffit walls

Now we can add several basic 2'x2' ACT ceilings.

65. Activate the *Ceiling* tool; set the *Type Selector* to **2' x 2' ACT System**.

66. Set the height to **9'-0"** in the *Properties Palette*.

67. In the Level 1 Ceiling Plan view, click within the two *Small Conference Rooms* (Figure 9-2.26).

FIGURE 9-2.26 2' x 2' ACT System added with a 9'-0" height.

Notice how the ceiling stopped at the edge of the soffit wall! Next you will adjust the grid position.

68. Click to select one of the grid lines.

69. Use the ***Move*** tool, on the *Ribbon*, to move the *Model Pattern*, which represents the ceiling grid, 1'-0" North.

70. Repeat this for the other *Small Conference Room*.

Next you will browse to the USG website and download a Revit file which contains one of their specialty ceilings. You will be downloading their **Celebration® Ceiling Panel** system. The file you will be downloading is a project template. This file is opened and the ceiling type will be copied to the *Clipboard* and then pasted into your project. Ceilings are *System Families* in Revit, which can only exist within a project, unlike *Loadable Families* such as doors.

71. Browse to the following URL:https://www.usgdesignstudio.com/

72. Click the **Revit Template** link on the page (Figure 9-2.27). **Note:** this website has changed recently, and the Revit files are not accessible, so the required file to download is provided. See inside front cover of this book for access instructions.

73. Save the file to your computer.

74. Double-click the file and copy or drag the Revit file (*USG Celebration Ceiling Panels.rte*) out of the compressed ZIP file.

75. Double-click the Revit file to open it.

76. In the *Project Browser*, expand Families → Ceilings → Compound Ceiling.

77. Select **1' x 5' USG Celebration…** ceiling type.

FIGURE 9-2.27
Revit link on USG website

Take a moment to open the *Type Properties* (via right-click) and notice all of the provided data; this is the "I" in BIM (Building Information Modeling). This ceiling would be pretty easy to create in Revit, but entering all this data would take a while.

78. Right-click and select **Copy to Clipboard** (Figure 9-2.28).

79. Switch back to your *Law Office* project.

80. Click within the drawing window, to make sure it is active, and then press **Ctrl+V** to paste the ceiling into your project.

The new ceiling type is now available in your *Law Office* model.

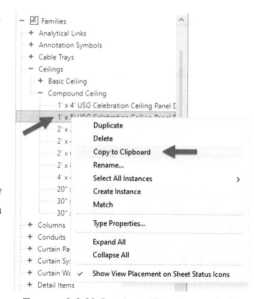

FIGURE 9-2.28 Copying ceiling type to clipboard

81. Switch back to the USG file you opened and close it without saving.

82. Zoom in on the *Impromptu Meeting* area between the two *Small Conference Rooms*, in the Level 1 Ceiling Plan view.

83. Activate the **Ceiling** tool, and set the *Type Selector* to the new **1'x5' USG Celebration** option.

84. Set the ceiling height to **9'-6"**.

85. Click to place the ceiling (Figure 9-2.29).

Your model should now look like Figure 9-2.29.

86. Adjust the grid pattern if necessary. Remember we laid out the soffits to accommodate the exact width dimension of this panel.

87. This would be a good time to **Save** if you have not done so in a while. Remember to save often!

Next you will specify the *Material* for this ceiling system.

88. With the *Celebration*® ceiling selected, click **Edit Type**.

89. Click **Edit...** *Structure.*

FIGURE 9-2.29 USG ceiling system placed in model

90. Click within the *Material* cell for the **Finish 2** row (which should be row 4).

91. Click the small icon that appeared to the right.

You are now editing the *Material* **Finishes - Interior - USG Celebration Metal Ceiling Panel 1' x 5'**.

FIGURE 9-2.30 Specifying finish for 1x5 USG Celebration ceiling panels

92. Click the **Replace Asset** icon on the *Appearance* tab name bar (Figure 9-2.30, Step #3).

93. Type **Maple** in the search box (Figure 9-2.30, Step #4).

94. For *Maple – Solid Natural Medium Gloss*, click the **Replace current asset** icon (Figure 9-2.30, Step #5).

You now have the wood finish selected, but the grid pattern still needs to be defined. You will use the same *Tile* pattern option used to define the ceramic tile pattern previously in the Materials chapter. Before changing the option to "tiles" you need to make note of the current image file defined for the maple wood appearance because it will be lost.

95. Click the material name "link" below the image preview (Figure 9-2.31, Step #1).

FIGURE 9-2.31 Applying grid pattern to ceiling

96. Select the file name listed at the bottom, and press **Ctrl+C** on the keyboard to copy the text to the *Clipboard*.

97. Click **Cancel** to close the *Select File* dialog.

98. Set the *Appearance Asset* type to **Tiles** (Figure 9-2.31, Steps 2 and 3).

The *Texture Editor* dialog is now open and can be used to define the wood ceiling tile and 1'x5' ceiling grid pattern. However, the image of maple wood has been replaced with a solid color option. The *Tile Appearance* can be changed to an image. Think of the *Tile Pattern* and the *Grout Appearance* as sub-components of the *Tiles* option.

99. Click **Edit…** below the preview and make the following changes in the *Texture Editor* (Figure 9-2.32):

 a. *Pattern:*
 i. 1 *per row*
 ii. 5 *per column*

 b. *Grout Appearance (gap width):*
 i. 0.25 Horiz.
 ii. 0.25 Vert.

 c. *Scale (sample size):*
 i. 5'-0" width
 ii. 5'-0" Height

 d. *Tile Appearance:*
 i. Set to **Image**.
 ii. Paste in the maple image file name via **Ctrl+V**.
 iii. See the paragraph below after selecting the new image file (before closing any dialogs).

The *Texture Editor* now shows a nested texture as highlighted in Figure 9-2.33. The image scale can be adjusted here, which is then applied to the tile area within the overall *Tiles* texture definition.

100. Click **Done** to close the *Texture Editor*.

FYI: Notice that materials from previous versions of Revit, which is the case with the ceiling we copied from USG, do not have physical or thermal assets. These features did not previously exist in Revit.

FIGURE 9-2.32 Applying grid pattern to ceiling

101. On the *Graphics* tab, click the **Texture Alignment** button to make sure the *Surface Pattern* and the *Appearance* asset align.

102. Click **OK** to close all open dialogs.

The USG Celebration material is now ready!

Next you will place a few more of these ceiling types before jumping on to the next task.

103. Place two more instances of the *1x5 Celebration* ceiling type as shown in Figure 9-2.34.

104. Adjust the grid lines as shown.

TIP: Use the **Align** *tool.*

FIGURE 9-2.33 Setting tile material for ceiling

FIGURE 9-2.34 Two more 1x5 ceiling types added

Next you will add several bulkheads to your model. These are different from a soffit wall in that both sides are partially visible and thus require gypsum board. You will create a new interior wall type for the bulkhead locations. This wall type will be thicker than the regular wall. This is for aesthetical reasons. We want the bulkhead to have a bit more girth to it, so the proportions are better. The typical 4⅞″ wall type could be used, and often is on the average project, but it can look a bit too slender.

Figure 9-2.35 shows an example of a bulkhead. Sometimes the lower ceiling aligns with the bottom of the bulkhead. However, both sides usually have gypsum board, at least up a few inches above the ceiling.

FIGURE 9-2.35 Bulkhead

As mentioned previously, the bottom of a wall cannot have a finish so you will add a sweep to the wall as you did with the soffit wall. Now you will add bulkheads at all the double-line locations: the *Mock Courtroom*, large conference room and the lunch room.

105. Create a 5/8" x 6" profile and load it into the project; name it **Gyp Bd Profile 4.rfa**.

106. Duplicate the typical 4⅞″ interior wall:

 a. *Name:*

 Interior – 6″ Bulkhead

 b. *Construction:*
 (2) layers ½″ gypsum bd.
 4″ Metal Studs
 (2) layers ½″ gypsum bd.

 c. Add a *Wall Sweep* using the profile just created.

 FYI: This wall construction will make a wall exactly 6″ thick. This will help with some modular building materials and the dimensional layout.

FIGURE 9-2.36 Bulkhead added to Mock Courtroom

107. Set the *Base Offset* to **9′-0″** and make sure the *Top Constraint* is set to **Level 2**.

108. Set the *Location Line* to **Finish Face: Exterior**.

109. Add the bulkhead in the *Mock Courtroom* as shown in Figure 9-2.36.

Do not forget to cut sections and add camera views occasionally to double check for issues and better visualize your progress.

110. Add a 2′x2′ ACT ceiling (at 9′-0″ AFF) and the 1′x5′ *Celebration* ceiling (9′-6″ AFF) in the *Mock Courtroom* as shown in Figure 9-2.37.

 a. The *Align* and *Move* tools are required to adjust the ceiling grid location.

 b. AFF = Above Finished Floor

111. Add the remaining ceilings and bulkheads as shown in Figures 9-2.37 and 9-2.38.

 a. Note: The center area in the *Large Conference Room* is currently a simple 2′x2′ ACT ceiling system.

 b. Do not add the ceiling height text at this time.

 c. The area below the lobby stairs is a soffit ceiling with soffit walls on <u>three</u> sides. The ceiling is the lowest you would ever want a ceiling: 7′-4″. This is just barely above a 7′-2″ door (7′ door plus 2″ hollow metal frame).

 d. Adjust the view's **Cut Plane** down to 7′-3″ so the low soffit shows up. **Reminder:** with nothing selected, edit *View Range* via *Properties Palette*.

FIGURE 9-2.37 Ceilings added to Level 1 ceiling plan – West side of plan

The last thing you will do to finish the level 1 ceiling plan is to add a few specialty ceilings in the *Large Conference Room*, the *Lunch Room* and the *Entry*. The specific systems to be used are available for download from the USG website, similar to the *Celebration* system.

However, they are not ideal Revit families because they consist of 3D AutoCAD geometry placed inside a Revit family. This creates extra lines and makes assigning *Materials* difficult. Therefore, this author has recreated these families for your use in this tutorial. This is very similar to what one might encounter when working on a real project. Many manufacturers do not even offer Revit content, so a general understanding of content creation is needed; see the chapter at the end of this book for an introduction to content creation.

FIGURE 9-2.38 Ceilings added to Level 1 Ceiling Plan – East side of plan

The 3D ceiling tile systems you are about to use are face-based families which will "stick" to the 2'x2' ceiling system previously placed AND cut a hole in it. The 2'x2' grid will be used to help align the 3D tiles. Visit https://www.usgdesignstudio.com/ and search for "topo" for examples.

 112. **Zoom** into the *Large Conference Room*.

113. Use the *Load Family* tool to load the following family from the provided online files:
DJS USG Topo 3D 4panel Down8.rfa

> *FYI:* See the inside-front cover of this book for instructions.

The *Family* just loaded is in the *Specialty Equipment* category and is placed using the *Component* tool.

114. Using the **Component** tool, place one instance of the *USG Topo* ceiling tile as shown in Figure 9-2.39 below.

FIGURE 9-2.39 Adding specialty ceiling system to *Large Conf. Rm.*

Notice this family actually represents four individual ceiling tiles. This is because the four tiles are always arranged together like this, so placement is faster.

115. Use the **Align** tool and reposition the tile in the upper left corner.

116. Use the **Copy** command to copy the 3D ceiling tile around until it fills the center area as shown in Figure 9-2.40.

FIGURE 9-2.40 USG *Topo* ceiling added to *Large Conf. Rm.*

The original 2′x2′ ceiling is still in the model but it is no longer visible. Try placing a *Camera* view in the *Large Conference Room* to check out how the ceiling looks (Figure 9-2.41).

FIGURE 9-2.41 Camera view showing USG *Topo* ceiling added to large conference room

117. Add the USG 3D *Topo* ceiling to the center feature area in the *Lunch Room*.

Next you will add the USG *Geometric* ceiling system to the *Entry* and *Lobby* areas. This *Family* consists of individual files due to the wide array of configurations possible.

118. Load the following *Family* from the provided online files: **DJS USG Geometric Wedge.rfa**.

119. Set the *Detail Level* to **Fine** in order to see an arrow in the family to indicate which direction the tile slopes.

120. Place the **Geometric Wedge** *Family* using the *Component* tool.

121. Use the ***Align*** tool to properly position the tile. Be sure to *Align* to the hidden reference plane just past the visible edge to properly center the tile in the 2'x2' ceiling grid.

122. ***Copy*** and ***Rotate*** the tile as shown in Figure 9-2.42. This is called the ***Pinwheel*** pattern.

FIGURE 9-2.43 Camera view

FIGURE 9-2.42 USG Geometric ceiling system

123. Add the ceilings as shown in the concept sketch to the Level 2 Ceiling Plan using the techniques thus covered (Figure 9-2.44).

 a. Notice the *Topo* ceiling above the *Lobby* needs the East and West soffit walls adjusted in because we laid the area out with an odd number of tiles. The *Topo* ceiling needs to have an even number to look correct.

FIGURE 9-2.44 Level 2 Ceiling Plan

124. Type **VV** and turn on the *Structural Framing* category in both the Level 1 and Level 2 Ceilings Plans. Both levels should have *Detail Level* set to **Fine**.

Having the *Structural Framing* category on, which makes the structure visible when not hidden by a ceiling, will help identify rooms with no ceilings and show the structure poking through the ceiling if the ceiling is too high.

Now you can delete the temporary detail lines and dimensions.

125. In your Level 1 plan view, select a window around your entire floor plan.

126. Click the ***Filter*** button on the *Ribbon*.

127. Click ***Check None*** (Figure 9-2.45).

128. Check ***Lines (<Hidden>)*** and ***Dimensions***.

129. Click **OK**.

130. Click the **Delete** key.

131. Repeat these steps for Level 2 *Detail Lines* and *Dimensions*.

> ***TIP:*** *In the camera views showing the* Geometric *panels, you can type* ***VV*** *and expand the* Ceilings *category and then turn off the* Ceiling Graphics *sub-category to hide the arrows associated with the geometric ceiling tiles.*

FIGURE 9-2.45 Using *Filter* tool to delete detail lines and dimensions

Self-Exam:

The following questions can be used as a way to check your knowledge of this lesson. The answers can be found at the bottom of this page.

1. Soffit ceilings and walls need to be joined to clean up properly. (T/F)

2. A wall's *Flip* arrow is always on the interior side. (T/F)

3. Detail lines are view specific. (T/F)

4. It is not possible to reposition ceiling grids. (T/F)

5. You can adjust a ceiling's height before or after it has been placed in the model. (T/F)

Review Questions:

The following questions may be assigned by your instructor as a way to assess your knowledge of this section. Your instructor has the answers to the review questions.

1. You can use *Filter* to narrow your selected elements down to just doors or furniture. (T/F)

2. What category are the special 3D ceiling tiles in: _____.

3. The *Ceiling Height* parameter is a type parameter. (T/F)

4. Revit cannot model soffits. (T/F)

5. Add a _____ to make a ceiling slope.

6. A bulkhead is just a wall with its base offset adjusted. (T/F)

7. The "Function" for a soffit wall should be set to soffit. (T/F)

8. Detail lines appear in multiple views. (T/F)

9. A ceiling can be either manually sketched or automatically placed. (T/F)

10. When just the ceiling grid is selected, the ceiling itself cannot be deleted. (T/F)

Notes:

Lesson 10
Break Room and Work Room:

This chapter will cover several concepts related to the development of casework using Autodesk® Revit®. Specifically, you will work on the *Break Room* and *Work Room*. The casework will be downloaded from the internet rather than using the content that comes with Revit. You will also learn about tagging and shared parameters.

Exercise 10-1:
Tagging

This section will study how **tags** work in Revit. This is an important feature in Revit, one used extensively by designers. This feature is often used on casework and ductwork as well as many other elements within the Revit model.

The basic premise of a *tag* is to be able to textually represent, in a drawing view, information found within an element, that is, an *Instance* or *Type Parameter*. One example found in nearly every set of construction documents is the door tag. This tag in Revit has been set up to list the contents of a specific door's **Mark** parameter, which is an *Instance Parameter*.

> The process to tag an element is simple:
> - Select the **Tag by Category** tool from the *Quick Access Toolbar*.
> - Adjust the settings on the *Options Bar* if needed.
> - Click the element(s) to be tagged.
>
> *FYI: Different tags are placed based on the category of the element selected.*
>
> - Click the *Modify* button or press the Esc key when done.

A *tag* is view specific, meaning if you add the tag to the **Level 1 Finish Plan**, it will not automatically show up in the **Level 1 Furniture Plan**. If you want the tag to appear in two Level 1 plan views, you need to add it twice. *(TIP: Use Copy/Paste.)* The advantage is you can adjust the tag location independently in each view. One view may have furniture showing and the other floor finishes. Each view might require the tag to be in a different location to keep it readable.

Tags are dependent on the element selected during placement. You cannot simply move a tag near another similar element and expect Revit to recognize this change. Similarly, if an element is deleted, the tag will also be removed from the model, even if the tag is in another view in the project.

The image below has several tags added to a floor plan view. All the listed information is coming from the properties of the elements which have been tagged. For example, the "M1" tag within the diamond shape is listing the wall type (i.e., *Type Mark*). Because the wall tag(s) is/are listing a *Type Parameter*, all wall instances of that type will report the same value, i.e., "M1." The door number "1" is reporting the element's **Mark** value, which is an *Instance Parameter*. Therefore, each door instance may have a different number. Notice some tags have the *Leader* option turned on. This is especially helpful if the tag is outside the room. Any 3D element visible in a view may be tagged, even the floor. In this case the *Floor Tag* is actually reporting the *Type Name* listed in the *Type Selector*. The leader can be modified to have an arrow or a dot.

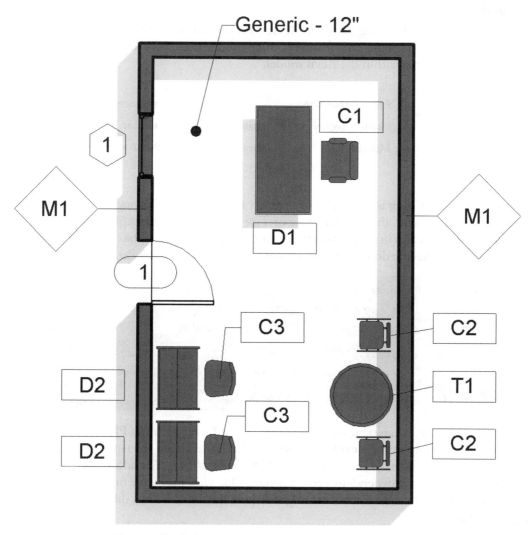

FIGURE 10-1.1 Floor plan with tags added

All of these tags were placed using the same tool: *Tag by Category*. Later you will learn how to specify which tag gets used for each category (e.g., walls, furniture, floors, windows, etc.).

The section / interior elevation below (Figure 10-1.2) shows many of the same elements tagged, as were tagged in the floor plan view on the previous page. If the wall's *Type Mark* is changed, the wall *Tag* will be instantly updated in all views: plans, elevations, sections, and schedules.

The tags had to be manually added to this view. They are not added here automatically just because you added one in the floor plan view, unlike when you add a section mark (where the section appears in other views automatically). If the view is deleted from the project, all these tags will automatically be deleted.

Some tags can actually report multiple parameter values found within an element. For example, the **Ceiling Tag** used in this example lists the *Type Name*, the ceiling height, and has fixed text which reads "A.F.F." Many tags can be selected and directly edited, which actually changes the values within the element. Changing the 8'-0" ceiling height will actually cause the ceiling position to change vertically.

It is possible to tag an element multiple times. You could add the same tag more than once. In the example below, you could place a **Wall Tag** on each side of the door. You can also add different tags which report different information from the same element. The **Wall Tag** "M1" and the **Material Tag** "PT-1" are two different tags extracting information from the same wall.

FIGURE 10-1.2 Section / interior elevation with tags added

Most design firms spend a little time adjusting the provided tags to make the text a little smaller and modify or delete the line work. This can easily be done by selecting a tag and then clicking **Edit Family** on the *Ribbon*. Revised tags can then be loaded back into the project and saved to a company template for future use.

With the more recent versions of Revit it is now possible to add tags in 3D views (Figure 10-1.3b). Before tagging a 3D view, you must **Lock** it (Figure 10-1.3a) to prevent changing the view and interfering with the position of the tags relative to the elements being tagged.

FIGURE 10-1.3A Locking a 3D view

When the *Leader* option is off, the tag is placed centered on the elected element. This initial position is not always desirable (e.g., note the D1 and C1 tags below) and you must move each tag so it is readable.

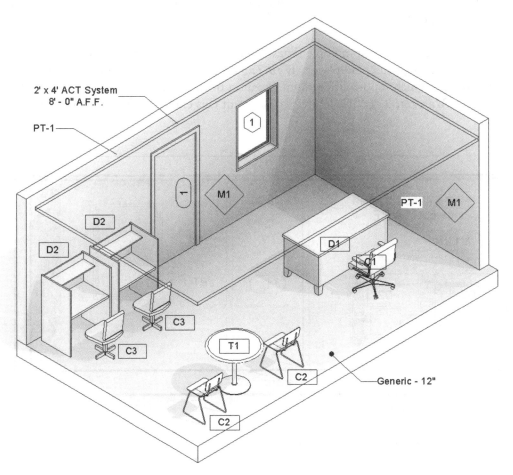

FIGURE 10-1.3B Locked 3D view with tags added

It is not possible to add *Room Tags* to 3D views yet.

It is helpful to understand where the information displayed in tags is stored. It is often necessary to change these values. In Figure 10-1.4 you can see that "PT-1" is the *Description*

for the *Material* assigned to the wall type.

> *TIP:* To find the Material *assigned to a specific wall type: Select the wall* → *Edit Type* → *Edit Structure* → *observe the Material column.*

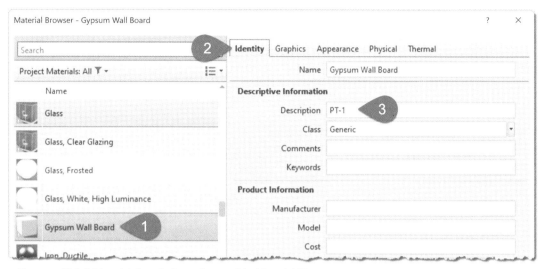

FIGURE 10-1.4 Material description value used in Material Tag

FYI: In the image above, the preview icons on the left with a gold triangle in the lower left corner are legacy materials, which are superseded by newer physically accurate materials.

A wall's *Type Mark* is found in the *Type Properties* dialog; select a wall → *Edit Type* from the *Properties Palette*. Notice in Figure 10-1.5 that the *Type Mark* is set to "M1." If you were to edit this value, all the tags for this wall will be updated instantly.

Seeing as all the information displayed in a tag is actually stored in the element being tagged, you will never lose any information by deleting a tag.

> *FYI: A tag will move with the element even if the element is moved in another view where the tag is not visible.*

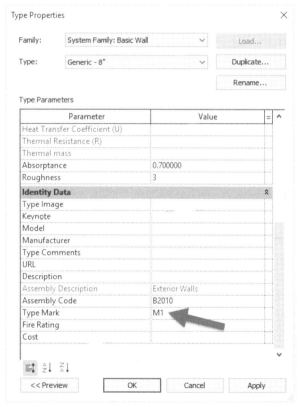

FIGURE 10-1.5 Selected wall's *Type Mark* value used in wall tag

The image below (Figure 10-1.6) shows the settings available on the *Options Bar* when the *Tag* tool is active.

- The tag can be ***Horizontal, Vertical, or Model*** *(the latter may be rotated)*.
- Clicking the **Tags…** button allows you to specify which tags are used when multiple tags for the same category are loaded in the current project.
- Checking the ***Leader*** option draws a line between the tag and the element.
 - *Attached End:* the leader always touches the tagged element.
 - *Free End:* allows you to move the end of the leader.
- **Length**: This determines the initial length of the leader.

FIGURE 10-1.6 Tag options during placement

When clicking the ***Tags…*** button you get the ***Loaded Tags*** dialog as shown in Figure 10-1.7.

Notice some categories do not have a tag family loaded. If you tried to tag a *Casework* element, Revit would prompt you to load a tag.

Clicking on a tag name in the right-hand column will allow you to select from multiple tag types/families, if available in the current project.

If you use one tag type in plan views and another in elevation views, you would visit this dialog just before tagging in plan views and then again adjust before tagging in elevation views.

FIGURE 10-1.7 Loaded tags dialog

> **FYI:** *It is possible to just let Revit place the default tag and then select the placed tag and swap it out via the* Type Selector.

> **TIP:** *If you want to swap several tags with another type: Select one of the tags* → *right-click* → **Select All Instances** *and then pick either* **Visible in View** *or* **In Entire Project**. *You now have all the tags selected and can pick something else via the* Type Selector.

The reader should be aware that some tags may be placed automatically when content is being added to the model. For example, when placing a window, the *Ribbon* has a toggle called ***Tag on Placement*** (Figure 10-1.8). When this is selected, which it is NOT by default, a door tag is placed next to every door. Note that the tag is only added to the current view, which could be an elevation, plan, etc.

You typically want to turn this off in the early design stages as the tags will just get in the way. Also, existing elements are not usually tagged, thus you would toggle this option off so you do not have to come back and delete the tag later.

FIGURE 10-1.8 Tag on Placement option; **off** (shown on left) and **on** (on right)

If you did not add tags, or some were deleted along the way, you will need to add them at some point. The easiest way to do this is by using the ***Tag All*** tool; this tool used to be called *Tag All Not Tagged*. When you use this tool, you can add a door tag to every door in the current view that does not already have a door tag. This is great on large projects with hundreds of doors and you want to make sure you do not miss any. You may still have to go around and rotate and reposition a few tags, but overall this is much faster!

Before concluding this introduction to tags we will take a brief look at how tags are created. If you select a tag and then click *Edit Family* on the *Ribbon*, you will have opened the tag for editing in the *Family Editor*. Keep in mind that your project is still open in addition to the tag family.

If you select a furniture tag and edit it you will see text (sample value shown) and four lines (Figure 10-1.9). Here the lines can be selected and deleted if desired. Also, the text can be selected and then you can click *Edit Type* to change the text properties (e.g., height, font, width factor, etc.).

If you select the text and then click the ***Edit Label*** option on the *Ribbon*, you can change the parameter being reported, or report multiple parameters (Figure 10-1.9).

TIP: *Open a tag family, edit it and then do a* Save-As *to make a new tag type.*

FIGURE 10-1.9 Editing a tag label in the family editor

In the *Edit Label* dialog above you can click the **New** icon in the lower left to load a *Shared Parameter* which allows you to create a tag which reports custom information. See the next section for more on this topic. Also, the last icon all the way to the right of the "New" icon allows you to override the formatting for numeric parameters (Figure 10-1.10).

This concludes the introduction to using tags in Revit. The next section continues this discussion with a more advanced concept called *Shared Parameters*.

FIGURE 10-1.10 Editing a label's format

Exercise 10-2:
Shared Parameters

FYI: You will not be using Shared Parameters in the tutorial, but this information is essential for a designer to be proficient and properly leverage Revit's power.

Introduction

Revit has many features which are unique when compared to other building design programs; one of these is *Shared Parameters*. The main idea with *Shared Parameters* is to be able to manage parameters across multiple projects, families and template files. This feature allows Revit to know that you are talking about the same piece of information in the context of multiple unconnected files. Here we will cover the basics of setting up *Shared Parameters*, some of the problems often encountered, and a few tricks.

The main reason for using **Shared Parameters** is to make custom information show up in tags; however, there are a few other uses which will be mentioned later. In contrast, a **Project Parameter** is slightly easier to make in a project but cannot appear in a tag; Revit has no way of knowing the parameter created in the family, info to be tagged, and the parameter created in the tag are the same bit of information. Both *Project* and *Shared Parameters* may appear in schedules.

The image below depicts the notion of a common storage container (i.e., the **Shared Parameter** file) from which uniquely coded parameters can be loaded into content, annotation and projects. This creates a connected common thread between several otherwise disconnected files.

Content Tags Projects

Shared Parameter File

Creating a *Shared Parameter* is fairly straightforward, but afterwards, managing them can be troublesome. Select **Manage → Shared Parameters** from the *Ribbon* to open the *Shared Parameters* dialog box (Figure 10-2.1). This dialog basically modifies a simple **text file** which should not be edited manually. If a text file has not yet been created, you will need to select the ***Create*** button and provide a file name and location for your *Shared Parameter* file.

It is important to keep in mind that this file will be the main record of all your *Shared Parameters*. From this file you create *Shared Parameters* within Projects, Templates and Families. Therefore, it

FIGURE 10-2.1 Edit Shared Parameters dialog

is ideal to only maintain one file for the entire firm, even if you have multiple locations. Of course, there is always an exception to the rule. This single file should be stored on the server and the software deployment should be set up to automatically point user computers to the *Shared Parameters* file. This can also be done manually by clicking the *Browse* button in the *Shared Parameters* dialog.

Once a *Shared Parameter* is loaded into a project, template or family, the text file is no longer referenced. So you technically do not need to send this file with your project file when transmitting to a consultant or contractor.

The name of the file should be simple and easy to find. You should create a new text file for each version of Revit you are using; include this in the file name. Some newer parameter types will cause older versions of the software to reject the text file.

Creating a Shared Parameter

Open the *Shared Parameters* dialog box. Here you can easily create *Groups* and *Parameters*. *Groups* are simply containers, or folders, used to organize the multitude of parameters you will likely create over time. When starting from scratch, you must first create one *Group* before creating your first *Shared Parameter*.

To create a new *Group*, click the ***New*** button under the *Groups* heading. In a multi-discipline firm you should have at least one group for each discipline. Parameters can be moved around later so do not worry too much about that at first.

Click the **New** button under the *Parameters* heading to create a new parameter in the current *Group*. You only need to provide three bits of information:

- Name
- Discipline
- Type of Parameter
- Tooltip

In this example, see image to the right (Figure 10-2.2); we will create a parameter called "Clear Width" which will be used in our door families (but can be used with other categories as this is not specified at

FIGURE 10-2.2 Setting Parameter Properties

this level). The other standard options, such as *Instance* versus *Type*, are assigned when the parameter is set up in the project or family.

The **Discipline** option simply changes the options available in the *Type of Parameter* drop-down. The **Type of Parameter** drop-down lets Revit know what type of information will be stored in the parameter you are creating. Many programming languages require parameters to be declared before they are used and cannot later be altered. Revit is basically a graphical programming language in this sense.

The **Edit Tooltip** button allows a description of the parameter to be entered. Anyone working in the project will see the description when they hover their cursor over the parameter in the Properties Palette or Edit Type dialog.

When a new *Shared Parameter* is created, a unique code is assigned to it. The image below (Figure 10-2.3) shows the code created for the "Clear Width" parameter just created. For this reason, it is not possible to simply create another *Shared Parameters* file a few months from now and have it work the same.

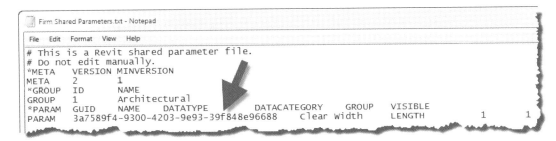

FIGURE 10-2.3 Shared Parameters text file

Creating a Shared Parameter in a Family

Now that you have created the framework for your *Shared Parameters*, i.e., the text file, you can now begin to create parameters within content; the next section covers creating *Shared Parameters* in project files.

Open a family file – in this example we will open the **Door - Single – Panel.rfa** file (download via Load Revit Family). Select the *Family Types* icon. Click the *New Parameter* icon on the lower-left (Figure 10-2.5). Now select *Shared parameter* and then the *Select* button (Figure 10-2.4). This will open the shared parameter text file previously created. Select "Clear Width" and then **OK**.

Now you only have two bits of information left to provide. Is the parameter *Type* or *Instance*, and what is the "Group parameter under" option? This controls which section the parameter shows up under in the *Properties Palette*.

Notice how the bits of information specified in the *Shared Parameters* file are grayed out here. They cannot be changed as this would cause discrepancies between families and project files. This information is hard-wired.

FIGURE 10-2.4 Parameter properties

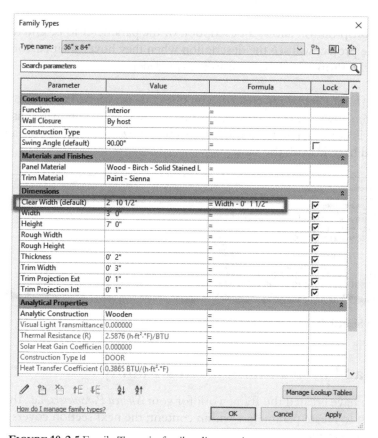

FIGURE 10-2.5 Family Types in family editor environment

Once the parameter has been created in the family it can be used just like a *Family Parameter*. In this example we created a formula to subtract the frame stops and the hinge/door imposition on the opening (Figure 10-2.5). Keep in mind that this value could now appear in a custom tag if desired, thus listing the clear width for each door in a floor plan. Maybe the code official has required this. Using *Shared Parameters* is the only way to achieve this short of using dumb text, that is, text manually typed which does not change automatically.

Next we will look at creating a door tag that lists the clear width. This is similar to the steps just covered for the door family as a tag is also a family. The only difference is this parameter will be associated to a *Label*.

Open the default **Door Tag.rfa** family (location: C:\ProgramData\Autodesk\RVT 2025\Libraries\US Imperial\Annotations\Architectural). Do a *Save As* and rename the file to **Door Tag – Clear Width.rfa**. Delete the linework, if desired. Select the text and click the *Edit Label* button on the *Ribbon*. In the *Edit Label* dialog you need to create a new parameter, by clicking the icon in the lower left; see image below. Your only option here is to select a *Shared Parameter*; as previously mentioned, only *Shared Parameters* can be tagged. Once the new parameter is created, click to move it to the right side of the dialog, in the *Label Parameters* column. Next, select the original *Mark* parameter and remove it. Finally, edit the *Sample Value* to something like 2'-10"; this is what appears in the family to give you an idea of what the tag will look like in the project.

> **FYI:** The Mark *parameter could be left in the tag if you wanted both the* Mark *and* Clear Width *to appear together. The other option is to have two separate tags which can be moved independently. Keep in mind it is possible to tag the same element multiple times. In this case you would have two door tags on the same door, each tag being a different type.*

Load your new door and door tag families into a new project file. In the next section you will see how these work in the project environment.

Using Shared Parameters in a Project

Now that you have your content and tags set up you can use them in the project. First, draw a wall and then place an instance of the *Single – Flush* door (turn off *Tag on Placement*). Select the door and notice the *Clear Width* parameter is showing up in the *Properties Palette*; it is an <u>Instance Parameter</u>. Add another door to the right of this one and change the width to 30″ via the *Type Selector*. Notice the *Clear Width* value has changed.

Next you will tag the two doors. Select the **Tag by Category** icon from the *Quick Access Toolbar*. Uncheck the *Leader* option and select each door. In the image below, the tag was also set to be *Vertical* via the *Options Bar*. If the tag placed is the door number, select the door tag and change it to the *Clear Width* option via the *Type Selector*.

> **TIP:** *If the text does not fit on one line you have to go back into the family and increase the width of the text box and reload the family into the project.*

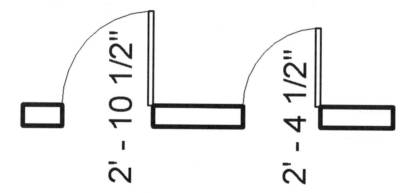

Things to Know

Only content which contains the *Shared Parameters* will display the parameter placeholder in a project. Try loading another door family (e.g., *Double – Flush*) into your test project without changing it in any way. Place an instance and then notice the *Clear Width* option does not appear in the *Properties Palette*. This means all content must have the *Shared Parameter* added to it. It is possible to create a *Project Parameter*, in the project/template file, but you have to specify *Instance* or *Type*. If this varies you cannot use a *Project Parameter*. The *Project Parameter* takes precedence and will change loaded content.

Selectively associating parameters with content is a great trick when it comes to certain categories having a variety of items, such as *Furniture* or *Mechanical Equipment*. Only loading

Shared Parameters into file cabinets or Air Handling Units (AHU) will make it so those parameters do not appear in your lounge chairs or VAV boxes.

> *FYI: This also works with* Family Parameters *but this information cannot be tagged or scheduled.*

Dealing with Problems

A number of problems can be created when *Shared Parameters* are not managed properly. The most common is when two separate *Shared Parameters* are created with the same name. This often happens when new users delete or otherwise lose a *Shared Parameter* file. They then try to recreate it manually, not knowing that the unique code, mentioned previously, is different and Revit will see this as a different *Shared Parameter*. The first place this problem typically shows up is when a schedule in the project has blank spaces even though the placed content has information in it. In this case, the schedule is using one specific version of the *Shared Parameter* and only the content using that same version of the *Shared Parameter* will appear in that schedule.

The fix for this problem is to open the bad content and re-associate the parameters with the correct *Shared Parameter*. It is not necessary to delete the bad parameter, which is good as this could cause problems with existing formulas.

If you get content from another firm you are working with and they have used *Shared Parameters* it is possible to export those parameters from the *Family Editor* into your *Shared Parameter* file. They will initially be located in the *Exported Parameters* group but can then be moved to a more appropriate location. This allows that unique code to be recreated in your file.

If you have a *Shared Parameter* with the same name as another firm's *Shared Parameter,* and you are working on the same model, that is a problem. You will have to decide whose version to use. There are some tools one can use to add and modify *Shared Parameters* in batch groups of families. One is found in the new *Extensions for Revit 2025*, which is available via the subscription website.

Conclusion

Using *Shared Parameters* is a must in order to take full advantage of Revit's powerful features. Like any sophisticated tool, it takes a little effort to fully understand the feature and its nuances. Once you have harnessed the power of the feature and implemented it within your firm's content and templates you can be much more productive and have less potential for errors and omissions on your projects.

Exercise 10-3:
Casework Content

In this chapter you will be downloading casework content to be used in your model. This section will discuss some of the added value one can benefit from when using well developed content. The image below shows the *Work Room* design you will create later in this chapter (Figure 10-3.1).

> *TIP: The image below was created from a camera view (see Chapter 2). All the casework was selected and then **Isolate Element** was selected from the Temporary Hide/Isolate fly-out on the View Control Bar at the bottom of the screen. Ambient Shadows is also turned on.*

FIGURE 10-3.1 Work room casework developed using content downloaded from manufacturer's website

The next several steps will walk you through downloading the content and then exploring the features. We will compare this content to the sample casework content provided by Revit. You will see many nice features and a lot of information provided in the downloaded content. This is not to say the downloaded content is perfect, but it is quite nice, especially for the price... free! When a manufacturer creates Revit content to represent their products, they will typically offer them to the public for free. They want to make it as easy as possible for a designer to use and specify their products. Charging a fee or requiring an online registration or special request is a hindrance that may cause the designer to keep looking for another product to download.

This exercise does not require you to make any changes to your law office project, but you will be instructed on how to download the content. This content will be required in the next exercise, so you must follow these numbered steps in order to complete the next section (i.e., Exercise 10-4).

Downloading Casework Content

1. Browse to the following website: http://www.hamiltoncs.com

2. Go to **Architects → Revit/CAD -- Casework**, select one of the Revit options, and then click **Download All Revit Blocks** to download this file (89+mb): **Modular Casework.zip**.

3. Save the file to your computer's hard drive.

4. Browse to the location you saved the file using *Windows Explorer* (not in Revit).

5. Double-click the ZIP file to view its content.

 a. A ZIP file is a compressed file which contains multiple files.

 b. Revit cannot directly access files within a ZIP file.

 TIP: You can create a ZIP file by selecting multiple files in *Windows Explorer* and then right clicking, and select Send to → Compressed (zipped) folder.

6. Select the two folders within the ZIP file.

7. Press **Ctrl + C** to copy the files to the *Clipboard*.

8. Browse to the folder you wish to store the casework content in.

9. Press **Ctrl + V** to paste the file into the current folder.

You now have the Hamilton Sorter casework downloaded and ready to use in your project! Before using it, we will review some of the features available within this content.

To better understand the options available with this manufacturer you may want to review their website.

As previously mentioned, we are not making any product or manufacturer recommendations. We are just trying to replicate the real-world process of accessing and reviewing content found online. This can save hundreds of hours, make your designs more accurate and allow for compelling renderings when compared to using the sample content provided with the software, or trying to create it yourself.

The image below (Figure 10-3.2) shows a 3D comparison between sample content provided by Revit and downloaded manufacturer created content on the right. Notice the difference in modeled 3D elements. The one on the right will look more accurate in sections and has nicer looking pulls. Keep in mind that more detail is not always desirable as this can decrease overall model performance. It takes experience to know where to draw the line, which is based on the size and complexity of the project.

FIGURE 10-3.2 Compare content: Revit sample content on left, downloaded manufacturer content on right

In addition to the differences in the way things are modeled, the downloaded content has a significant amount of information and options contained within the individual families. For example, if you try to change the width of the cabinet to a value larger than provided by this manufacturer, a conditional formula will replace your value with the maximum size.

Some Revit content also offers multiple handle, or pull, options by changing a setting in the families *Type Properties* (Figure 10-3.3).

FIGURE 10-3.3 Multiple handle options

The images on the next page, Figure 10-3.4, show the vast difference between the sample Revit content parameters and the downloaded content. These images are from the *Family Editor* environment, where the formulas are visible. Notice in the downloaded content that not all the parameters are even visible and some of the formulas run off the screen. When you open the downloaded content, you can explore and dissect it to get a better understanding of how the content works and how it was created.

Family Types

Type name: 36"

Search parameters

Parameter	Value	Formula
Construction		
Construction Type		=
Materials and Finishes		
Cabinet Material	<By Category>	=
Door/Drawer Material	<By Category>	=
Handle Material	Cabinets - Handles	=
Finish		=
Dimensions		
Depth	2' 0"	=
Height	2' 10 1/2"	=
Toe Depth	0' 3"	=
Toe Height	0' 3 1/2"	=
Width	3' 0"	=
Identity Data		
Assembly Code	C1030410	=
Type Image		=
Keynote		=
Model		=
Manufacturer		=
Type Comments		=
URL		=
Description		=
Cost		=

FIGURE 10-3.4 Compare content; Revit sample content on above, downloaded mfr. content on right

Family Types

Type name: MCBE302434

Search parameters

Parameter	Value	Formula
Width (default)	30.0000	= Overall Width
Top Draw Opening Model (default)	0.0000	= if(Top Drawer Opening > Overa…
Shelf Elevation (default)	16.0013	= if(Shelf Adjustability, Shelf Eleva…
Second Drawer	2.0000	=
Right Door Opening Placement (def	0.0000	= Wood Thickness * sin(Right Do…
Overall Height	34.0000	=
Open Door Width (default)	15.0000	= Overall Width / 2
Model Width (default)	30.0000	= if(or((Special Size Width < Widt…
Shelf Elevation Model (default)	9.1617	= if(User Shelf Elevation < (Kickpl…
Model Depth (default)	24.0000	= if(or((Special Size Depth < Dept…
Model Code Width (default)	30.0000	=
Model Code Depth (default)	24.0000	=
Lock Position	1.5000	=
Left Door Opening Placement (defa	0.0000	= Wood Thickness * sin(Left Door…
Kickplate Height	4.6020	=
Kickplate Depth	2.6870	=
Height (default)	34.0000	= Overall Height
Handle Position (default)	12.9000	= Open Door Width * 0.86
Drawer Width (default)	29.7500	= Model Width - (2 * Gap)
Gap	0.1250	=
Drawer Box Depth (default)	23.0000	= Model Depth - 1"
Drawer Height 2	6.5995	= (Overall Height - Kickplate Heig…
Door Top (default)	7.4120	= Drawer Height + Gap + Centre…
Depth (default)	24.0000	= Overall Depth
Centre Gap (default)	0.0625	= Gap / 2
IFC Parameters		
MasterFormat Section Title	Casework	=
MasterFormat Section Number	12 30 00	=
Other		
Handle Elevation	3.6748	= (Overall Height - Kickplate Heig…
Handle Elevation B (default)	12.6620	= Door Top + 5.25"

As with most content, a family represents many sizes, but once the geometry changes a new family is used. For example, when a base cabinet goes from a single door to a double door it is easier to use a separate family rather than try and set up complex visibility settings within a single family, although it can be done. The image below is a partial list of the downloaded casework content. It includes base and wall cabinets, tall cabinets, and accessories (Figure 10-3.5).

FIGURE 10-3.5
Loading content

When an individual family has many *types* (i.e., options and/or sizes) the content creator will usually employ a ***Type Catalog***. This appears while loading content into your project and allows you to select only the types needed in the current project. This helps to reduce errors by limiting the options within a project and also makes the *Type Selector* much more manageable. Autodesk does this with many of the structural steel shapes such as wide flange beams.

The image below (Figure 10-3.6) shows a *Type Catalog* which appears when loading one of the downloaded base cabinets. Notice the different depths and heights available. Here we are selecting all the 30″ deep and 2′-11″ tall cabinets; you hold down the **Ctrl** key to select multiple types. In elevation you would not be able to see whether you selected the correct depth and in plan you would not immediately notice the wrong height being selected. So it is better to only load what you need.

Specify Types

Family: HS_MCBH.rfa

Types:

Type	Model	Width	Depth	Height	Description
(all)		(all)	(all)	(all)	(all)
MCBH242429	MCBH	2′ 0″	2′ 0″	2′ 5″	242429
MCBH242433	MCBH	2′ 0″	2′ 0″	2′ 9″	242433
MCBH242435	MCBH	2′ 0″	2′ 0″	2′ 11″	242435
MCBH242437	MCBH	2′ 0″	2′ 0″	3′ 1″	242437
MCBH242441	MCBH	2′ 0″	2′ 0″	3′ 5″	242441
MCBH243029	MCBH	2′ 0″	2′ 6″	2′ 5″	243029
MCBH243033	MCBH	2′ 0″	2′ 6″	2′ 9″	243033
MCBH243035	MCBH	2′ 0″	2′ 6″	2′ 11″	243035
MCBH243037	MCBH	2′ 0″	2′ 6″	3′ 1″	243037
MCBH243041	MCBH	2′ 0″	2′ 6″	3′ 5″	243041
MCBH302429	MCBH	2′ 6″	2′ 0″	2′ 5″	302429
MCBH302433	MCBH	2′ 6″	2′ 0″	2′ 9″	302433
MCBH302435	MCBH	2′ 6″	2′ 0″	2′ 11″	302435
MCBH302437	MCBH	2′ 6″	2′ 0″	3′ 1″	302437
MCBH302441	MCBH	2′ 6″	2′ 0″	3′ 5″	302441
MCBH303029	MCBH	2′ 6″	2′ 6″	2′ 5″	303029
MCBH303033	MCBH	2′ 6″	2′ 6″	2′ 9″	303033
MCBH303035	MCBH	2′ 6″	2′ 6″	2′ 11″	303035
MCBH303037	MCBH	2′ 6″	2′ 6″	3′ 1″	303037
MCBH303041	MCBH	2′ 6″	2′ 6″	3′ 5″	303041

Select one or more types on the right for each family listed on the left

[OK] [Cancel] [Help]

FIGURE 10-3.6 Type catalog

Large Number of Loaded Family Types ✕

You have selected many family types to load. Loading family types that are not going to be used takes up additional memory and should be avoided. Do you want to continue?

[Yes] [No]

FIGURE 10-3.7 Type catalog related warning

If you try to load a lot of types from a *Type Catalog*, Revit will give you a warning telling you there are some drawbacks to loading "many" types (Figure 10-3.7). You can do this if you really want to but with consequences.

Another aspect of Revit content is the **Detail Level** options in each view. Each view has a *Detail Level* setting; it can be **Coarse**, **Medium** or **Fine**. Each option progressively reveals more detail. The *Coarse* option is often used for early phase schematic drawings, *Medium* is typically adequate for construction documents and *Fine* is used for high level details and renderings. The *Coarse* option also helps with increasing computer/model performance. When a view does not have hundreds of handles on doors and cabinets, the model can be spun via the *Orbit* tool much more smoothly.

Notice in the following examples (Figures 10-3.8 and 10-3.9), additional detail is revealed as the *Detail Level* is increased. It is possible to set things up so certain elements never show in a plan or elevation view. However, all 3D elements will show up in a 3D view regardless of *Detail Level*, except when visibility is specifically turned off in the *family* (e.g., the hidden handles types not in use).

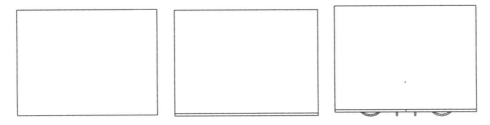

FIGURE 10-3.8 Plan view for same base cabinet with different detail level settings: *Coarse*, *Medium* and *Fine*

FIGURE 10-3.9 Elevation of base cabinet; *Coarse* (left) and *Fine* (right)

It should be mentioned that some content does not have any *Detail Level* control. In this case, all of the 3D geometry is visible in plan or elevation view.

An important aspect to developing functional casework is using accessories. These are mainly fill and back panels that close off leftover gaps and allow for scribing against irregular walls, especially in existing conditions. The image on the next page (Figure 10-3.10) shows a few samples of the accessories from the content pack you downloaded. Notice the base cabinet filler on the left, with a built-in toe-kick, and the end-panel/countertop support legs to name a few.

FIGURE 10-3.10 Accessories available for downloaded casework content

Some content downloaded will provide a file specifically pertaining to available finishes (Figure 10-3.11). This example has several *Materials* created in it, similar to what you did in Chapter 4. Additionally, small solid elements have been added and assigned a material. These elements can be selected, copied to the *Clipboard* (Ctrl+C) and then pasted into your project (Ctrl+V). The element pasted can then be deleted, but the imported *Material* will remain. This new *Material* can be assigned to your casework or anything else in your model.

There are many things we would continue to look at regarding casework, but time and space will not permit at this time. In the next two exercises, you will add casework to the law office.

FIGURE 10-3.11 Material options

Exercise 10-4:
Work Room

Introduction

This exercise covers the basic techniques used to add casework to your model. Casework is typically any built-in cabinets, countertops, desks, bars, etc., and is usually installed by a carpenter. This does not include systems furniture, or cubicles, which are sometimes attached to the wall.

The architectural floor plans typically show the built-in items (i.e., casework) and the FFE Plan (furniture, fixtures and equipment) typically shows the moveable items. Furniture placement is covered in the next exercise.

Cabinets rendered in Revit

The words "typically" and "usually" are used a lot because there is no exact science to how a set of drawings is put together. They can vary by country, state, city and even office.

The items dimensioned in Figure 10-4.1 are not actually dimensioned in an interior elevation, except the cabinet height, because this is typically a manufactured item, so the dimensions are what they are. They still need to be modeled correctly, but the dimensions do not need to be added because they would not help the contractor and would "muddy up" the drawings. The dashed lines represent shelving within the cabinets and door swing.

FIGURE 10-4.1 Nominal base cabinet dimensions

Work Area Cabinets:

Cabinets in a work area can vary quite a bit depending on the work being performed there. The picture on the next page is similar to what you will be drawing next. The **base cabinets** are on the floor and are typically 24″ deep but come in a variety of depths. The **wall cabinets** are attached to the wall and also come in various depths, 12″ and 14″ being the most common.

As you can see, the **electrical outlets** need to be coordinated with the electrical engineer; they need to be installed above the countertop and below the wall cabinets. When preparing the program, the designer should get a list of all the equipment required in each area. This will determine the number of outlets and locations. Often, items are overlooked or added later that can create an unsafe strain on the electrical system or require extension cords. For

example, it is possible that the microwave shown in the image below was added by the users long after the design of the building; if the electrical system did not have enough capacity, it could make for an unsafe condition. If that were the case, a new outlet on another circuit would need to be added.

Notice two things about the wall cabinets:

Example of typical work room cabinets

1) The cabinets have **"under-cabinet lights"** across the entire length. If this is desired by the client/designer, the electrical drawings need to be coordinated and a valance added or the bottom of the cabinet recessed so the light fixture is not directly seen.

2) For a clean, dust-free finish, the space between the top of the cabinet and the ceiling is filled in with a soffit. This is constructed of studs and gypsum board. This is not always in the client's budget.

Placing Casework Content:

You will be prompted to load just the components needed for this *Work Room*. All content should be loaded from the location you saved the downloaded content to in the previous exercise.

1. **Open** your Law Office model.

2. Switch to the **Level 1 Floor Plan** view.

3. Set the *Detail Level* to **Medium**.

With the *Detail Level* set to *Medium* you will be able to see additional lines which will help determine which side is the front. You would not want to set the *Detail Level* to *Fine* because too much detail would be shown, e.g., the handles.

4. Zoom into the *Work Room*, which is located near the front *Lobby*.

5. Load the following base cabinet family (Insert → Load Family):
 HCS_MCBH.rfa (which is a Base Cabinet w/ 2 Door 2 Drawer)
 FYI: See page 10-17 for custom content download instructions.

 a. Select the following *Types* from the **Type Catalog**:

 - MCBH243035
 - MCBH303035
 - MCBH363035
 - MCBH423035

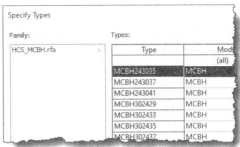

The model number for most cabinet manufacturers includes the width, depth and height. That is the case here. Notice we have four widths (24″, 30″, 36″ and 42″) and all are 30″ deep and 35″ tall. The deeper cabinet is ideal in a work room to accommodate equipment such as printers, postage machines, etc.

6. Using the *Component* tool, place an instance of **MCBH363035** in the middle of the room as shown in Figure 10-4.2.

 a. The cursor is at the middle-back and the side with flip-arrow icon (when cabinet is selected) is on the front of the cabinet.

 b. Before clicking to place the content, you can press the **space bar** to rotate the item if necessary.

 FYI: If the Detail Level *were set to* Coarse *you would not be able to tell which direction was front without looking at an elevation or 3D view.*

FIGURE 10-4.2 First base cabinet placed in workroom

7. Use the *Align* tool to reposition the cabinet flush with the wall in the Northeast corner.

 TIP: You may want to click the **Thin Lines** *option on the Quick Access Toolbar in order to more clearly see the wall lines and properly position the cabinet.*

8. **_Copy_**, or **_Array_**, the cabinet three times to the West (Figure 10-4.3).

Later, when viewing these cabinets in elevation, you can quickly select one and swap it out with another _Type_ via the _Type Selector_.

To prepare for the remaining cabinets to be used, you will load all needed content now.

FIGURE 10-4.3 Four base cabinets added

9. **Load** the following _Families_ and _Types_ into your model:

 a. **HCS_MCBL.rfa** (Base Cabinet - 2 Door 9 Shelf Sorter)
 - _Type:_ MCBL363035

 b. **HCS_MCSA.rfa** (Storage Cabinet - 2 Door)
 - _Type:_ MCSA361884 (or, use corrected file provided with book)

 c. **HCS_MCWA.rfa** (Wall Cabinet - 2 Door)
 - _Type:_ MCWA361430

 d. **HCS_MCWF.rfa** (Wall Cabinet - Sorter - 2 Door Legal Shelves)
 - _Type:_ MCWF361830

 e. **HCS_MCCT.rfa** (Accessories Misc - Work Surface - 24-30 Deep)
 - _Type:_ MCCT7231 Elevation 35

10. Place two instances of the **HCS_MCSA.rfa** (Storage Cabinet) as shown in Figure 10-4.4.

FIGURE 10-4.4 Two tall storage cabinets added

Next you will place the countertop. The content you downloaded has a somewhat limited selection of countertops. We will have to use two pieces to cover the entire row of base cabinets. This will cause a line to show in plan and 3D views. If you did not want this, you could use one of the countertops provided with Revit or adjust the family.

The countertop you will be using does not have a backsplash, or side splashes, but that is not usually a problem in a work room.

11. Place the HCS_MCCT.rfa (Work Surface - i.e., countertop) in the middle of the room and use **Align** to properly position it in the Northeast corner of the room.

 FYI: Note that the surface is positioned at the correct height off the floor as should be apparent by the Type Name *containing* "Elevation 35."

12. **Copy** this element **6'-0"** to the West.

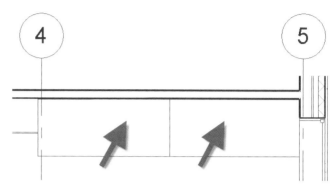

You now have a countertop, which hides the visibility of the cabinets below it (Figure 10-4.5). They are 3D elements which cannot be seen through unless the view is set to *Wireframe* or the visibility of the elements has been overridden; e.g., select one, right-click → Override Graphics in View → By Element.

FIGURE 10-4.5 Countertop added

Next up are the wall cabinets. Some wall cabinet content is *wall hosted*. Thus, they can only be placed on a wall and will be deleted if the wall is deleted. The good thing is that they move with the wall, but the base cabinets do not (unless *Aligned* and *Locked*). The cabinets you downloaded are not wall hosted and therefore may be placed anywhere in the model, for better or worse.

The wall cabinets have a default height off the floor that puts the top of the cabinet at 7'-0". This is pretty common and is what will be used. Be sure to notice the *Cabinet Elevation* instance parameter, in the *Properties Palette*, while placing the wall cabinets.

13. Place two instances of **HCS_MCWA.rfa** (Wall Cabinet - 2 Door**)** in the Northeast corner of the room; this will be over the base cabinets.

14. Next to that place two **HCS_MCWF.rfa** (Wall Cabinet - Sorter - 2 Door Legal Shelves); see Figure 10-4.6.

You may want to add a *Camera* view to track your progress in 3D. You will be instructed to create an interior elevation in this section.

You have now placed all the cabinets in the *Work Room* (Figure 10-4.6). Notice we made the two wall cabinets closest to the window only 14" deep due to the location near the window.

FIGURE 10-4.6 Wall cabinets added

Creating a Filter and applying it to a View:

Next you will add a *Filter* to make all of the wall cabinets, which are above the cut plane, a dashed line. Some content you download is set up to be dashed automatically. In this case we want a smaller dash and we also want the side/end lines dashed.

15. Type **VV** on the keyboard.

16. Click the *Filter* tab.

17. Select the *Edit/New...* button near the bottom (Figure 10-4.8, Step 1).

18. In the *Filters* dialog, click the *New Filter* icon in the lower left (Figure 10-4.7).

19. Enter the name **Wall Cabinets**.

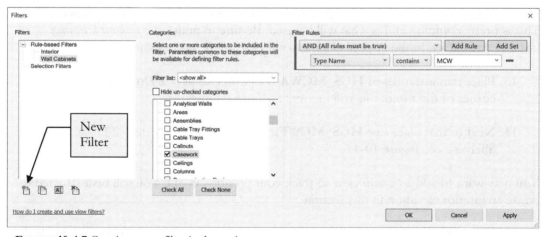

FIGURE 10-4.7 Creating a new filter in the project

20. Make the following edits for the new filter:

 a. *Categories:* **Casework**

 b. *Filter by:* **Type Name**

 c. **Contains**

 d. **MCW** *FYI: This is case sensitive.* **Also:** *If using additional content, all wall cabinets must follow this naming convention for this filter to work properly.*

21. Click **OK** to close the *Filters* dialog.

You now have a new *Filter* created in your project which can be used in any view. Keep in mind the filter you just created is not tied to the view from which you just created it (i.e., Level 1 Plan View). The *Filter* dialog is accessible directly via the *View* tab.

Big picture: a *Filter* is defining the criteria used to search (i.e., filter) for elements. Next you will associate this *Filter* with your plan view and then tell Revit what to do with the elements it finds.

> *FYI: It is possible to have different views do different things with the elements it finds. For example, in this view we will make the cabinets dashed, but in another view we could make them change to the color magenta so they stand out.*

22. Click the **Add** button to add the new *Wall Cabinet* filter to the plan view (Figure 10-4.8; Step 2).

FIGURE 10-4.8 Adding a filter to the floor plan view

23. Set the *Transparency* to **100%** and *Projection/Surface* lines to **Dash 3/32"** as shown in Figure 10-4.8; Steps 3 and 4, respectively.

Notice a few of the other options: you could make everything meeting the *Filter* criteria not visible, halftone and more.

> **FYI:** *The cabinets do not pass through the view's cut plane so that is why you edited the* Projection *line column rather than the* Cut.

24. Click **OK** to close the *VV* dialog.

Your wall cabinets are now dashed and can be more easily distinguished from the tall and base cabinets (Figure 10-4.9). Because you added a *View Filter* rather than overriding the individual elements, any new wall cabinets added to this view will automatically become dashed. You will have to apply the "Wall Cabinet" *Filter* to the Level 2 Floor Plan view if needed.

FIGURE 10-4.9 Wall cabinets are now dashed in the floor plan view

Creating an Interior Elevation:

Next you will create an interior elevation, learn how to change types and add tags to your cabinets.

25. Using the **Elevation** tool from the *View* tab, and with the *Type Selector* set to **Interior Elevation**, place an elevation as shown in Figure 10-4.10.

FIGURE 10-4.10 Interior Elevation added in Work Room

26. Find the new elevation view in the *Project Browser* and rename it to **WORKROOM ELEVATION - NORTH**.

27. Double-click the view title in the *Project Browser* to open the view.

 a. You can also double-click the pointer part of the elevation in plan to open the view. The elevation symbol cannot be selected for this to work.

You should now have an interior elevation view similar to that shown in Figure 10-4.11. Next you will adjust the *Crop Region* bottom so the floor structure does not show, change the *Detail Level* and turn off the grids and levels.

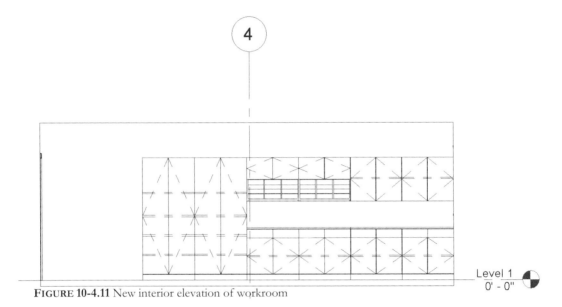

FIGURE 10-4.11 New interior elevation of workroom

28. To hide the *Grid*, do the following:

 a. Select the grid

 b. Right-click

 c. Hide in View (fly-out menu)

 d. Category

29. Hide the *Level* line as well.

30. Set the view's *Detail Level* to **Fine**.

31. Select the ***Crop Region*** and drag the bottom grip up until it snaps to the top of the floor slab. Be careful not to drag it up higher than the floor line.

Your view should now look like Figure 10-4.12 below. Notice the cabinet pulls are now visible.

FIGURE 10-4.12 Elevation view adjusted

Adding Tags:

You are now ready to add tags, notes and dimensions to this view. If this were a real project, we would not add annotation at this time. It is best to wait until the project is in the construction documents (*CD*) phase to avoid redoing the annotation if (more like "when") things change.

32. Click the ***Tag by Category*** tool from the *Quick Access Toolbar* (or *Annotate* tab).

33. Select the tall storage cabinet on the far left.

When a tag is not loaded in the model for the category of a selected element you will get a prompt to load one (Figure 10-4.13). You will load a casework tag next.

FIGURE 10-4.13 Load tag prompt

34. Click **Yes**.

35. Browse to the download/provided content with this book.

36. Open the **Familes** folder to view its contents.

37. Select **Casework Tag** and click **Open**. FYI: This family can also be downloaded via the Load Autodesk Family command.

Before clicking to place the tag, you will turn off the *Leader* option on the *Options Bar*. Notice the tag changes and moves with your cursor as you hover around the model (without clicking).

38. Uncheck the **Leader** option on the *Options Bar*.

39. Click the tall storage cabinet on the far left again.

When the *Leader* option is turned off, the tag is placed at the center of the element as seen in Figure 10-4.14. Notice the tag shows a question mark when selected and is blank when not selected. This means the parameter the tag is pointing to is blank. This tag is looking at the *Type Mark* parameter.

You will move the tag below the floor line and edit the tag *Family* to show the model number and remove the box around it.

FIGURE 10-4.14 Tag placed at center of cabinet

40. If needed, select the tag and move it below the cabinet.

41. With the tag still selected, click the ***Edit Family*** button on the *Ribbon*.

Edit
Family
Mode

FIGURE 10-4.15 Tag opened in the family editor environment

You are now in the *Family Editor* (Figure 10-4.15). Your *Law Office* project is still open. These next steps are just meant to show you how easy it is to adjust a tag family. This is not meant to be an in-depth overview of creating families.

42. Select the four lines which define the rectangle and delete them.

Edit
Label

Label

43. Select the text **1i** and then click the ***Edit Label*** button from the *Ribbon*.

44. <u>Remove</u> the ***Type Mark*** label (i.e., parameter) and <u>add</u> ***Description*** per the steps shown in Figure 10-4.16 below.

45. Click **OK** to commit the changes.

> ***FYI:*** *Notice the sample value changed, but this is no problem. You would set this to the worst case when you have a box to make sure the value fits.*

FIGURE 10-4.16 *Edit Label* dialog for tag family being modified

The last thing you will do is adjust the text to be *Arial Narrow* as well as reduce the width factor. This will make the information fit better under each cabinet.

46. Select the label, which should now read "This is a sample description."

47. Click ***Edit Type*** from the *Properties Palette*.

48. Change the *Text Font* to **Arial Narrow**.

49. Change the *Width Factor* to **0.9** (Fig. 10-4.17).

50. Click **OK**.

Text	
Text Font	Arial Narrow
Text Size	3/32"
Tab Size	1/2"
Bold	☐
Italic	☐
Underline	☐
Width Factor	0.900000

FIGURE 10-4.17 Label font

You are now ready to load the tag back into the project. But before you do you need to make a decision: do you want to overwrite the original tag in the project? If you do, you can just click the **Load into Project** button and you are done. Any tag previously placed gets updated and the old tag looking at the *Type Mark* no longer exists. If you want to use both tags, you need to save your tag family with a new name (via *Save-As*) and then load it into your project. Saving it as a named file also gives you the ability to use the new tag in future projects.

We will just overwrite the original tag for this tutorial. You may save it to a file as well if you wish.

51. Click the **Load into Project** button on the *Ribbon*.

Load into Project

You are now taken back to the project and prompted as to whether you want to overwrite a family with the same name.

52. Click the top option: ***Overwrite the existing version*** (Figure 10-4.18).

FIGURE 10-4.18 Overwrite family prompt

The tag is now updated to show the *Model* number (Figure 10-4.19)

53. **Tag** the remaining cabinets as shown in Figure 10-4.20; do not copy the tag.

> *TIP: Once the wall cabinet tags are placed, select them all while holding the **Ctrl** key and then use the **Up** arrow key to move them into the desired position.*

Because the model number provides the dimensions of the cabinet, you do not need to add but one dimension. Even this dimension is not really needed because the top of the wall cabinets align with the tall storage cabinets. But if we did not have the tall storage cabinets the installer would need to know where to mount the wall cabinets relative to the floor.

54. Add the 7'-0" vertical **Dimension** shown in Figure 10-4.20.

361884

FIGURE 10-4.19 Updated tag showing model number

FIGURE 10-4.20 Tags, notes and dimensions added to interior elevation

> *FYI: The door frame on the adjacent wall is showing on the left side of the elevation – notice the arrow pointing this out in the image above (Figure 10-4.20).*

Next you will add a line to represent the resilient wall base. This is not actually modeled in Revit due to the time and reduced model performance it would require. The room finish schedule lists what each room requires. So you will just add a 2D line when needed. If you end up doing a photorealistic rendering in an area, you would have to add a 3D version of the wall base. But that is only needed in a few locations usually.

55. Select the **Detail Line** tool from the *Annotate* tab.

56. Set the *Line Style* to **Thin Lines** on the *Ribbon*.

57. Draw a horizontal line from the tall cabinet to the adjacent wall.

58. Select the **Text** tool.

59. Pick **3/32" Arial** from the *Type Selector* and then click **Edit Type**.

60. Make these changes:

 a. *Leader Arrowhead:* **Arrow Filled 15 Degree**

 b. *Width Factor:* **0.9**

61. Select the leader option shown in the image to the right.

 Add the notes shown in Figure 10-4.20 above. *FYI: All text in the drawings is typically uppercase.*

62. **Save** your project; you do not need to save the *Tag* family that is still open.

Exercise 10-5:
Lunch Room

Now you will develop the *Lunch Room*, making your model look like the one shown below. You will utilize many techniques you have already learned, such as loading content, creating an interior elevation, adding soffit walls and tagging. In addition, you will look at how the cabinet family dimensions can be changed and what the ramifications of doing so are. You will also be working around a structural column which will require the use of filler panels and a custom countertop and backsplash.

FIGURE 10-5.1 Lunch Room design

The first step is to cover the exposed column with gypsum board. Some designers may wish to shift the wall or fur it out behind the casework to eliminate the bump out. In this case we will use this anomaly to learn a few new things in Revit. Something like this may occur in the design of an existing building, so it is good to know about the techniques needed to deal with these situations. This involves using an *Architectural Column*, casework filler panels (i.e., accessories) and creating an *In-Place* family to deal with the countertop wrapping around the column.

1. Zoom into the *Lunch Room* in the Level 1 floor plan.

2. Create an interior elevation and set the *Detail Level* to **Fine** (Figure 10-5.2).

Notice in the image below (Figure 10-5.2) that the structural HSS column is thicker than the stud wall and therefore is visible in the room and interrupts the otherwise flat or smooth wall. You will load an *Architectural Column* to handle this situation.

lunchroom staff (the café)

254

321 SF

FIGURE 10-5.2 Lunch Room layout – floor plan view

Adding Column Furring:

Architectural Columns will automatically take on the properties of the wall they are placed on; thus, they will show the gypsum board wrapping around the column. *Structural Columns* do not take on the properties of elements they are joined to like *Architectural Columns*.

3. Select **Architecture → Build → Column → Column: Architectural**

4. You are prompted to load an *Architectural Column* because none exist in the current model; click **Yes**.

5. Open the ***Columns*** folder, <u>not</u> *Structural Columns*.

6. Select **Rectangular Column.rfa** and then click **Open**.

Looking at the *Type Selector*, you see there are three sizes available:
- 18″x18″
- 18″x24″
- 24″x24″

You will create a new type within the *Rectangular Column* family that is 8″x8″, which is the size we want to wrap around and conceal the column.

FIGURE 10-5.3 Enter type name

7. While still in the *Architectural Column* tool, select ***Edit Type*** from the *Properties Palette*.

8. Click ***Duplicate*** in the upper right to create a new *Type*.

9. Enter **8″ x 8″** for the type *Name* (Figure 10-5.3).

You now have a new type within the *Rectangular Column* family, but you still need to adjust the new type's parameters to be the correct size.

10. Change the *Width* and *Depth* parameters to **8″** (Figure 10-5.4).

11. Click **OK** to save the changes.

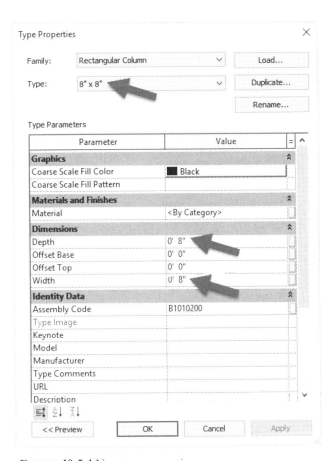

Now you are ready to place the architectural column.

12. **Zoom** in to the structural column in Figure 10-5.2.

13. Click at the intersection of the two Grids, <u>not</u> at the center of the wall.

FIGURE 10-5.4 New type properties

As you can see (Figure 10-5.5), the outline of the column is not really visible. It has blended in with the adjacent walls and now you have gypsum board that continues from the walls around the column. Notice how little of the column appears when placed at a corner (Figure 10-5.6).

In reality, this column may need to be larger to accommodate the metal furring which can range from ⅝″ channels to full 3⅝″, or larger, metal studs.

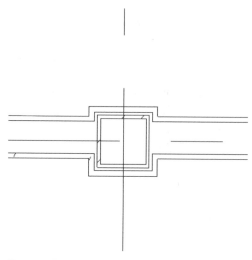

FIGURE 10-5.5 Architectural column added

Keep in mind the *Architectural Column* you just placed has *Base Offset* and *Top Offset* parameters just like walls. In our case the default setting is from floor to floor. In some cases the column furring may just need to go a few inches above the ceiling. We will leave it as is to better deal with acoustics, minimizing any noise coming from the adjacent mechanical room.

> *FYI: This may also be part of the wall's fire rating which separates the mechanical room from the rest of the building.*

Next you will be given a list of content to load from the local library or from provided content. You might want to note the size of your Revit model before and after loading the content, via *Windows Explorer*, to see how loading content affects project file size. File size is not directly related to model performance, but it is good to keep an eye on the overall file size.

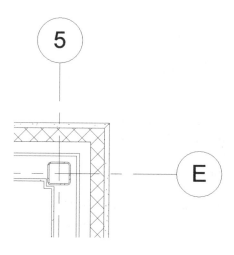

FIGURE 10-5.6 Architectural column added

Loading Needed Content:

14. **Load** the following *Families* and *Types* into your model:

From the provided content:

a. **Refrigerator**

b. **Microwave**

c. **Vending – Candy**

d. **Faucet-Kitchen_Sink-Zurn-Z871B1**

e. **Sink-Five_Hole_Drilling-Kohler-Hartland-5818_5u**

TIP: See the inside-front cover, of this book, for instructions on downloading provided files.

From the Hamilton casework library previously downloaded:

 a. **HCS_MCBE.rfa** (Base Cabinet - 2 Door 1 Drawer)
- *Type:* MCBE302433

 b. **HCS_MCBI.rfa** (Base Cabinet – Sink)
- *Type:* MCBI362433

 c. **HCS_MCDD.rfa** (Base Cabinet – 4 Drawer)
- *Type:* MCDD242433

 d. **HCS_MCWA.rfa** (Wall Cabinet - 2 Door)
- *Type:* MCWA301418
- *Type:* MCWA301436
- *Type:* MCWA361424

 e. **HCS_MCWB_C.rfa** (Wall Cabinet - 1 Door)
- *Type:* Right MCWC231436

 f. **HCS_MCPF.rfa** (Accessories - Base End Filler)
- *Type:* MCPF33

 g. **HCS_MCPH.rfa** (Accessories - Wall Cab Filler)
- *Type:* MCPH36

FYI: *Model numbers change occasionally. In that case, select a similar style and size.*

Placing Content:

15. Place the *Casework* and *Specialty Equipment* as shown in Figure 10-5.7, plus:

 a. The cabinets are positioned relative to the column; the sides of the base cabinets and wall cabinets are aligned with the 8″ architectural column.

 b. The top of the wall cabinets is at 7′-6″ above the floor.

 c. Place the filler panels in the floor plan view and *Align* with adjacent cabinets as needed (temporarily change the view to wireframe if needed).

 d. Verify vertical positions via the interior elevation and camera views.

16. Set the casework finishes:

 a. Load the family: Standard Finishes\HamiltonCaseworkSolutions-Materials.rfa

 b. Select each casework item, one at a time, just placed and set the following material instance parameters:
- **Front Finish** *Material* to **HS-Laminate-Grey Moss**
- **Cabinet Body Finish** *Material* to **HS-Laminate-Grey Moss**
- **Kick Plate Finish** *Material* to **HCS Kick Plate**

Because the *Material* parameter is an *Instance Parameter* and not a *Type Parameter*, you are able to vary the finish for each space (or even within the same space).

FIGURE 10-5.7 Placing Casework, Refrigerator and Vending Machine (type name and description shown)

Next you will modify the base cabinet dimensions. In our case we want to use a thicker 2″ countertop (1½″ thick is standard), but we want the overall height to be 3′-0″. Therefore, we need to increase the cabinet height by one inch. This can easily be done via the cabinet parameters; we do not need to edit the family. Making a change like this requires the following consideration: can the manufacturer make this custom size or am I just going to use this family to represent custom casework?

We will assume the manufacturer is able to make a custom size. To maintain the integrity of the information we will need to rename the *Type Name* and the *Model* number as well.

17. Select each of the **base cabinets** in the *Lunch Room* and do the following:

 a. **Edit Type**

 b. **Duplicate…** (change the 33 at the end to 34)

 c. Edit the **Description** number (change the 33 at the end to 34)

 d. Change the *Height* to **2′-10″**

 e. See **Figure 10-5.8**

Remember the description number is the Width, Depth and Height of the cabinet. In this case we have a standard 24″ deep base cabinet. We do not need as much depth here as we did in the workroom.

MCBE302434

MCBE302434

MCBI362434

MCDD242434

MCPF34

FIGURE 10-5.8 Editing cabinet height, type name and model number

Notes:

Creating a Custom Countertop:

Now you will create a custom countertop like the one shown in Figure 10-5.9 below. There are a number of families which represent simple countertops but none that wrap around a column or have a curved end and a unique sink hole. You will create this custom countertop using the *Model In-Place* tool.

FIGURE 10-5.9 Custom countertop to be created

Use of the **Model In-Place** tool should be limited to unique conditions relative to model context, like the unusual column condition in this example. It is better to create a family in the family editor so it can be copied and shared between projects. An In-Place family can be copied, but each copy becomes a unique element, and they cannot be shared between projects.

18. Switch to the floor plan view.

19. Select **Architecture → Component → Model In-Place** from the *Ribbon*.

Next you need to select the *Category* so Revit knows how to control the visibility and cutting of the element you are about to create.

20. Select **Casework** and then click **OK** (Figure 10-5.10).

21. Enter the family name: **Lunch Room Countertop**.

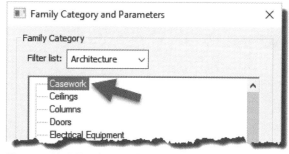

FIGURE 10-5.10 Selecting family category

Now you are in the *Family Editor* but within the project. The *Family* you are about to create will only exist within the current project. While in this edit mode you will create two separate simple extrusions: one for the countertop and one for the backsplash.

22. Click the **Extrusion** tool on the *Create* tab (Figure 10-5.11).

FIGURE 10-5.11 Selecting the *Extrusion* tool while in the *Model In-Place* mode

Before sketching the perimeter of the shape to be extruded you will enter the thickness. You are currently working relative to the Level 1 floor plane, so the bottom of the countertop will be 2'-10" and the top will be 3'-0".

23. In the *Properties Palette*, set the following (Fig. 10-5.12):

 a. *Extrusion End:* **3'-0"**

 b. *Extrusion Start:* **2'-10"**

24. Sketch the countertop per the following:

 a. See Figure 10-5.13.

 b. Sketch lines directly on the wall and column bump out.

 c. Sketch a line directly on the East side of the cabinet.

 d. Front edge to be **2'-1"** away from back wall.

 e. Make the West side extend **3'-2"** 'past the cabinet.

 f. Use the **Spline** option to sketch the curved West end (Figure 10-5.14).

FIGURE 10-5.12
Setting extrusion thickness

FIGURE 10-5.13 Sketching outline of custom countertop

25. Click the **green checkmark** to finish the extrusion.

You are still in the *Edit Family Mode*. Before finishing the family you will also create the backsplash. Having both elements in one family is ideal, and moving one will move the other.

26. Create the backsplash per the following:

a. Select **Extrusion**.

b. *Extrusion End:* **3'-4"**

c. *Extrusion Start:* **3'-0'"**

d. Follow the back wall and column bump out (Figure 10-5.15).

e. **Offset** the wall/column lines out 1".

f. **Trim** and add lines to each end to close off the outline.

FIGURE 10-5.14 *Spline* option within *Extrusion* tool

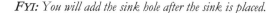

FIGURE 10-5.15 Sketching outline of backsplash

27. Click the **green checkmark** to finish the *Extrusion*.

28. Click ***Finish Model*** to complete the custom countertop.

> **FYI:** *You will add the sink hole after the sink is placed.*

You have now created a custom countertop! This is a very simple process, especially once you have done it a few times. Before moving on you will do one more thing to the countertop to make it easy for you or someone else to edit the *Material*. Currently the only way to edit the *Material* is to select the countertop and click *Edit In-Place* and then select each extrusion and change the *Material* via the *Properties Palette*.

In the next few steps you will create a *Material Parameter* and then "link" the material assignment for the two extrusions to this new *Material Parameter*. This will allow anyone working on the project to simply select the countertop and change its *Material* via the *Properties Palette*.

29. Select the countertop and click **Edit In-Place** from the *Ribbon*.

30. Select the countertop extrusion.

31. Click the **Associate *Parameter*** button for *Material* in the *Properties Palette* (Figure 10-5.16).

FIGURE 10-5.16 Associate parameter button

You are now presented with a list of parameters of type "material." However, the list is empty, so you need to create one. Fortunately, a button is provided to make this easy.

32. Click the **Add parameter...** button (Figure 10-5.17).

33. Enter the name: **Countertop Material** (Figure 10-5.19).

34. Leave the other settings as is and then click **OK**.

You now have a *Material Parameter* to select.

35. Select the **Countertop Material** parameter and then click **OK**.

36. Link the same *Material Parameter* to the backsplash:

 a. **Select** the backsplash.

 b. Click the **Associate Parameter** button (Figure 10-5.18)

 c. Select **Countertop Material**.

 d. Click **OK**.

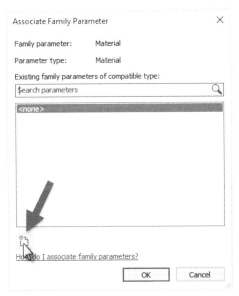

FIGURE 10-5.17 Creating a new parameter

Notice how you did not need to create the *Material Parameter* a second time. You can tell when an element parameter is mapped to a family parameter because the parameter is grayed out and the *Map Parameter* button has an equal sign next to it.

FIGURE 10-5.18 Map parameter

37. Click *Finish Model* to exit the *Edit In-Place* mode.

Next you will see how we can now edit the material via the *Type Properties* without needing to go into edit mode.

38. Select the countertop.

39. Click **Edit Type**.

40. Click within the ***Countertop Material*** parameter's field (Figure 10-5.21).

Note that this parameter is the one you created back in Step 29.

41. Click the small button that appeared to the right.

42. Select **Counter Top** from the list of *Materials*.

43. Click **OK**.

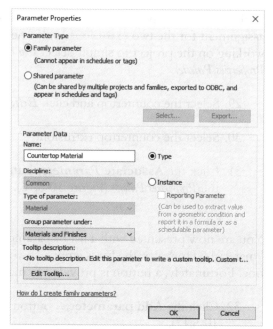

FIGURE 10-5.19 Creating a new parameter

You have now set the *Material* for the countertop, both the surface and the backsplash, without needing to edit the *Model In-Place*.

Before completing this exercise, you will add the sink, microwave and soffit walls above the wall cabinets.

44. Using the *Component* tool, place the **Microwave** in floor plan view, using family *Type* **30″x16″x18″**.

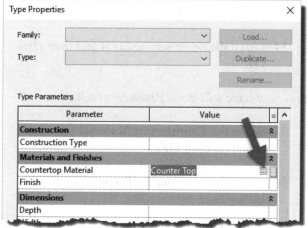

FIGURE 10-5.20 Changing the material for countertop

45. Set the proper elevation in elevation view (Figures 10-5.1 and 10-5.21).

46. Add the sink and faucet centered on the sink base cabinet (Figure 10-5.21).

 a. The sink is a face-based family, so you have to click on the countertop in order to place it correctly. ***Note:*** *Visual Style* cannot be set to *Wireframe*.

 b. Adjust the Offset value for the faucet, so it aligns with the countertop.

47. Edit the countertop to add a hole for the sink (Figure 10-5.22).

Both the sink and the faucet have MEP connectors associated with them. These are highlighted when the family is selected. This allows the mechanical engineer to connect piping to these families using Revit MEP. Also, when you loaded the sink, several materials were loaded. If you select the sink, and go to *Edit Type*, you can change the *Material* (i.e., *Finish*) to one of the other manufacturer provided materials.

FIGURE 10-5.21 Final lunchroom layout

48. In the Level 1 ceiling plan, add the **Soffit Walls** with the base set at 7'-6". Verify the proper placement via the interior elevation and a camera view.

 a. This will consist of three walls: one along the front and one on each end.

FIGURE 10-5.22 Edit countertop family to add hole for sink

49. Add the tags, notes and dimensions shown in Figure 10-5.23.

 a. Use *Detail Lines* to represent the countertop support bracket and baseboard on the adjacent walls.

 b. Note: It is important to minimize the use of detail lines in favor of true 3D elements, so they appear in 3D views and MEP and structural models.

When finished your interior elevation should look similar to Figure 10-5.23.

FIGURE 10-5.23 Edit countertop family to add hole for sink

Conclusion

You should now have the knowledge required to work through most casework designs using Revit. Occasionally you will come across a cabinet you need which cannot be found anywhere. This is when things get a little tricky for those who do not know how to create custom families. You either need to learn how to create custom content or find someone who can make it for you.

It is great that so many manufacturers are now creating Revit content for designers to download and use. Most of the content downloaded for this chapter was not available just a few years ago.

Self-Exam:

The following questions can be used as a way to check your knowledge of this lesson. The answers can be found at the bottom of this page.

1. An _____ was used to wrap the steel column.

2. Tags are view specific. (T/F)

3. Wall cabinets must all be the same depth. (T/F)

4. Cabinets are in the *Casework* category in Revit. (T/F)

5. *Shared Parameters* are stored in a text file. (T/F)

Review Questions:

The following questions may be assigned by your instructor as a way to assess your knowledge of this section. Your instructor has the answers to the review questions.

1. *Tags* report information found with drawing elements. (T/F)

2. Manufacturer created content can be downloaded from the internet. (T/F)

3. When families have many types, a _____ is used to select specific options.

4. The countertop is usually part of the base cabinet family. (T/F)

5. Deleting a tag can remove information from tagged elements. (T/F)

6. _____ were used to make wall cabinets dashed.

7. Tags cannot be edited in the *Family Editor*. (T/F)

8. When the handles do not appear in elevation, try increasing the

 _____.

9. *Model In-Place* can be used for custom countertops. (T/F)

10. Shared parameters are used to coordinate parameters between Families, Projects and Schedules. (T/F)

Notes:

Lesson 11
Toilet Room Design:

This chapter will cover the development of the toilet rooms using Autodesk® Revit®.

Exercise 11-1:
Toilet Room Layout Introduction

Introduction

There are several important factors in laying out a toilet room. The primary consideration typically is accessibility; that is, complying with local, state and federal accessibility codes so that people with various handicaps can easily access and use the toilet room facilities, and for that matter the entire building in general.

Other considerations include sight lines. It is undesirable for a person in the hallway to see any of the toilet fixtures when the door is being opened by someone entering that space. It should be obvious that privacy, especially between males and females, should be expected in a toilet room design. Seeing the sinks while people wash their hands is not too bad, but the designer needs to make sure the sight lines are not extended by the mirror above the sink. The design techniques employed to impede improper sight lines often take up a fair amount of floor space.

Another consideration has to do with the plumbing fixtures. It is important to have the correct type of fixture drawn and have it drawn the correct size. Some of the code requirements are based on the clear space in front of the fixture; thus, the fixture needs to be the correct size in order to verify compliance. Also, whether the fixture is wall-hung or a floor mounted tank-type fixture is important; the mechanical engineer needs to know as they specify the fixture in the **Project Manual**. If the toilet fixture is wall-hung the wall needs to be thick enough to conceal the support bracket and the plumbing. This wall is even thicker when two toilets are back-to-back. From a designer standpoint, wall-hung toilets are often desirable in commercial buildings because they are easy to clean under and can be equipped with automatic flushing to help ensure a better "space experience" for everyone!

Downloading Content from the Internet:

In this chapter you will again go to the internet to download Revit content to be used in your project. The availability of Revit content is continually growing, as is the quality of said content.

A WORD OF WARNING:

Do not assume all Revit content found on the internet is worthy of using in your models. There are many locations on the web where individuals have uploaded files that are less than perfect, so use caution. On the other hand, if you are downloading content from a product manufacturer's website, you can be fairly confident that they have accurately drawn their own product and want you to have a successful project using both their product and Revit files! However, some content is over-modeled and can have a negative impact on your project's performance and stability. It takes some trial and error to develop the experience needed to know which content is worth using.

In this chapter you will use some provided content. In an office setting you might want to save downloaded plumbing fixture families to a "Plumbing Fixtures" folder on your hard drive for convenient access in the future.

Space Design:

The **toilet partitions** offer privacy for toilets and urinals. They come in a few styles and materials. Styles range from floor mounted to ceiling hung; the latter offers better access to cleaning the floors, like wall-hung toilets. Materials range from prefinished metal to solid phenolic core (i.e., plastic) to stainless steel.

The divider panels, pilasters and doors come in various thicknesses. However, each is typically modeled 1″ thick, in plan view, and is dimensioned to the center of the panel so the thickness is not too important.

If you want to see images and more information on toilet partitions, you should visit the Bradley Corporation website list below:

https://www.bradleycorp.com/product-category/toilet-partitions

The example to the left shows a floor mounted, overhead-braced toilet partition system.

Also notice the urinal screen at the end of the countertop.

Floor drains are usually positioned beneath a toilet partition, so users do not have to walk over the slightly uneven surface.

Notice the accent floor tile that also takes into consideration the location of the toilet partitions.

Accessibility codes offer several layout options for toilet partitions. For example, if the door swings into the stall, more room is required in the stall to maintain the required clear floor space for the toilet fixture. Having a cheat-sheet of several different layout options is a big help (Figure 11-1.1). Consult your local code for specifics.

Grab Bars are an important part of the design of accessible toilets. The **Americans with Disabilities Act** (ADA) and state/local building codes require grab bars. Most building codes are very similar in the quantity, location and dimensions required for grab bars (e.g., Figure 11-1.2). However, some municipalities vary as can be seen in Figure 11-1.3 below, which shows a vertical grab bar was originally only required in Minnesota. As such, the designer needs to be aware of the building codes applicable for any given project. It is interesting to point out that the ADA is not a building code, but rather a federal civil law. So, architects' and interior designers' projects are not reviewed based on ADA, but rather the state and local building accessibility codes; for example, Chapter 1341 in the Minnesota Building Code. Those state and building accessibility codes are based on the ADA but are often added to. The designer still needs to meet the ADA codes and must do so to protect their client who would be liable for any ADA violations in their facility.

FIGURE 11-1.1 Reference drawing showing the minimum required dimensions for several possible toilet room layouts.

FIGURE 11-1.2 Grab bar dimensions; 1½″ diameter pipe

FIGURE 11-1.3 Grab bar picture; codes vary by location. Minnesota requires a vertical bar as shown above.

Sight lines are typically explored by drawing a line from the adjacent hallway (or space), through the door opening, into the toilet room. Looking at Figure 11-1.4 you can see the sight lines are quite acceptable. As previously mentioned, this design solution takes up a fair amount of floor space but is necessary.

FIGURE 11-1.4 Dashed line shows sight line

An unacceptable design solution is shown in Figure 11-1.5. The toilet room is much larger and can even have another toilet fixture without the door modification for sight lines; also, the door is right off the circulation path which can be problematic. Notice two lines are shown to visualize the full visual range. Granted, some of this view is obstructed by a person's body as they pass through the door opening; it is still an undesirable design solution.

This is the full view range in this design solution.

FIGURE 11-1.5 Dashed lines reveal unacceptable sightlines into the toilet room

Placing Plumbing Fixtures:

You will start this exercise by loading several components to be placed into your project. Oftentimes, the architectural team places plumbing fixtures as they are doing the initial toilet room design. However, the content between the *MEP* users and the architectural users is, unfortunately, not the same. The *MEP* content has special connectors on them that allow the *MEP* designers to connect water and sanitary piping to the element. You will actually be using manufacturer created content provided with this book. This content has the Revit MEP connectors.

Placing Plumbing fixtures,
Image credit: Stabs, Wingate

Standard Mounting Height Diagram:

A *Standard Mounting Height Diagram* (Figure 11-1.6) is a drawing that lists all the mounting heights for fixtures and equipment. This helps to reduce the need to dimension every item in an interior elevation; instead, a note is added to the interior elevations sheet that refers to the *Mounting Height Diagram* for more information. It cannot be repeated too many times that this information can vary by location, so the designer must verify these requirements when working in multiple locations.

FIGURE 11-1.6 Standard mounting height diagram example

Exercise 11-2:
Toilet Room Layout

In this exercise you will lay out the *Toilet Rooms*. There are several design considerations to be made in this space. You will need to download some provided content from SDC Publications. You will learn how to use a few more features in Revit in this section.

You will create a *Toilet Room* similar to the one shown in the image below. This will include creating a new wall type with a partial height tile pattern built in. You will also use the *Split Face* tool to "cut in" accent tile and rotate the pattern 45 degrees. The plumbing fixtures are manufacturer created content with automatic flush valves. Finally, the countertop family has the sink and faucets built in and when the overall length is adjusted, additional sinks automatically appear.

> *FYI: The flooring will come in the next chapter.*

In the image below the toilet partition is selected; notice how it is transparent when selected.

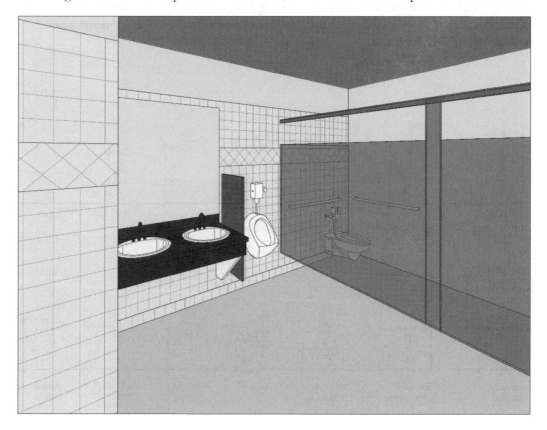

1. Open the **Law Office.rvt** file.

2. Open the **Level 1 Floor Plan** view if needed.

3. Load the following families into your *Law Office* project from the provided content:

 a. **Counter Top w Sinks-3D**

 b. **Grab Bar-3D**

 c. **Low_Consumption_Washout_Urinal_-_TOTO_-_UT104**

 d. **Bradley-Washroom_Accessories-Mirrors-748(SafeCare)**

 e. **Toilet Stall-Accessible-Side2-Braced-3D**

 f. **Toilet Stall-Braced-3D**

 g. **Urinal Screen-3D**

 h. **Wall_Hung_Flushometer_Toilet_-_TOTO_-_CT708E**

Now that the content is loaded you can start placing the toilet partitions. All of the content you are about to place is done using the *Component* tool. This is the same tool used to place all the casework in the previous chapter. Any element that does not have its own tool, e.g., Door, Wall, Ramp, etc., is placed using the *Component* tool. Unfortunately, this can make the *Type Selector* very long on a large project. A good naming convention helps a lot! For example, have all the plumbing fixtures start with "PF" and casework with "CW." This will place the different categories together in the *Type Selector* list.

4. Click to activate the **Component** tool.

5. Select **Toilet Stall-Accessible-Side2-Braced-3D: <u>60″ x 96″ Clear</u>** from the *Type Selector*.

6. Zoom in to the *Toilet Room* (East of the northern stair).

7. **Place** the toilet stall as shown (Figure 11-2.1).

 a. This is a *wall hosted* family, so it must be placed on a wall to exist in the project.

 b. Notice when the toilet partition is selected, you have two *Flip* controls which can be used to quickly change the orientation of the stall.

 c. Click in the corner of the room as the insertion point for the family is 5′-0″ away to allow for proper placement.

You will address the issue with the extra space on the left in a moment.

FIGURE 11-2.1 Accessible Toilet stall

With an accessible toilet stall placed, you will place a standard size toilet stall adjacent to it.

8. Place toilet stall **Toilet Stall-Braced-3D: <u>36″ x 60″ Clear</u>** as shown in Figure 11-2.2.

As with most projects, you will need to modify the model as you develop the design. In this case, we notice that toilets which are back-to-back and stacked on each floor will require a thicker wall to accommodate the fixture brackets, as toilets are not supported by light gauge metal studs, and larger piping. You will make this adjustment next.

For information on wall thicknesses behind double-hung toilets visit www.pdionline.org and review the publication *Minimum Space Requirements for Enclosed Plumbing Fixture Supports.*

9. Select the middle wall and **Move** it **5 15/16″** to the West.

Notice how the toilet partitions moved with the wall because they are wall-hosted families. The ceilings also adjusted.

10. Add an additional **4⅞″** gyp. bd. wall as shown in **Figure 11-2.3**.

FIGURE 11-2.2 Std. Toilet stalls added

In this case, you would not want to delete the original central wall and add two new walls. If you deleted the center wall, the wall-hosted content would also be deleted and the ceilings would no longer update.

FIGURE 11-2.3 Toilet room wet-wall locations

Next, you will adjust the *Location Line* of the two "wet walls" (i.e., walls that have plumbing behind them). This will help keep the finished face of the wall in the proper location when you adjust its thickness in the next few steps.

11. Select the two wet walls centered on the two toilet rooms and set the *Location Line* to **Finished Face: Exterior** via *Properties*.

12. Adjust the *Flip* control for each wall so they are positioned, in relation to the wall, as shown in Figure 11-2.3.

 FYI: You only see the Flip *control when a single wall is selected.*

13. Reposition the walls, if needed, so they match the dimensions shown in Figure 11-2.3; use the *Measure* and *Move* tools plus the temporary dimensions.

Creating a Custom Wall Type

Now you will create a new wall type within your BIM project. The new wall type will be similar to the 4⅞" stud wall, but it will only have gypsum board on one side of the wall.

FIGURE 11-2.4
Toilet room wet wall

Looking at Figure 11-2.4, you see the wall has gypsum board in the plumbing cavity, which is not appropriate or even possible. This would never be constructed this way, and if the model were used for quantity takeoff you would have too much gypsum board in the total.

14. Select one of the two center *Toilet Room* wet walls.

15. Select *Edit Type* from the *Properties Palette*.

16. Click **Duplicate**.

17. Type: **Interior - 4 1/4″ Furring Partition with Tile**.

18. Click **OK**.

19. Click the **Edit** button next to the *Structure* parameter.

20. Click on Row 5 to select it, the gypsum board *Layer* on the *Interior Side*.

21. Click the **Delete** button to remove the selected *Layer*.

Your *Edit Assembly* dialog should now look like Figure 11-2.5.

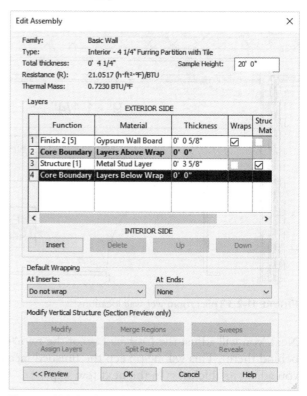

FIGURE 11-2.5 Edit wall configuration

22. Click **OK** twice to close the open dialog boxes.

23. Select the other wet wall and change it to **Interior - 4 1/4″ Furring Partition with Tile** via the *Type Selector*.

Your two furring walls should now look like Figure 11-2.6. If the *Location Line* was not set properly, or the exterior side not identified properly, the wall would have moved out of its proper position when you changed to a wall with a different thickness.

24. Use the *Measure* tool to ensure the walls are in the proper position; see Figure 11-2.3.

The exterior wall can have continuous gypsum board as shown. We will not modify that.

FIGURE 11-2.6
Toilet room wet-wall revised

Now that the walls are in the proper position you can adjust the size of the accessible toilet stall to extend the full length of the room.

25. Select the accessible toilet stall.

26. Click **Edit Type** via the *Properties Palette*.

You will make a new *Type* and then adjust the *Length* parameter.

27. Click **Duplicate**.

28. New name: **60″ x 109.25″ Clear**.

29. Adjust the **Length** parameter from 8′-0″ to 9′-1¼″.

Public Restroom

226

143 SF

FIGURE 11-2.7 Accessible stall length adjusted

Your toilet stall now properly fits within the room (Figure 11-2.7).

> *FYI: The toilet partition panels come in various lengths and with their adjustable trim/brackets, they are able to conform to custom room sizes like ours.*

> *FYI: There is a nice set of toilet compartment families available from the Bradley Corporation (Figure 11-2.8). It has an array to set the number of standard stalls next to the accessible stall and check boxes, via the* Properties Palette, *for things like door swing and code requirements such as the need for an ambulatory stall. This family also has more detail which is good for visualization but may slow down the model.*

FIGURE 11-2.8 Toilet stall family offered by Bradley Corp. via the internet

Now you will place the countertop with the built-in sink. As mentioned at the beginning of the exercise, you can adjust the length of the countertop. When the length is adjusted, the number of sinks automatically changes. The sink has been set up to be arrayed within the family. This is a nice way to begin to understand how complex families can be. It should be pointed out that this sink does not have Revit MEP connectors so it would not be appropriate for projects where the mechanical design is being done in Revit, but it is a good early design tool!

30. Use the **Component** tool to place **Counter Top w Sinks-3D: <u>48″</u>**.

> **FYI:** *This is a wall hosted family, so you need to select the wet wall first and then move/align the adjacent wall.*

When the countertop family is selected, you see grips appear at each end (Figure 11-2.9). These grips allow you to quickly stretch the counter to any length. Give it a try and notice how the number of sinks changes. Be sure to *Undo* any changes before moving on. One thing that is not good about the way this family is set up is that when the length is stretched manually in the model, the length no longer matches the *Type* name.

Now you will place most of the remaining items in the model for the Level 1 restrooms.

FIGURE 11-2.9 Countertop with sink

31. Add the additional components as shown in **Figure 11-2.10**.

 a. Place the content loaded at the beginning of this exercise.

 b. The height (hgt) can be changed via the *Instance Properties*.

 c. *Wall hosted families* do not always mirror very well, so you will have to manually place all the elements on the men's side.

 d. To center the toilet in the standard stall: add a dimension string, then click the *EQ* option to center the toilet, and then delete the dimension.

 TIP: Use the Align *tool to make the plumbing fixtures aligned between the two restrooms.*

Now that the Level 1 plan is laid out you will create an interior elevation of the men's restroom and begin to develop the wall finishes.

FIGURE 11-2.10 Toilet Room layout

Creating an Interior Elevation

32. From the *View* tab on the *Ribbon*, click the **Elevation** tool.

33. Select **Interior Elevation** from the *Type Selector*.

34. Click to place the elevation tag as shown in Figure 11-2.11.

If you clicked too far to the left, your elevation line, shown in plan when the elevation is selected, might not extend the full length of the room. In some cases that might be all you need to show. In our case we want to show the entire room as we have a tile pattern with accents to depict.

If the line landed in the wrong location, you can click and drag to reposition it and adjust the length. These types of adjustments will affect what is seen in the elevation view.

FIGURE 11-2.11 Adding elevation

35. Rename the new elevation view to **MEN'S PUBLIC RESTROOM ELEVATION - WEST**.

36. Open the elevation view.

37. Hide the *Level* and adjust the bottom edge of the *Crop Region* to hide the floor.

38. Set the *Detail Level* to **Fine**; notice the additional detail that appears on the Toto plumbing fixtures.

Your elevation should now look like Figure 11-2.12.

FIGURE 11-2.12 Adding elevation

Adding Wall Tile

You will look at two ways of showing wall tile, which goes up the wall about 7'-0". The first way is to build it into the wall type. This will cover new features in Revit as the material does not extend the full height of the wall. This will work great for the wet wall as we will want every instance of this wall to have tile.

Another way in which tile is often represented is by using the **Split Face** and **Paint** tools. This technique allows the designer to define an area, within the larger wall, to receive a different material.

There are additional ways to represent tile which we will not be covering. For example, you could draw in a thin wall, similar to the floor finish technique covered later, and then *Join* the main wall and the thin finish wall. This has some issues with the joins at the corners. Another example is just using *Fill Pattern* and/or *Detail Lines*, but this limits the ability to

visualize/render and to do quantity takeoffs. This method is utilized by many firms starting to learn Revit but should generally be avoided.

39. Select the wet wall; it may be easier to do this from a plan view.

40. Go to **Edit Type** → **Edit** *Structure*.

41. Make sure the preview pane is open; click the **<<Preview** button in the lower left if needed.

In the next few steps you will make a new *Layer*, *Split* an existing *Layer* and then *Assign a Layer* to one side of the split. This can only be done when the preview pane is set to section, rather than plan view.

42. Create a new *Layer* and *Split* the gypsum board *Layer* per the following (see Figure 11-2.13 on the next page):

 Step 1 - Select the **Section Preview** mode.

 Step 2 - Create a new *Layer* for the **CT1** *Material* created previously.

 Step 3 - Click the **Split Region** button.

 Step 4 - Click on the outside edge of the gypsum board *Layer* in the preview pane; you may need to spin the wheel on your mouse to zoom in; you can also *Pan*.

 Step 5 - Adjust the dimension: (Steps numbered in Figure 11-2.14):

 a. Click the **Modify** button (1).

 b. Zoom in and select the horizontal line (2).

 c. Edit the dimension to **7'-5 1/4"**.

Your gypsum *Layer* is now split, but you still need to assign the tile *Layer* CT1 to one of the split *Layers*. You will do that next.

FIGURE 11-2.13 Adding wall tile wainscot to wet wall

FIGURE 11-2.14 Selecting split line

43. Assign the CT1 *Layer* per the steps below and Figure 11-2.15:

 a. Select the *Layer* row CT1.

 b. Click the ***Assign Layer*** button.

 c. Zoom in and click the lower part of the gypsum board *Layer*; it should highlight before selecting it.

The CT1 *Layer* should now automatically have the thickness **0′ 0 5/8″** set to match the gypsum board *Layer*. This is a big limitation with this technique – that is, the two *Layers* must be the same thickness. In our case here, the tile and backer board are not too much thicker than the gypsum board so this will not be a problem.

Now you can look at your elevation and see the wall has a ceramic tile wainscot up to 7′-5¼″ (Figure 11-2.16). This is true for the women's restroom as well; seeing as we edited the *Type* properties, all instances will have this change. That means we must be careful when making changes like this, to be sure this will not affect another area of the project which should not change. That is why we added "with Tile" to the wall type name.

FIGURE 11-2.15 Adding wall tile wainscot to wet-wall

FIGURE 11-2.16 Ceramic Tile wainscot added to wall type

Looking at our interior elevation (Figure 11-2.16), we decide the fill pattern for the tile is too dark in terms of what is to be printed on paper. We want the plumbing fixtures and countertop to stand out more. You will edit the *Material* for CT1, making the projection fill pattern gray rather than black.

> *FYI:* The actual line thickness is controlled by Line Weight #1. This is found here: **Manage (tab) → Additional Settings → Line Weights**. *However, changing this may change other elements within the model, so we will not touch this. Some firms develop their content and template to not use* Line Weight #1 *so it can be changed and only affect fill patterns.*

44. Open the *Materials* dialog: **Manage → Materials.**

45. Select *Material* **CT1** and change the Foreground *Surface Pattern* color per Figure 11-2.17.

FIGURE 11-2.17 Materials dialog; editing color for surface pattern

The ceramic tile surface pattern is now more subdued in the interior elevation (Figure 11-2.18). This pattern is a *Model Pattern*, rather than a *Drafting Pattern*, as discussed in the *Material* chapter; this pattern will not change size when the *View Scale* is adjusted. We can also move the pattern similar to the ceiling tile pattern, without moving the wall itself.

FIGURE 11-2.18 Ceramic Tile pattern color and location adjusted

46. Use the ***Align*** tool to make the tiles align with the wall on the left. (Compare Figures 11-2.16 and 11-2.18.)

The tile on the left is much more visible than the tile in the toilet stall. So, the full tile will be on the left and any remaining tiles needing to be cut will be less visible in the toilet stall. The image below is a rendering of the toilet room; rendering is covered in Chapter 17.

Exercise 11-3:
Toilet Room Wall Finishes

Using the Split Face Tool

Looking back at the elevation (Figure 11-2.19), we want to continue the ceramic tile around the room. But the remaining walls, interior and exterior, occur all over the building so we cannot edit the wall type as we just did. This would cause tile to show up everywhere. We could create new wall types, and then split the walls up in plan and swap parts out with the new type, but this gets a little messy so we will use another option.

To add the CT1 *Material* to the remaining walls, you will use a tool called **Split Face**. This allows you to define an area on the face of a wall to receive another *Material*. The new *Material* is applied using the **Paint** tool. Both of these tools are found on the *Modify* tab.

1. Select **Modify → Split Face** (Figure 11-3.1).

2. Move your cursor over the edge of the wall to highlight it – click to select it (Fig. 11-3.2 – Step #1).

 a. Press **Tab** if needed to cycle through the selection options below your cursor.

FIGURE 11-3.1 Split Face tool

FIGURE 11-3.2 Split Face tool – selecting surface to edit

Notice the entire profile of the wall to be split appears; this includes the portion above the ceiling in this elevation which extends to the roof.

3. Select the **_Line_** option from the _Ribbon_ (Fig. 11-3.3).

FIGURE 11-3.3 Split Face tool – line option

If you are splitting across the entire surface, you only need to sketch one line; you do not need to sketch an outline.

4. Draw a horizontal line from side to side and then align with the adjacent tile pattern (Figure 11-3.2 – Step #2).

5. Click the **green check mark** to finish the _Split Face_ tool.

FIGURE 11-3.4 Results of Split Face tool

Your elevation should now look like Figure 11-3.4. Depending on how your walls join at the outside corner, you may have an issue where you are seeing the end of the wall around the corner. Notice how this situation is emphasized in Figure 11-3.5.

In this case you simply have to use the _Split Face_ tool again on this smaller, adjacent wall end.

FIGURE 11-3.5 End of wall is exposed in West elevation

6. Per the previous steps, use the **Split Face** tool on the end of the wall, if needed, to add the wall base as shown in Figure 11-3.6.

FIGURE 11-3.6 Using Split Face on end of wall in west elevation

Next you will use the *Split Face* tool to add the wall base on the wet wall and also to add an accent tile band. You may split the same wall multiple times. You can also split the *Material* which is part of the wet wall, added at the end of the previous section.

How to delete a split surface: If needed, you can select the edge of a split face, not in *Edit* mode, and press the **Delete** key to remove a split.

7. Use the ***Split Face*** tool to split the walls as shown in Figure 11-3.7 below, plus:

 a. Lines 4 and 7 have already been added.

 b. Each numbered line is a separate horizontal line.

 c. Each wall face requires a separate use of the *Split Face* tool.

 d. The lines are heavier and dashed in the image below to make them stand out; yours will be the same as the previous steps.

FIGURE 11-3.7 Using Split Face to define wall base, accent tiles and wainscot

Now that the walls have their surfaces split, you are ready to change the *Materials*. Each area has the *Material* originally defined on the wall. You will use the *Paint* tool to make this change.

Using the Paint Tool

The first thing you will do is refresh your memory of how *Materials* are created. You will rename *Material* CT2 to CT4, as we will not be using CT2 right now, and *Duplicate* CT1 to CT2. The new material will be the same tile, but a different color.

8. **Manage → Materials**.

9. Select *Material* **CT2** and then click the ***Rename*** icon in the lower left.

10. Change the name to **CT4**.

11. Edit the *Mark* value for CT4 to match the new name.

 a. This is found on the *Identity* tab.

12. Select *Material* **CT1** and click the ***Duplicate*** button in the lower left.

13. Enter **CT2** for the new name (no spaces).

14. Make the following edits to the new **CT2** *Material*:

 a. <u>Identity Panel:</u>
 i. *Comments:* **WALL TILE**
 ii. *Model:* **0744 ARTISAN BROWN**
 iii. *Mark:* **CT2**

 b. <u>Appearance Panel:</u>
 i. Duplicate the appearance asset first!
 ii. Set ceramic color to: RGB **100, 82, 68**

 c. <u>Graphics Panel:</u>
 i. Check: **Use Render Appearance of Shading**

Now you are ready to use the *Paint* tool.

15. Select **Modify → Paint** (Figure 11-3.8).

A dialog listing *Materials* now appears. The process is this: select a *Material* from the list and then pick a surface to receive the *Material*. The surface does not have to have a split applied to it. But as mentioned in the *Materials* chapter, the painted *Material* will then be applied to the entire wall, even if the wall extends into another room, part of which is not visible in the current interior elevation. This does not apply to our situation.

FIGURE 11-3.8 Paint tool on Modify tab

If you need to remove a *Material* later, you use the ***Remove Paint*** option shown in Figure 11-3.8. This makes the area revert back to the original *Material* assigned to the wall type being used.

16. Select the *Material* **CT1** and then pick the area shown in Figure 11-3.9.

 a. Be sure the correct area highlights before picking to paint.

FIGURE 11-3.9 Using the Paint tool to change a material

As new *Materials* are painted, the fill pattern position may shift out of alignment with adjacent *Materials*. You will use the *Align* tool to fix that.

17. Use the ***Paint*** tool to apply *Materials* as shown in Figure 11-3.10, and:

 a. The accent band and wall base are to be **CT2**.

 b. The remaining areas are to be **CT1**.

 c. The accent band tiles will be rotated in the next step.

18. Use the *Align* tool to align the CT1 tile patterns with the outside corner as shown in Figure 11-3.10.

19. Use the *Align* and *Move* tools to make the wall base offset a half tile.

 a. First, ***Align*** wall base tile with CT1 pattern above.

 b. ***Tab/Select*** one of the tile pattern lines.

 c. **Move** to the right **2.125″**.

20. ***Tab/Select*** the tile pattern for the accent bands one area at a time.

21. Use the ***Rotate*** tool to make the tile 45 degrees as shown in Figure 11-3.10.

22. With a line in the pattern selected, use the arrow keys to move the pattern vertically to match the image shown below.

FYI: The tips of the tile will be cut off. You would have to adjust the accent band height and the overall wall tile height to make them fit exactly, which would be ideal, but we will not do that now.

Edit
Boundary
Mode

a. ***Editing a Split Face Boundary:*** You can select the edge of a *Split Face* area and then click **Edit Boundary** on the *Ribbon*.

23. Use ***Align*** as needed to make the angled tiles align with each other.

TIP: You may have to use Align *on both line angles.*

FIGURE 11-3.10 Materials painted on surface and repositioned / rotated

You now have the materials properly placed in this view. You will need to create elevation and/or section views of the other walls to properly place tile throughout the space. The painted materials will appear in all views, including 3D views and renderings. You can also schedule the total square footage of materials, which include regular element materials and painted materials.

Tagging Materials

With the *Materials* applied you will see how they can be tagged. As with other tags, these **Material Tags** will automatically update if a different *Material* is selected or *Painted* onto a tagged surface. Also, as with other tags, they are view specific.

In our case, we will need to edit the default ***Material Tag*** to report the information we want, that being the ***Type*** value we entered on the *Identity* tab of the *Materials* dialog.

24. Select **Annotate → Material Tag**.

25. Click **Yes** to load a *Material Tag* (if prompted to do so).

26. Select **Annotations\Architectural\Material Tag.rfa**.

27. Un-check **Leader** on the *Options Bar*.

28. Click in the area pointed out in Figure 11-3.11.

You should now see a *Material Tag* which reads "Interior finish" as shown here.

Take a moment to open the *Materials* dialog, select CT1 and then look at the *Identity* tab. Notice the ***Description*** field is what the *Material Tag* is pointing at (Figure 11-3.12).

Next you will edit this tag to make it point at the ***Type*** value, which is CT1.

FIGURE 11-3.11 Material tag added to elevation

29. Select the *Material Tag* which reads "Interior finish" in your elevation view.

30. Click ***Edit Family*** from the *Ribbon*.

You are now in the *Family Editor*, where you can make changes to what the tag looks like and what information it reports.

31. Select the text ***Material Description***.

FIGURE 11-3.12 Material tag added to elevation

32. Click the *Edit Label* button on the *Ribbon*.

33. Make the following changes in the *Edit Label* dialog (Figure 11-3.13).

 a. <u>Remove</u> **Description** from the right (Steps 1 and 2)

 b. <u>Add</u> **Mark** (Steps 3 and 4)

34. Click **OK**.

FIGURE 11-3.13 Editing the Material Tag

35. With the *Label* text still selected, click *Edit Type* from the *Properties Palette*.

36. Set the *Width Factor* to **0.9**.

 a. Notice the **Background** is set to **Opaque** here in *Edit Type*. This is what hides the line work under the text. If this is not desired, one can change this setting to **Transparent**.

This last change will make the overall width of the text a little narrower so we can fit more information in the drawing.

Before loading this tag back into the model, you need to decide if the old tag type should be preserved. That is, will you ever want to tag the *Material's* description? If so, you will need to do a **Save-As** to give the tag a different name before loading it back into the project. In this case you would have access to both tags. Keep in mind the *Description* could be changed to say something like "Ceramic Tile over backer board." The *Material Tag* could then be used to add notes to your drawings. In our case we will just overwrite the old tag.

37. Click the **Load into Project** button.

38. Click **Overwrite the existing Version**.

Your tag is now updated to display the contents of the *Material's* **Mark** field, which is "CT1" (Figure 11-3.14).

FIGURE 11-3.14 Revised Material Tag

The modified *Material Tag* could be saved to a file for use in other projects, or even loaded into a template making it readily available in new projects.

39. Add seven additional *Material Tags* as shown in Figure 11-3.15 below.

FIGURE 11-3.15 Remaining Material Tags added

Things that are wall hosted, like the mirror, can be placed in elevation. You will do that next.

40. While in the interior elevation view, select the *Component* tool and then select **Mirror: 72″ x 48″** from the *Type Selector*.

41. Create a duplicate *Type*: **48″ x 48″**. Change the width accordingly.

42. Place an instance of the **48″ x 48″** mirror on the wall above the countertop.

43. Use the *Align* tool to align the mirror with the edge of the countertop – both vertically and horizontally (Figure 11-3.16).

44. Add the notes and dimensions per Figure 11-3.16.

 a. Add three angled lines in the mirror. This is often done to represent glass.

FIGURE 11-3.16 Updated interior elevation

Adjusting the Reflected Ceiling Plan:

Because you added a wall in the East *Toilet Room*, the definition of the room that the reflected ceiling plan uses is incorrect. You will adjust that next.

45. Switch to the Level 1 Ceiling Plan (Figure 11-3.17).

46. **Hide** the interior elevation tag from this view. Select the two components, right-click and select **Hide in View → Elements**.

47. Select the ceiling in the *Men's Toilet Room*, and then click **Edit Boundary** on the *Ribbon*.

 a. You may have to tap the *Tab* key to select the ceiling.

48. **Adjust** the left side of the ceiling to align with the new wall; when finished, click *Finish Ceiling* on the *Ribbon* (Figure 11-3.18).

Remember, deleting an element and then replacing it can cause problems!

49. Type **VV** and make the following changes:

 a. Set *Model Categories\Structural Framing* to **Halftone** (check box on right).

 b. **Turn off** *Annotation Categories\Elevations* to hide them.

50. **Save** your project.

FIGURE 11-3.17 Level 1 RCP **FIGURE 11-3.18** Revised Level 1 RCP

Consider loading in additional toilet room accessories, such as a paper towel dispenser, toilet paper dispenser, etc. Also, edit the remaining walls to have the tile wainscots.

Keep in mind, when looking at VV, that the *Specialty Equipment* category does not cut in elevations or sections. If an item in this category touches a cut plane, the entire element appears. This creates a problem in the women's restroom as the stall doors hide the tile pattern information on the walls. You can either hide the stall or make it transparent. Otherwise the family would need to be edited to make the doors and front supports hidden in elevation views; this is the best option. This requires an understanding of how families are created.

Exercise 11-4:
Level 2 Private Toilet Rooms

Single Use Rooms

The Level 2 restrooms only accommodate one person at a time. Therefore, toilet partitions are not required. Often, the men's side will still have a urinal in addition to the toilet.

1. Open the **Level 2 Floor Plan** view.

2. Load **Lavatory - Wall Mounted.rfa** from the provided content.

3. Add content per Figure 11-4.1, and:

 a. Create a new mirror type: **24″ x 36″**.

 b. Add the wall tile pattern to all walls.

 c. Add the two wet walls and update the ceiling.

4. **Save** your project.

FIGURE 11-4.1 Level 2 private restrooms

You should now have a good understanding of how to lay out a toilet room and how to apply materials to the walls. In the next chapter you will look at the flooring for the entire building, and how to manage different materials in various rooms.

Self-Exam:

The following questions can be used as a way to check your knowledge of this lesson. The answers can be found at the bottom of this page.

1. Specialty equipment does not cut. (T/F)

2. The *Paint* tool only changes the material for the visible portion of a wall. (T/F)

3. The tile pattern can be repositioned. (T/F)

4. The sink in countertop arrays with length of counter. (T/F)

5. Materials can be tagged. (T/F)

Review Questions:

The following questions may be assigned by your instructor as a way to assess your knowledge of this section. Your instructor has the answers to the review questions.

1. All content downloaded from internet is high quality. (T/F)

2. The countertop length is _____.

3. The color of the tile material surface pattern was changed to make it lighter in elevation. (T/F)

4. The painted material on a wall cannot be removed. (T/F)

5. Which field does the default material tag point to? _____.

6. It is considered bad design to see public toilet stalls from the hallway. (T/F)

7. Ceilings always expand to conform to space – like rooms. (T/F)

8. The *Split Face* tool is used to define smaller areas on a wall. (T/F)

9. Painted tile patterns can be aligned with each other. (T/F)

10. To edit the vertical structure of a wall, the preview "view" setting has to be _____.

Lesson 12
Floors:

In this chapter you will begin to develop the floor finishes using Autodesk® Revit®. We will cover all the basics related to modeling floors plus offer a few tips and tricks to make your design ideas a reality.

Exercise 12-1:
Introduction to Revit Floors

In Revit, the *Floor* tool is used to model the horizontal surfaces people walk on. A floor is very similar to a wall in how one defines the structure of the floor assembly, except walls are vertical and floors are horizontal. In this exercise you will look at how floor assemblies are defined, the various properties associated with floors and how to create them within a model.

This introduction section on Revit floors will use examples from a "fresh" copy of the **Commercial-Default.RTE** template file. The *Law Office* was started from this file but has been modified. If you want to explore some of the concepts and techniques mentioned in this first section, you should start a new project file based on this template. This way you will not make any unintended changes to the *Law Office* model.

Floors are typically created in a floor plan view but can also be created in 3D views. They cannot be created in elevation, section or perspective views. Here is an overview of the process:

- Select the *Floor* tool.
- Define the perimeter of the floor.
 - Sketch lines and arcs.
 - Use the *Pick Lines* option to create floor sketch lines based on line work already in the model (can even be from a linked in CAD file).
 - Use the *Pick Walls* option to create floor sketch lines based on walls previously added to the model.
- Define any openings within the floor:
 - Two story lobby
 - Mechanical shafts
 - Stairwells
 - But not for small items such as individual pipes.
- *Finish* the Floor Sketch.

In Figure 12-1.1 below you see a simple floor in *Sketch* mode. The *Boundary Lines* are magenta and must form a perfectly enclosed area: no gaps or crossing lines. It is possible to have multiple enclosed *Boundary Lines* within the outer perimeter – these will define holes in the floor. Finally, the *Span Direction* symbol defines which direction the floor spans structurally, usually perpendicular to the joists. This mainly relates to which direction the flutes go for metal decking (see the Level 2 floor in Figure 12-1.2). Structural engineers will use *Span Direction* for additional analytical needs as well as adding the *Span Direction* tag. Clicking the *Span Direction* button on the *Ribbon*, while in *Sketch* mode, allows you to pick another line to move the *Span Direction* symbol as well.

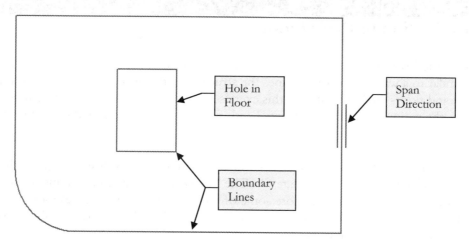

FIGURE 12-1.1 Floor sketch example with terms

By default, when a floor is created its top surface is aligned with the *Level* on which it was created (Figure 12-1.2). The level is tied directly to the floor plan view in which you are creating the floor element. No matter what the thickness of the floor, the top is aligned with the *Level*; you will see later how this can be offset up or down via *Instance Properties*.

Early on in a project, a generic floor might represent the overall thickness of all the components of a floor; e.g., concrete, metal deck, floor joists. As the project develops, the floor might be swapped out, not deleted, with something more descriptive.

FIGURE 12-1.2 Floor relative to level

Keep in mind that floors only represent horizontal planes, such as wood sheathing, rigid insulation, concrete, etc. A Revit floor element cannot represent bar joists or a concrete pan and joist system. It is possible to add a *Layer* to the floor which represents the total thickness of the structure. Sometimes this is done as a placeholder, and then removed later in the project as the actual structure is modeled by the structural engineers.

During **Schematic Design** it is not usually necessary to even add a floor. Revit plan views stop at the floor level by default, so elements beyond, for example, walls on the floor below, will not show. Floors are needed when sections and camera views are used.

Figure 12-1.3 shows a thick generic floor added during the ***Design Development*** (*DD*) phase of a project. This floor element represents the floor plane AND the structure below it. Usually the exact thickness of the floor is not known, but the structural engineer has provided their best guess. When the *DD* phase floor includes the overall thickness, it helps in determining how high ceilings can be. However, the mechanical engineer does not know where some of the "nooks and crannies" are where they might be able to run the main or branch ducts in and allow you to raise a ceiling higher in a specific location.

During the ***Construction Documents*** phase the generic floor is changed to represent the actual floor plane, for example, just the metal deck and concrete (Figure 12-1.4). The structural beams and bar joists are added separately below using the structural tools.

Figure 12-1.5 shows how the generic floor represents the depth of the structure. Like most elements in Revit, a *Floor* has both *Instance* and *Type* properties.

FIGURE 12-1.3 Placeholder floor in design development

FIGURE 12-1.4 Floor with structure added separately

FIGURE 12-1.5 Generic floor represents structure

When a floor is selected, its **Instance Properties** can be seen in the *Properties Palette* (Figure 12-1.6). Notice a few things:

- **Level:** By default this is set relative to the floor plan view in which it was created. But you can change that here by clicking on the current level and then selecting from the drop-down list available.

- **Offset:** Here you can move the floor up or down relative to the referenced level. Add a minus sign in front to offset down. If the level data is moved, the floor will parametrically move with it.

- **Room Bounding:** When this is checked, the *Room* elements will stop when a *Floor* element is encountered.

 FYI: Because most floors align with the Level*, the bottom of the* Room *does not typically ever extend into a* Floor *anyway. But it is still best to keep this option checked.*

- **Structural:** This option helps control what elements appear in a view with the discipline set to *Structural*. This option is really the only difference between the *Floor* tool and the *Structural Floor* tool.

- **Dimensions:** Notice these read-only values provide information about the selected floor element.

- **Phasing:** this allows for the management of existing, demolished, new and future elements.

 Every modeled element in Revit has these two settings: *Phase Created* and *Phase Demolished*. Additionally, every view also has two phase

FIGURE 12-1.6 Floor Instance Properties

settings: *Phase* and *Phase Filter*. Between these two groups of settings it is possible to do things like have the existing elements be gray and solid filled in a new view, and also have items to be demolished, hidden or appear dashed.

At the top of the *Properties Palette* is the ***Type Selector***. With the floor selected, it is possible to swap it out with another floor system in the list. If the one you want is not there, you click *Edit Type*, and then *Duplicate* to create a new one. Generally speaking, a floor should always be swapped rather than deleted and redrawn.

When a floor is selected and you click the *Edit Type* button on the *Properties Palette* you see its **Type Properties** (Figure 12-1.7). Any changes here will be applied to all instances of this floor type in the model, not just the selected one.

You can also access the *Type Properties* by right-clicking on the family name in the *Project Browser*. This is helpful when a floor of this type has not yet been added to the model.

Take a minute to note the following points about a floor's *Type Properties*:

- **Construction:** The *Structure* option is where you define the horizontal planes within the floor assembly. We will look at this more closely in a minute. The *Function* option can be set to "Interior" or "Exterior." This can be used in scheduling and filtering a view.

- **Graphics:** This is similar to what was already covered for walls. The fill and color of the floor can be set here. This only applies to the floor when in section and the view's *Detail Level* is set to *Coarse*.

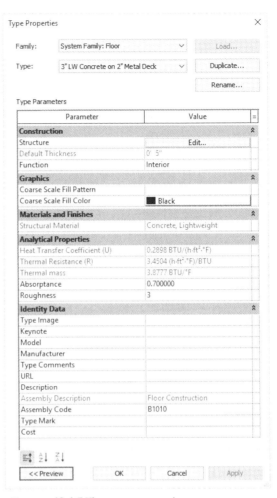

FIGURE 12-1.7 Floor type properties

- **Identity Data:** Most of the *Identity Data* parameters are straightforward. The *Assembly Code* can be used by some external specification writing programs to automatically add specification sections based on what elements have been added to the model. *URL* stands for Universal Resource Locator, which just means internet address. Here you can add the web address for a specific product webpage or product cut sheet, which might be a PDF file located online.

Changing the *Type* from the drop-down list will change the type of the selected floor in the model. If you need a new floor type, set the *Type* option to one that most closely matches the one you want to make to reduce the effort required to define the new floor type.

Next you will take a look at a floor's **Structure Properties**. This relates more to the *Layers*, or planes, of material within the overall floor assembly than to its actual physical structural characteristics. In a floor's *Type Properties*, click **Edit** next to *Structure* to see the *Edit Assembly* dialog box (Figure 12-1.8).

FIGURE 12-1.8 Edit Assembly dialog for floor type "3″ LW Concrete on 2″ Metal Deck"

As you can see, the *Edit Assembly* dialog is very much like the one for walls covered in Chapter 6; notice the main difference being the options near the bottom. The walls have options for reveals and sweeps, and floors have an option to control how the *Structural Metal Deck* appears in section views.

Refer back to Chapter 6 for more information on adding *Layers*, what *Function* does and how to set the *Material*. Note that even if a "finish" *Layer* is added above the *Core Boundary*, the entire floor is still below the level datum.

The options for *Structural Metal Deck* only become active when a *Layer* whose *Function* is set to **Structural Deck [1]** is selected (Step #1 in Figure 12-1.8 above). When active, you can select from the various *Deck Profiles* loaded in the current project, and you can also select the *Deck Usage*: *Bound Layer Above* or *Standalone Deck* (Step #2 in figure above).

The template file you start from may already have one or more deck profiles loaded. You can load a deck profile using the *Load Family* button on the *Ribbon*. The profile families are found in the following location: **Profiles\Structural**. A profile is simply a series of connected 2D lines which gets extruded within the model.

When the *Deck Usage* is set to **Bound Layer Above**, the *Layer* above, usually concrete, fills the top flutes of the deck. The thickness of the metal deck is included within the overall thickness of the *Layer* above. The cross section, showing the flutes, is always visible in the *Edit Assembly* dialog box. However, in the model it depends on the *Span Direction* setting.

When the *Deck Usage* is set to **Standalone Deck**, the thickness of the metal deck occurs completely below the *Layer* above (Figure 12-1.9). The material from the *Layer* above does not fill the flutes either. Notice the *Thickness* for the metal deck is not editable as this is based on the dimensions within the *Profile*. If a different dimension is needed, simply create a new *Profile*, **New → Family → Profile** template, or open an existing deck profile, edit it and then *Save-As*.

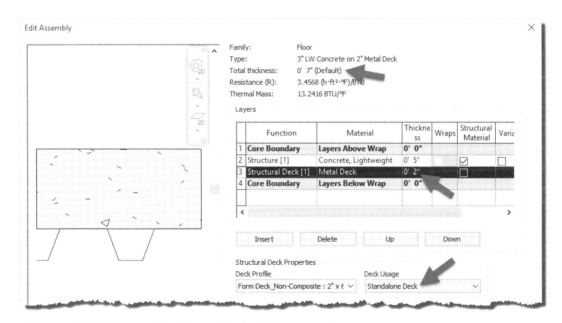

FIGURE 12-1.9 Edit assembly dialog – adjusting desk usage option

It is possible to include a *Layer* for the floor finish, such as carpet or tile. However, this finish would be applied to the entire building. Additionally, it is best to separate the primary floors (i.e., what we have just covered) created by the architect or structural engineer from the interior designer's floor finishes. This prevents unintended edits or movements being made by another discipline, and allows for better visibility control, renderings and scheduling of interior finishes. Plus, creating separate floors to represent floor finishes is the only way possible when the primary floor is coming from a linked structural model.

Now we will take a look at how to model and manage separate floors, in addition to the primary floors, for the interior finishes. The *Law Office* model already has all of the primary floors modeled. On top of those, you will model ⅛″ thick floor elements throughout the building (Figure 12-1.10). This will be done on a room-by-room basis. You will have one ⅛″ floor type for each finish you wish to represent, for example, carpet, ceramic tile, porcelain tile, terrazzo, sealed concrete, etc.

These separate floor elements can be scheduled to determine quantities and cost for the various floor finishes.

> *TIP: Turn off* Room Bounding *for the finish floor element so it does not cause any conflicts with volume calculations for the space or coordination issues for* Revit MEP Spaces *when they are linking your model.*

FIGURE 12-1.10 Floor finish floor element added on top of the primary floor element (scale exaggerated)

Here is how the "floor finish" *Floors* are set up:

- **Duplicate** an existing floor type (preferably a simple one with one *Layer*).

- Add "**Floor Finish**" as a prefix to all of these floor types.

- Set the *Function, Material* and *Thickness* for the single **Layer** (Figure 12-1.11).

 o Set the thickness to ⅛″ regardless of the material's actual thickness – this is meant to be representative.

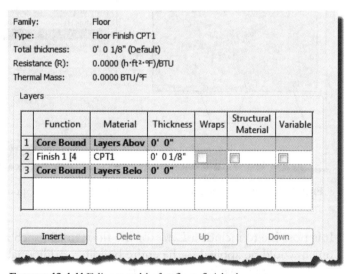

FIGURE 12-1.11 Edit assembly for floor finish element

- When placing the floor, set the *Instance Parameter* **Offset** to ⅛″. This will move the floor up ⅛″, the thickness of the finish floor element, so it sits on top of the primary floor.

Most of the time, the primary floor and the finish floor elements are drawn as large as possible. In the *Law Office* model, the primary floor element represents the entire *Level 1* floor area. Sometimes the floor needs to be broken up into several elements. This may be due to varying thicknesses, deck span direction, etc. When multiple floors are used, a line appears between them. Often, it is not desirable to see these lines in plan or section. Using the ***Join Geometry*** tool provides the solution (Figure 12-1.12). This tool will make multiple elements appear to be monolithic, such as floor to floor, floor to wall and roof to roof conditions, as long as the *Material* is the same between the two elements.

FIGURE 12-1.12
Join tool on Modify tab

The first image below (Figure 12-1.13) shows three separate floor slabs. The smaller square shape is a different thickness, but all three are the same *Material.*

FIGURE 12-1.13 Three separate floor elements – **not joined**

The next image (Figure 12-1.14) shows how the three floors appear monolithic after using the *Join* tool on them. They are still separate elements which can move and be edited, or even deleted. They even appear continuous in sections; this setting is not view specific. The join will remain until the ***Unjoin Geometry*** tool is used.

FIGURE 12-1.14 Three separate floor elements – **joined**

Floors Accents:

Within a room it is not always necessary to sketch each ⅛" floor finish separately when there are multiple styles/ colors. You can, for example, create a floor finish for the larger area, the field, and then superimpose any accents within the field. The last, but very important, step is to *Join* the floor elements.

FIGURE 12-1.15 Carpet field and accent band created separately

In Figure 12-1.15 we have a **150 S.F.** carpet area, the *field*. Adjacent to it we also have a **38 S.F.** *accent band*.

If we move the accent band within the larger carpet field, the areas do not change automatically. You would still have **188 S.F.** of carpet listed in the *Properties Palette*: the field and accent band totals combined.

FIGURE 12-1.16 Carpet field and accent band joined together

To properly represent the area of the field we need to *Join* the two carpet areas together. Once this has been done, the carpet field will report **112 S.F** (Figure 12-1.16).

One variation on the ⅛″ finish floor workflow previously mentioned is to not offset the floor up and just use the *Join* tool. A limitation to this is it cannot be done when the primary floor is coming from a linked model. We will be using the previous method, so the finish is always above the *Level* datum and the primary floor is not reduced in thickness by this technique.

> *FYI: This works great for recessed floor mats in vestibules.*

The image below shows a ⅛″ floor element representing a carpet field. The floor has not been offset up and the view's *Shade* mode is on (Figure 12-1.17). Notice the finish graphics do not appear correctly because Revit is confused by the two *Materials* occupying the same plane.

FIGURE 12-1.17 Two floors overlapping – **not joined**

When the ⅛″ floor is *Joined* things clean up properly.

FIGURE 12-1.18 Two floors overlapping - **joined**

The last thing we will have time to cover is **Slab Edges**. This tool allows you to modify the default flat edge of a floor. A *Slab Edge* is based on a *Profile*; recall we just talked about *Profiles* in structural floor decking a few pages back.

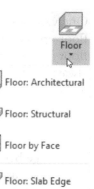

> *FYI: Loading a* Profile *into your project does not make it available from the* Type Selector *when using the* Slab Edge *tool. You need to* Duplicate *an existing* Slab Edge **Type** *and change the* Profile *specified.*

This feature can be used for many things, such as a thickened slab shown in the icon to the right. It could be for a floor transition from

your "finish floor" to the primary floor in an open area. It can even be for a continuous bench along the edge of a floor seeing as the *Material* can be set independent of the floor hosting the floor edge. Plus, the position can be adjusted via the *Instance Properties*: **Horizontal Offset** and **Vertical Offset**.

FIGURE 12-1.19 Slab Edge tool

When in the *Family Editor*, you create a simple outline as shown in Figure 12-1.20. This is then saved to a file with a specific name. That *Profile* is then loaded into a project and associated with a *Slab Edge* type.

> *FYI: This 6" high profile will be shown added to a 4" thick floor.*

The *Slab Edge* may be applied to a *Floor* in plan or 3D views. In a 3D view you can pick the top or bottom edge of the slab to host the *Profile*.

> *TIP: In a 3D view, select the* Floor *and then click the* **Isolate Element** *on the* View Control Bar *(i.e., the eye-glasses icon).*

When you begin selecting the edges of a floor, the profile extends the entire length of the edge. Inside and outside corners are automatically mitered as shown in Figure 12-1.21. Even our tapered edge turns the corner properly.

FIGURE 12-1.20 Profile sketch

Once you are done adding the *Slab Edge* to your floor, click *Modify* to end the command. Later, if you select one of the *Slab Edges*, you can drag one of the end grips to reposition or change the length of it. Notice two lengths of the *Slab Edge* in Figure 12-1.21 have been adjusted inward.

FIGURE 12-1.21 Slab Edge added to four sides of floor – two lengths have been dragged inward

Notice how the floor and slab edge are automatically joined when viewed in section (Figure 12-1.22).

The *Slab Edge* feature allows just about any type of common floor to be modeled. When something more elaborate is needed, you can create a *Mass Element* and then use the *Floor by Face* tool.

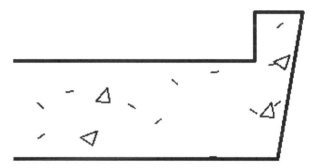

FIGURE 12-1.22 Slab edge automatically joins with floor

Floors have the ability to be sloped using a ***Slope Arrow*** while in *Sketch* mode. You may refer back to Chapter 9 on *Ceilings* for a review of *Slope Arrows*. *Slope Arrows* work the same for *Floors, Ceilings* and *Roofs*. Another thing you may wish to explore on your own is **Shape Editing**. This feature is available from the *Ribbon* when a floor is selected, as shown below.

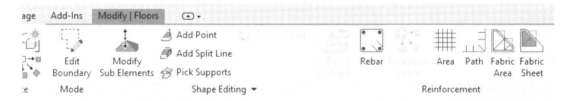

Exercise 12-2:
Modeling Floor Finishes – Carpet

There will certainly be plenty of opportunities for the interior designer to model primary floors, but sometimes the floor has already been modeled by the architect or the structural engineer. The latter situation will be the case with your *Law Office* model; the primary floors already exist. Of course, you are aware of this because a previous exercise had you modify the Level 2 floor for the stairs.

The first thing you will do is create the required materials and then a floor type to apply them to. You will then create the ⅛" thick floors, a.k.a., sliver floor, room-by-room.

Floor Finish Materials:

Below is an overview of the floor finishes you will be using in this chapter. This information will be referenced in subsequent steps in this exercise.

CPT1
Style: Float Tile 59484
Color: Mink 83761
Size: 24"x 24"
Install: Quarter Turn
Mfg: Shaw

CPT1 12"x12" TILE

CPT2
Style: Vertical Edge Tile 59114
Color: OchreVerge 67556
Size: 18" x 36"
Install: Monolithic
Mfg: Shaw

CPT3
Style: Fade Tile 59597
Color: Washed 95106
Size: 18" x 36"
Install: Brick
Mfg: Shaw

CPT2 18"x36" TILE

CPT4
Style: Overlay Tile 59598
Color: Washed 95106
Size: 18" x 36"
Install: Brick
Mfg: Shaw

CPT5
Style: Colour Plank Tile 59595
Color: Orange 95675
Size: 18" x 36"
Install: Brick
Mfg: Shaw

CT3
Product Type: Colorbody Porcelain
Style: Mont Blanc
Color: D'Alpe P236
Size: 18" x 18"
Mfg: Daltile
Install: Brick (running bond)
https://www.daltile.com/

EM1
Entry Mat
Style: Normad Modular matting 9200
Size: 8" x 8" Nominal
Mfg: 3M
https://www.3m.com/3M/en_US/company-us/

RF1
Tile goods
Style: Noraplan Mega
Color: Crystal Blue 2667
Mfg: Nora
Tile: 24" x 24"
https://www.nora.com/united-states/en/product-lines/flooring

RF2
Sheet goods
Style: Linoleum
Color: Pollen LP170
Mfg: Armstrong
http://www.armstrong.com/commflooringna/products/linoleum

Please keep in mind that flooring styles and products are constantly changing and some of the specific styles and/or colors are no longer available. However, the image file is provided with this book; see inside front cover to download.

Floor Finish Design:

The first two images below show the flooring design to be modeled. Be sure to look at the colored images, provided with the online files, for this section.

Floor Finish Plans:

The primary architectural floor plans, which you have been working in thus far, do not typically show the floor finishes. This helps to make the drawings clearer and less cluttered. Therefore, you will create a set of "Finish" floor plans.

1. **Right-click** on **Level 1 Floor Plan** in the *Project Browser*.

2. Select **Duplicate View → Duplicate with Detailing** from the pop-up menu.

 FYI: The Duplicate with Detailing *option creates a copy of the view and also copies any view specific annotation such as notes, dimensions and tags. We mainly want to do this for the room name and number tags.*

3. Rename the new view to: **LEVEL 1 FINISH FLOOR PLAN**.

4. Repeat these steps to create a Level 2 finish plan:

 a. **Duplicate with Detailing** Level 2

 b. Rename view to **LEVEL 2 FINISH FLOOR PLAN**

 WARNING: It is not possible to duplicate a Level 1 plan view *and make it a* Level 2 view. *You can only duplicate a plan view for the same level. Otherwise, you need to select* **View → Plan Views → Floor Plan**.

Next you will turn off a few unneeded categories.

5. Type **VV** and turn off, for both finish plan views:

 a. **Grids**

 b. **Elevations**

 c. **Sections**

6. Set the *Detail Level* to *Coarse*, for both finish plan views:

You now have two clean plan views you can use to define your finishes. We will only be looking at floor finishes at this point. Later you will apply a filter to your primary floor plan views to hide the ⅛″ finish floors.

Modeling Floors – Carpet:

Each room or area will need a separate floor element added. You would not want to have one floor element span multiple rooms. The overall area of the floor would include the footprint of all the walls it overlaps. This can add up and throw off your quantity schedule used for estimating. Of course, there is always extra material included for cutting. So, as long as you understand how materials are counted it is possible to have a floor span multiple rooms, but we will create separate floors per room. Some rooms, with multiple carpet styles, will have multiple floor elements within that single room; e.g., note the *Small Conference Rooms* on Level 1.

An example of individual floor elements by room can be seen in the image below. Exposed concrete floors do not have a finish; you just see the primary structural floor in this case. If the concrete floor had a special sealer or finish, you might even create a floor type for this so the exposed concrete area may be calculated as well.

FIGURE 12-2.1 Individual floor element added to each room – Level 1

New Floor Types:

You will create a floor type for each floor finish. This floor will be ⅛" thick and will be modeled such that it sits on top of the primary structural floor as mentioned in the beginning of this chapter.

7. Open the **LEVEL 2 FINISH FLOOR PLAN** (if needed).

8. Select the **Floor: Architectural** tool from the *Architecture* tab.

9. Select the **4″ Concrete Slab** type from the *Type Selector*.

 FYI: This is a simple floor type from which to start your new floor.

10. Select *Edit Type*, and then *Duplicate*.

11. Enter name: **Floor Finish – CPT1**

All of the sliver floors will have the same prefix "Floor Finish." This will help later when we want to filter the main plans so the finish floors do not appear.

12. Select *Edit Structure*.

13. Change the floor thickness to **⅛″** and uncheck *Structural Material*.

14. Click **OK**; you will change the *Material* later.

15. In the **Choose Assembly Code** dialog, change the *Assembly Code* to **C3020520 Carpet Tile** (Figure 12-2.2).

16. Click **OK**.

Selecting the correct *Assembly Code* is helpful for estimating and specification programs that can import Revit data.

17. In the Level 2 Finish Plan, zoom in on the **First Year Associate/ Paralegal** room (near the restrooms).

18. In the *Properties Palette* (Figure 12-2.3):

 a. Height Offset: **1/8″**

 b. Un-check **Room Bounding**

19. Select the **Rectangle** option in the *Draw* panel on the *Ribbon*.

20. Pick two opposite corners in the office space.

21. Click the **green checkmark** to finish the creation of your first floor finish.

 a. Click *No* if prompted to attach walls below to floor.

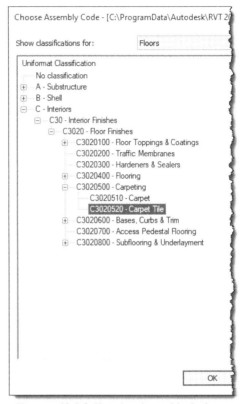

FIGURE 12-2.2 Changing Assembly Code

FIGURE 12-2.3 Properties Palette

Based on the *Material* currently assigned to the new floor element just placed, we do not see anything except a line at the door opening (Figure 12-2.4). This line is the edge of the ⅛″ floor. The other edges coincide with adjacent walls and are not readily visible. This can make selecting floors a bit tricky. The *Tab* key is needed to select a floor when an edge is not visible.

Next you will create a new *Material* and then assign it to the new floor type you just created and placed in the model. Once this is done, the previously placed floor will automatically be updated to show a 2′x2′ grid representing carpet tile.

FIGURE 12-2.4 First floor finish floor added

22. Select the ⅛″ finish floor element in the *First Year Assoc. Room*; use the **Tab** key if needed to highlight the floor before clicking to select it.

23. Select *Edit Type*.

24. Select *Edit Structure*.

25. Click the *Material* field and then click the button which appears to open the *Materials* dialog.

It turns out you do not need the original CPT1 material previously created. But just in case you change your mind, we will rename it rather than delete it.

26. Select *Material* CPT1 and **Rename** it to CPT 8.

 a. Remember to change the **Mark** value on the *Identity* tab.

27. Duplicate CPT8 to a new *Material* named **CPT1**.

Now you have a new material you can modify for the CPT1 material listed at the beginning of this exercise.

28. Make the following entries on the **Identity** tab:

 a. URL: https://www.shawcontract.com/

 b. Post-Consumer Content: click **spec → Eco Specs** via URL listed above.

 c. Source (miles): Click **LEED Calculator** via URL link above, enter your zip code and then click submit. Notice the manufacturer's zip code is **30120** (which

is Cartersville, GA). Use an internet map program to determine the distance in miles between your location and where the carpet is made.

 d. See Figure 12-2.5 and the *Materials* list at the beginning of this exercise for the remaining information.

FIGURE 12-2.5 New CPT1 information added

29. Switch to the **Graphics** tab in Revit's *Material Browser*.

30. To set the *Foreground Surface Pattern*, click the preview to open the **Fill Patterns** dialog.

31. Click on the **Model** *Pattern Type* option.

 FYI: Remember, model type patterns will not change size when the View Scale *is changed.*

There is no 24″ x 24″ tile pattern, so you will need to make one.

32. Click the **New** button to create a new fill pattern.

33. Create a new pattern named **24″ Tile** per Figure 12-2.6.

The pattern name does not need to have the word "carpet" in it, just *24"Tile* helps keep the overall number of patterns down. This pattern may be used for several other materials such as ceramic tile or ceiling tiles.

34. Set the color to the darker gray option (Figure 12-2.7).

FIGURE 12-2.7 Adjusting new foreground fill pattern color

35. Click **OK** to close all the dialogs and notice the 2'x2' carpet tile pattern now appears in your plan view.

36. Hover your cursor over one of the fill pattern lines,

 press *Tab* to highlight it, select it and use the ***Move*** tool to move the pattern **1'-0"** to the right to center better in the room (Figure 12-2.8).

FIGURE 12-2.6 New fill pattern

The next step is to download the image file for the carpet. This is a bit more complicated than what you have already done in the previous *Materials* chapter. Here we want to use multiple carpet tiles and have some of them rotated a ¼ turn. We can actually get this image from the Shaw website, but the added text at the bottom of the image needs to be removed from the image. You will need an image editing program for this – we provided the modified image file for you, with the online files, if you need it.

Keep in mind that this step of picking a render appearance does not need to occur until you need to create a rendered image. We will do it now because we know what we want.

FIGURE 12-2.8 New fill pattern in plan-view

37. **Save** your project file.

Due to a change in available products, the next steps are for reference only, if you want to select a similar product. Otherwise, the required image and information are provided with this book.

38. **Ref only:** Browse to the following URL: https://www.shawcontract.com/

Notice that the last number is the model number/color for the Shaw carpet we are using. This can be changed to another known model number to quickly access similar information for another Shaw product (including the color number is optional).

39. **Ref only:** Follow the steps shown in Figure 12-2.9 to download an image file for the specified carpet.

40. **Ref only:** Select a location on your hard drive to save the file.

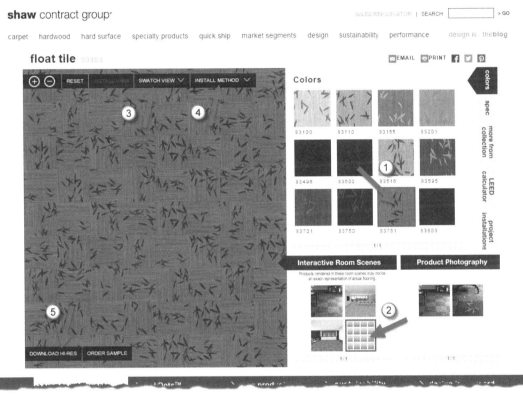

FIGURE 12-2.9 Downloading Shaw carpet image file

If you open the file you downloaded, it should look similar to what is shown in Figure 12-2.10. Notice the text across the bottom gives a disclaimer and information about the carpet in the image. In our case we need to crop this image down from a 9x9 tile pattern to 8x8. We will select our pattern from the upper left corner to get a pattern which will repeat properly in Revit.

FIGURE 12-2.10 Downloaded image file

41. Open the image file in an image editing program such as Adobe Photoshop and crop the image down as shown in Figure 12-2.11 (8 tiles x 8 tiles).

 a. If you do not have access to an image editing program, you can use this file from the online files: **59484_83761_ninexniesb_Cropped.jpg**

Now you will switch back to Revit and associate the new image with your CPT1 material. Rather than take the extra steps of selecting the floor, we will go directly to the *Materials* dialog and edit CPT1 given we know the name and that it is already assigned to the floor.

FIGURE 12-2.11 Cropped Shaw carpet image file

42. In Revit, go to **Manage → Materials**.

43. Select the *Material* **CPT1** and then click the *Appearance* tab.

44. Click the file name, which is a link, below the image preview.

45. Browse to the cropped Shaw image file.

The next step is very important to ensure the image renders properly in the project. This step adjusts the scale of the image to correspond to the 2′x2′ tile size needed.

46. Click directly on the image preview to access the *Texture Editor*.

47. In the *Texture Editor* dialog, under *Transforms*, set the **Scale** to **16′ x 16′** (Figure 12-2.12).

 FYI: The image has 8 tiles which are 2′-0″ each, thus 16′-0″ total.

48. Click **OK** to close the open dialog boxes.

FIGURE 12-2.12 Adjusting image size in model

You have now successfully set up CPT1 in your project. If this is a material you use often, you could save it to your firm's template to save time. Even if

the color changed, most of the settings for the material will still be correct.

The image below (Figure 12-2.13) is a result of using the **Section Box** feature in a 3D view to crop the top of the building. The *Visual Style* has been set to **Realistic** which causes the rendered appearance image to show in the model without using the *Render* tool; notice the carpet image and the wood grain on the doors. Another feature utilized in this image, and many others in this book, is **Show Ambient Shadows**. This gives the image a deeper and more interesting feel.

FIGURE 12-2.13 Cropped 3D view set to realistic mode and Ambient Shadows turned on

You are now ready to create the remaining floors and materials. We will move through these at a faster pace, assuming you can refer back to the steps just covered if you forget something. We will mainly be focusing on unique aspects of the remaining materials.

Next you will create CPT2 for the *Law Office* project.

49. In the ***Materials*** dialog, rename the existing CPT2 to **CPT9**, and change its *Mark* value on the *Identity* tab to CPT9.

50. ***Duplicate*** CPT1 and name it **CPT2**.

51. Modify the fields on the *Identity* tab for CPT2.

52. On the *Graphics* tab, create a new foreground ***Surface Pattern*** (Figure 12-12.14).

 a. In order for this pattern to align with the render appearance, the spacing values must be entered in the order shown.

53. On the *Appearance* tab:

 a. Duplicate the *Appearance Asset* and *Rename* it to CPT2 (right-click on the appearance name to duplicate and then expand the *Information* section below to rename).

 b. Load the proper image file. The one provided online is 10 tiles x 5 tiles. (Thus, the image scale should be 15′x15′.)

FIGURE 12-2.14
18x36 tile pattern created

You have three more carpet *Materials* to create. The only thing different about them is the pattern is not monolithic; it is staggered like a brick pattern.

54. Rename CPT3 to **CPT10** and change the *Mark* value.

55. Duplicate CPT2 to **CPT3**, and then update the *Identity* and *Appearance* information.

56. Switch to the *Graphics* tab to create a new *Surface Pattern*.

57. In the *Surface Pattern* area, click the button to open the ***Fill Patterns*** dialog.

58. Click **New**.

Because the standard grid pattern will not work, you will need to create a custom pattern. A custom pattern will be imported from a pattern file provided by Autodesk. This pattern file is a series of commas and numbers which define repeating patterns.

> *FYI: This is the same format AutoCAD uses. There are many patterns available online for AutoCAD.*

59. Click Custom and then Import.

60. Browse to the following file:
 C:\Program Files\Autodesk\Revit 2025\Data\revit.pat.

61. Select **Block 8x16** from the *Custom* pattern list (Figure 12-2.15).

62. Enter the name:
 18x36 Carpet Tile - Running Bond

Next you will enter the scale factor of 2.25, to go from 8x16 to 18x36.

63. Set the *Import Scale* to **2.25**.

Once you set the import scale and click **OK**, it cannot be changed. If you need to change it, you have to delete the pattern and re-import it.

64. Click **OK** to get back to the *Materials* dialog.

65. Create the last two carpet materials, CPT4 and CPT5, by *Duplicating* CPT3.

 a. Edit the *Identity* tab.

 b. Duplicate and edit the *Appearance* tab.

 c. Adjust the appearance image *Scale* to **18″ x 36″** for CPT5 only.

66. **Save** your project.

67. Under *Families\Floors*, in the *Project Browser*, right-click on **Floor Finish - CPT1**.

68. Click **Duplicate**.

69. Type **Floor Finish – CPT2.**

70. Right-click again and pick *Type Properties*.

71. Change the *Material*, in *Edit Structure*, to CPT2.

 TIP: You can just type CPT2; you don't have to open the Materials dialog.

72. Repeat the previous 5 steps to create floors for the remaining carpet types: CPT3 – CPT5 (Figure 12-2.16).

FIGURE 12-2.15 Importing a custom fill pattern

FIGURE 12-2.16 Project Browser with new floors created

The last step in this exercise is to create the floor elements in the model. This design involves extending the corridor finishes 1'-0" into the adjacent rooms. This will be apparent to the occupants due to the large expanse of glass between the corridors and adjacent rooms.

73. Switch to the **Level 1 Finish Floor Plan**.

Next you will remove an override which is currently hiding the floor patterns.

74. Type **VV**, click twice on the word *Hidden* next to the *Floors* category, check *Visible* (if not already checked,) and then close the open dialog boxes.

75. Add the four **Floor Finish - CPT1** areas and the one **Floor Finish - CPT2** area as shown in Figure 12-2.17.

 a. The carpet edge is **1'-0"** from the corridor side of the wall.

 b. Align tiles as shown using the *Align* tool.

FIGURE 12-2.17 Level 1 carpet added

76. Switch back to **Level 2 Finish Floor Plan**.

77. Add the carpet as shown in Figure 12-2.18. All areas shown are **Floor Finish – CPT1**, except for the open office area which is **Floor Finish – CPT2**.

 a. Use *Align*, *Move* and *Rotate* as required making the tile patterns match.

 b. Corridor and open office carpet extend **1'-0"** into adjacent rooms.

 c. Carpet from corridor into rooms can extend under walls to save time.

FIGURE 12-2.18 Level 2 carpet added

Next you will create a multi-color flooring pattern which does not follow the boundaries of the room. This will be done in the *Small Conference* and *Impromptu Meeting* area on Level 1. The pattern is somewhat irregular, so you will load a group from the disc to make it match what is presented here. If the online files are not readily available you can just make it as close as possible.

78. Switch back to the **Level 1 Finish Floor Plan** view.

Load as
Group

79. From the *Insert* tab, select **Load as Group**.

80. Browse to the *Materials* folder provided with the online files.

81. Select the following file: **Law Office Carpet Pattern Lines.rvt**.

82. Right-click on the new *Detail Group* in the *Project Browser* and select **Create Instance** (Figure 12-2.19).

FIGURE 12-2.19 Project Browser, Groups

83. For the insertion point, click the center of the column shown in Figure 12-2.20 below.

You now have a group of 2D detail lines you can use to create your carpet areas.

FIGURE 12-2.20 Imported lines placed as a Detail Group

84. Select the *Floor* tools; select **Floor Finish – CPT4** from the *Type Selector*.

85. Click the *Pick Lines* option from the *Ribbon* (Figure 12-2.21).

FIGURE 12-2.21 Pick Lines option for Floor tool

86. Click the four lines that define the upper left area (Figure 12-2.22).

87. Use *Trim* to clean up the corner conditions.

88. Click the **green checkmark** to finish the floor.

89. Click **No** when/if prompted to extend walls to floor.

You now have an irregular floor area defined for CPT4 (Figure 12-2.23).

FIGURE 12-2.22 Sketch lines trimmed

FIGURE 12-2.23 New floor added

90. Using the same steps, add the carpet as shown in Figure 12-2.24.

 a. Use *Rotate* and *Align* to adjust the pattern.

91. Select the *Detail Group* and delete it.

92. Select **Annotate → Material Tag**; tag the floors as shown (no leader). FYI: The material tag has been modified to report material Mark, rather than material Description parameter.

 TIP: The material tag tool does not work when the view is set to wireframe.

93. **Save** your project.

FIGURE 12-2.24 Carpet added

Exercise 12-3:
Modeling Floor Finishes – Ceramic Tile, Sheet Goods, Etc.

In this exercise you will add the remaining floor finishes. You will begin with the ceramic tile.

1. Open the **Level 1 Finish Floor Plan**.

2. Open the *Materials* dialog.

3. Select **CT1** and then click *Duplicate*.

4. New name should be **CT3**.

 FYI: If you have a CT3 from Chapter 4, rename it first, to Old_CT3.

5. Edit the *Identity* tab (Fig. 12-3.1).

 a. Note the product origin is Italy and has no *Post-Consumer Content*.

6. On the *Graphics* tab, create a new pattern named **18x18 Tile - Running Bond**.

 a. Note the new pattern definition from Figure 12-3.2 below.

FIGURE 12-3.1 CT3 Identity data

7. Check use **Render Appearance** for *Shading*.

8. On the *Appearance* tab, Duplicate the appearance asset and then select an image downloaded from the Dal Tile website.

 a. Click *Image* from the drop-down pointed out in Figure 12-3.3.

 b. For a better representation, use the image provided online:
 CT3 Dal Tile DT_P236_18x18_D_Alpe.jpg.

 FYI: It has been adjusted for the running bond pattern in Adobe Photoshop.

FIGURE 12-3.2 CT3 fill pattern

9. Be sure to adjust the image scale on the *Appearance* tab; goal is 18" x 18" individual tiles.

10. Uncheck *Relief Pattern*.

11. Click **OK** to close the *Materials* dialogs.

Next you will create the entry mat material for the vestibule.

12. In the *Materials* dialog, select CPT1 and **Duplicate** to **EM1**.

13. Adjust the *Identity* tab as needed; be sure to set *Mark* to **EM1**.

14. On the *Graphics* tab, set the *Surface Pattern* to **8″ Tile**.

15. On the *Appearance* tab, duplicate the *Appearance* (name it EM1) and then set the image to **EM1 3MFloorMat.jpg** from the online files provided. This file has been cropped so there is no extra white space.

FIGURE 12-3.3 CT3 fill pattern

16. Click on the image preview and set the scale/size to **8″ x 8″**.

Next you will create the flooring material for the exercise room.

17. Select EM1; *Duplicate* to **RF1**.

18. Edit the *Identity* tab; set the *Mark* parameter to **RF1**.

19. On the *Graphics* tab, set the *Foreground Surface Pattern* to **24″ Tile**.

20. On the *Appearance* tab, *Duplicate* and *Rename* the appearance asset, select the Nora image from the online files for this book or download it from their website.

21. Set the image scale to **2′ x 2′**.

Now you will create the last flooring material.

22. Select RF1; *Duplicate* to **RF2**.

23. Adjust the *Identity* tab as required.

The linoleum is installed as sheet goods rather than tiles, so you do not need to have a grid pattern for this. You could set the *Surface Pattern* to *None*, but we want to see where the flooring occurs. You will select a stipple drafting pattern.

24. On the *Graphics* tab, click the button to launch the *Fill Patterns* dialog in the *Surface Pattern* area.

25. Select the **Drafting** option at the bottom (Figure 12-3.4).

26. Select the **Sand** option.

27. On the *Appearance* tab, set the image file to an image downloaded from the internet or use the one provided online for this book.

28. Leave the image scale at 2′ x 2′.

29. Close the *Materials* dialogs.

30. **Save** your project.

You have now created all the *Materials* needed for the floor finishes. The last "setup" step is to create one ⅛″ floor *Type* for each *Material*. You will just *Duplicate* the CPT1 *Type* and change the *Material* assigned to it.

FIGURE 12-3.4 RF2 fill pattern

31. Under *Families\Floors*, in the *Project Browser*, right-click on **Floor Finish - CPT1**.

32. Click **Duplicate**.

33. Type **Floor Finish – CT3.**

34. Right-click again and pick *Type Properties*.

35. Change the *Material*, in *Edit Structure*, to **CT3**.

36. Repeat the previous 5 steps to create **EM1**, **RF1** and **RF2**.

FIGURE 12-3.5 Project Browser

When finished creating the new floor types, your *Project Browser* should look like Figure 12-3.5.

37. Add the remaining Level 1 flooring as shown in Figure 12-3.6:

 a. Extend the linoleum (RF2) under the cabinets.

 b. To save time, let the tile (CT3) extend under the corridors' walls.

 c. Extend the tile (CT3) below the plumbing fixtures and countertop.

FIGURE 12-3.6 Remaining Level 1 flooring to be added

38. Add the remaining Level 2 flooring as shown in Figure 12-3.7.

 a. Extend floor into door openings.

 b. *Align* and *Join* floor patterns at door openings.

You now have a highly accurate model of your project's flooring!

FIGURE 12-3.7 Remaining Level 2 flooring to be added

Next you will switch to your main architectural floor plans and add a *Filter* to hide the finish floors. This is the same feature you used to make the wall cabinets dashed.

First you will create a *Filter* and then you will apply it to the views.

39. Click **View → Filters**.

40. Click the **New** icon to create a new *Filter* (lower left).

41. Name it: **Floor Finish.**

42. Set the dialog per Figure 12-3.8.

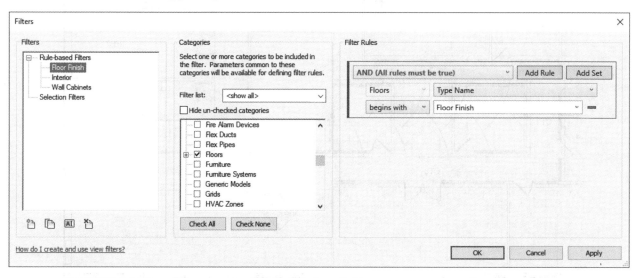

FIGURE 12-3.8 Filters dialog – creating filter to hide floor finishes

We will use this filter to hide all the floors whose *Type Name* begins with "Floor Finish."

43. Switch to your **Level 2 Floor Plan** view.

Notice the floor finishes appear and that they also hide the *Room Separation Lines*.

44. Type **VV**.

45. Switch to the *Filters* tab.

46. Click the *Add* button.

47. Select **Floor Finish** from the list.

48. Uncheck the *Visibility* options (Figure 12-3.9).

FIGURE 12-3.9 Level 2 Visibility / Graphic Overrides for Level 2 Floor Plan – Filters tab

49. Click **OK** to finish adding the *Filter* and see the results.

Your floor finishes should now be hidden and the *Room Separations Lines* visible again. Keep in mind that *Filters* are case sensitive, so **Floor Finish** is different than **FLOOR FINISH**. Therefore, you need to make sure you name all the floor *Types* correctly and create the *Filter* correctly as well.

50. Switch to your **Level 1 Floor Plan** view and apply the **Floor Finish** *Filter* as well.

Conclusion

You have now added all the floor finishes and have created special floor plan views to present and document them. The process of setting the *Materials* up can be a bit involved, but there are many benefits as you will see when we cover scheduling and rendering later on. Keep in mind that you can create several floor types and save them in your firm template, so they are ready to use when you need them. Be sure to save the image files in a common place so every computer has access to them. You could also create "resource files" which are project files with a bunch of 1'x1' floors created. The designer can then open the file and just Copy / Paste what they need into their project.

Self-Exam:

The following questions can be used as a way to check your knowledge of this lesson. The answers can be found at the bottom of this page.

1. A floor system cannot include bar joists. (T/F)

2. It is not possible to add a thickened slab to a floor. (T/F)

3. When a floor is created, the bottom aligns with the *Level* on which it was created. (T/F)

4. While sketching a floor, use a _____ to define a slope direction of the floor.

5. Use the *Join* tool to make two or more floor elements appear monolithic. (T/F)

Review Questions:

The following questions may be assigned by your instructor as a way to assess your knowledge of this section. Your instructor has the answers to the review questions.

1. A floor material can be displayed with the visual style set to *Realistic*. (T/F)

2. In a floor's properties, the _____ can be used by some external specification writing programs.

3. What is the *Assembly Code* for Carpet Tile? _____

4. A 2D profile is used to define a slab edge on a floor. (T/F)

5. The _____ defines which way the floor spans structurally.

6. Floors cannot contain custom information such as the amount of post-consumer content. (T/F)

7. A thin floor is modeled on top of another floor to represent various floor finishes. (T/F)

8. A floor does not have to align with a *Level*. (T/F)

9. What is the *Layer Function* for metal roof deck? _____

10. Some materials need an image edited in an image editing program to show rotated tiles. (T/F)

Lesson 13
Custom Reception Desk and Sign:

This chapter will cover creating a custom reception desk for the *Lobby* area using Autodesk® Revit®. Additionally, you will learn how to edit the profile of a wall to follow the underside of a stair and create 3D text on the wall adjacent to the desk.

A good reception desk design starts out with a few, if not many, sketches. The image below shows a sketch developed for our project early on in the design. This allows the designer to quickly explore several concepts before committing time to the more involved process of creating a 3D model in Autodesk Revit. When the floor plans are initially being developed in Revit, it may be convenient to simply use 2D *Detail Lines*, via the *Annotation* tab, to represent custom elements such as this reception desk. Eventually it needs to be modeled in 3D for sections, elevations, details and MEP coordination.

Sketch by Jill Isola Johnson, CID, IIDA, LEED® AP

Exercise 13-1:
Creating a Custom Reception Desk

The receptionist and reception desk are often the first impression a visitor has of a business. Therefore, the reception desk is often custom built and unique. This requires the designer to leverage the modeling tools provided in Revit. One can create an ***In-Place*** family or a separate ***Loadable*** family when modeling the desk. There are pros and cons to each method.

You will use the *In-Place* method as it is appropriate to one-off items like a reception desk or built-in bookcase. The main drawbacks are 1) the family cannot be easily saved to a file and loaded into another project, and 2) if it were copied around, the family would be duplicated for each instance; you cannot have multiple instances of an *In-Place* family.

Edit Wall Profile:

The first thing you will do is add walls below the open stair in the lobby. Most building codes require the underside of a stair either be closed off or protected by a curb or railing. This requirement is primarily meant to prevent seeing impaired or blind individuals from running into the stair.

Unfortunately, a wall cannot conform to the underside of a stair as it can the underside of a sloped floor or roof. However, Revit has a feature related to walls called ***Edit Profile***. When looking at the wall in a straight on elevation or section view, you can modify the outline of a wall. A few examples of this are shown below (Figure 13-1.1). These three walls have been modified using *Edit Profile*, each from the same original wall. Notice, when the wall is selected, the original perimeter, or profile, is shown with dashed lines.

FIGURE 13-1.1 Edit wall profile examples

1. Switch to your **Level 1 Floor Plan** view (not the Finish Plan).

2. Activate the *Wall* tool; select **Interior – Partition 4⅞″ (1 hr)**.

3. Duplicate the wall, with the name **Interior – 4¼″ Furring Partition**.

4. Edit the wall structure to only have gypsum board on the "exterior" side of the wall.

 FYI: This is different than the restroom wall with tile on it.

5. Verify the *Base Offset* for the wall is **0′-0″**.

6. Draw the **three** walls under the stair (Figure 13-1.2).

UP

Waiting / lobby

219

303 SF

FIGURE 13-1.2 Walls added below stair

The three walls you just added, plus a few from earlier, extend up through the stair and need to be fixed.

7. Create a *Section* in plan as shown in Figure 13-1.2. The section line must be perfectly parallel with the stair for the next step to work properly. Revit should snap to align the section with the stair as you are drawing the section.

 a. Adjust the view depth line (i.e., dashed line when section is selected) so Revit does not "look" too far into the model.

 FYI: *This would also work within an elevation view.*

8. Switch to the section view and select the facing wall below the stair.

9. With the wall selected in elevation/section, click **Edit Profile** on the *Ribbon*.

Edit Profile

10. Using the *Edit Profile* sketch tools and *Trim*, redefine the wall profile as shown in Figure 13-1.3 below.

 a. The final result must be an enclosed perimeter with no extraneous line work.

FIGURE 13-1.3 Editing wall profile to conform to bottom of stair

Again, notice the original wall perimeter appears dashed wherever it has been relocated.

11. Repeat these steps for the wall on the other side of the stair.

12. Select the wall which is shortest in length and is completely below the landing, then:

 a. Set the *Top Constraint* to **Unconnected**.

 b. Set the *Unconnected Height* to **7'-5"**.

13. Create the three sections shown in Figure 13-1.4 and use them to edit the profile of wall previously drawn so they are below the stair.

 a. Add a wall along grid line C to hide the Level 1 ceiling area adjacent to the landing from level 1 ceiling to level 2 floor.

FIGURE 13-1.4 Add sections to aid in adjusting walls to occur below the stairs

When finished editing the walls, you should create a camera view to inspect your work. Be sure to rename the new camera view, in the *Project Browser*, so you can find it again later, or delete the view when finished with it. The model should look similar to Figure 13-1.5 below. There are lots of little issues which need to be addressed in the model, but overall the design is really coming together.

FIGURE 13-1.5 Camera view showing walls aligned below sloped stair stringer

Before you begin modeling the desk, we will create two *Reference Planes* and a *Section* view to help us in the process.

14. Using the **Reference Plane** tool, from the **Architecture** tab, create the two *Reference Planes* shown in Figure 13-1.6. Be sure to draw the line in plan view from right to left. This affects which direction solids are created in elevation and section views.

 a. The dimensions are **1'-1½"** and **3"** in the image below.

FIGURE 13-1.6 Creating two reference planes

We will name these reference planes so they can be selected from a list, later when we are in an elevation/section view.

15. Select the longer *Reference Plane* (1'–1½" away from the wall).

16. In the *Properties Palette*, enter the *Name*: **Recpt Desk Front** (Figure 13-1.7).

17. Name the other *Reference Plane*: **Recpt Desk Sign Wall**.

FIGURE 13-1.7 Naming the reference plane

Now, whenever the *Reference Plane* is selected, the name will appear at one end in the view. Keep in mind that *Reference Planes* print, so you have to turn them off in **VV** or uncheck **Hide ref/work planes** in **Print → Settings** dialog.

You will create a section and then adjust the windows so you can see the plan and section views at the same time. This will help you see that things are being created properly while sketching extrusions in plan.

18. Create the ***Section*** as shown in Figure 13-1.8.

19. Close all views except the Level 1 Plan and the new *Section* view created in the previous step.

20. On the keyboard type **WT** (Figure 13-1.9).

FIGURE 13-1.8 New Section created

FIGURE 13-1.9 Tiled views

To give you a sense of what you will be creating the image below is provided.

FIGURE 13-1.10 Reception desk to be created in this exercise (3D view with Section Box applied)

FIGURE 13-1.11 Reception desk section with dimensions

You will use the *Model In-Place* tool to create the reception desk. The steps are very similar to those used to create the custom countertop previously. The entire desk will be created in a single *In-Place* family. When the family is finished, selecting any part of the desk will select the entire desk. This means once you enter *Sketch* mode for the desk, you will create all the solids and assign materials before completing the *Sketch* mode. You can exit *Sketch* mode any time and then return to make additional modifications.

The section to the left shows the dimensions you will refer to when creating the desk (Figure 13-1.11). It is made with plastic laminate, solid surface tops and 1½″ x 3″ solid maple trim on the front.

21. Select **Component → Model In-Place** from the *Architecture* tab.

22. Set the *Category* to **Casework**.

23. Enter the *Name*: **Reception Desk**.

You will create several extrusions, starting at the floor, and working your way up. The first extrusion will be the 3″ wide by 4″ tall toe space (Figure 13-1.11).

24. Select ***Extrusion*** from the *Architecture* tab.

25. Set the *Extrusion* constraints (Figure 13-1.12):

 a. *Start:* **0′ 0″**

 b. *End:* **0′ 4″**

26. Sketch the 3″ wide outline, 2″ back from *Reference Plane* as shown in Figure 13-1.13. Note the return, on the East side, aligns with the wall; use *Trim* and *Offset*.

FIGURE 13-1.12 Extrusion start and end

27. Click the **green check mark** to finish the extrusion, but do not finish the *In-Place* family. This would be done by clicking the *Finish Model* button which also has a green check mark.

> ***TIP:*** *If you finish the* In-Place *model and need to get back into* Sketch *mode, select any part of the desk and then click* **Edit In-Place** *from the* Ribbon.

70 SF

 FIGURE 13-1.13 First extrusion on floor – toe space

Notice in the section view that the 4″ tall block appears after finishing the extrusion.

28. In the plan view, select ***Extrusion*** to create another 3D solid.

29. Set the *Extrusion* constraints in the *Properties Palette*:

 a. *Start:* **0′ 4″**

 b. *End:* **2′ 5″**

30. Sketch the **5″** wide privacy wall (Figure 13-1.14).

 a. The back side aligns with the back side of the toe base.

 b. The main front side aligns with the *Reference Plane* previously added.

FIGURE 13-1.14 Second extrusion – privacy wall

31. Finish the extrusion via the **green check mark**.

Again, notice the section view continues to update (Figure 13-1.15). Next you will create the thickened portion of the front edge of the main work surface. This and the work surface will be *Joined* to look monolithic.

32. Create another **Extrusion** for the following:

 a. *Start:* **2′ 3″**

 b. *End:* **2′ 5″**

 c. **4½″** wide in plan view

 d. Sketch per Figure 13-1.16.

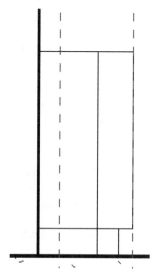

FIGURE 13-1.15 Section view of reception desk

A portion of the counter will extend along the wall (Figure 13-1.10) to the east exterior wall.

33. Finish the *Extrusion*.

FIGURE 13-1.16 Thickened front edge of countertop

Now you will create the main countertop/work surface at 30″ above the floor.

34. Create another **_Extrusion_** for the following:

 a. _Start:_ **2' 5"**

 b. _End:_ **2' 6"**

 c. **3' 0"** wide in plan view

 d. Sketch per Figure 13-1.17 (do not add dimension)

FIGURE 13-1.17 Main work surface / countertop

35. Finish the _Extrusion_.

Looking at the section view, you can see a line between the thickened edge and the work surface. You will use _Join_ to change that.

36. In the section view, use the **_Join_** tool (_Modify_ tab).

 a. Select the work surface.

 b. And then the thickened edge.

 c. See Figure 13-1.18.

FIGURE 13-1.18 Before and after Join

You will add a transaction surface at 42" above the floor. This is a convenient surface for clients/public to fill out a form or sign a document without bending over.

37. Create another **_Extrusion_** for the following:

 a. _Start:_ **2' 6"**

 b. _End:_ **3'-5"**

 c. **5"** wide in plan view

 d. Sketch per Figure 13-1.19 (do not add dimensions)

FIGURE 13-1.19 Transaction wall element

If you want to see the transaction wall in section, you may have to move your section mark in the plan view so it cuts through it. You first need to finish the *In-Place* model to move the section.

38. Create another **Extrusion** for the following:

 a. *Start:* **3' 5"**

 b. *End:* **3' 6"**

 c. Width varies; see Figure 13-1.20 (do not add dimensions).

FIGURE 13-1.20 Transaction surface

39. Finish the *Extrusion*.

You have the main parts of the reception desk modeled. Next you will add the wood trim to the front of the desk. This will be done in elevation. In this case you will use the *Extrusion* tool to create solids on the face of the *Reference Plane*, which will extrude horizontally rather than vertically, as has been done thus far via floor plan view. Drawing the trim in elevation will be easier because there are several pieces stacked on top of each other and they are not continuous.

In order to create an elevation or section view of the front of the desk, we need to finish the *In-Place* model.

40. Click **Finish Model** from the *Ribbon*.

41. **Save** your project.

42. Create a **Section** as shown in Figure 13-1.21.

43. Switch to the new *Section* view; zoom into the reception desk.

44. Select the reception desk and click **Edit In-Place** from the *Ribbon*.

You are now ready to model the 1½" thick wood trim on the face of the desk. The *Reference Plane* will be very helpful here.

FIGURE 13-1.21 Section added to view/edit front of reception desk

45. In the section view looking at the front of the desk, select the **Extrusion** tool.

Because Revit does not know where to start and stop an extrusion in elevation or section views, you are prompted to specify a *Work Plane*. You will select your *Reference Plane* from a list.

FIGURE 13-1.22 Selecting a work surface for extrusion

46. Select *Recept Desk Front* as shown in Figure 13-1.22; click **OK**.

47. Create the Extrusion:

 a. *Start:* **0′ 0″**

 b. *End:* **0′ 1½″**

 c. Trim is **3″** wide in elevation

 d. See Figures 13-1.11 and 13-1.23.

FIGURE 13-1.23 Sketching trim on face of desk

The first row of trim is continuous, but several rows will not be for aesthetic affect. Each piece of trim will be a separate extrusion within the *In-Place* family. It is possible to create multiple pieces of trim within the save extrusion but making edits later gets a bit more difficult.

48. Finish the *Extrusion* if you have not done so already, but stay in the *In-Place* family.

49. Select the trim element just drawn.

50. **Copy** it up **4″** (3″ trim plus 1″ space will properly position the next row).

Notice how the selected extrusion has grips on each side (Figure 13-1.24). These can be modified without needing to get back into *Sketch* mode for the extrusion. So now you can *Copy* and drag the end grips around to quickly create the remaining pieces of trim.

FIGURE 13-1.24 Wood trim added to face of desk

51. Using *Copy* and the edge **grips**, create the trim as shown in Figures 13-1.11 and 13-1.25.

FIGURE 13-1.25 Wood trim added to face of desk

Notice how the left edge of the trim is not able to die-into the angled stair wall as the desk itself was able to do. This is due to the fact that the extrusion only extends perpendicular to the view in which the sketch was created. The trim could have been extended into the angled wall and then a **Void** could have been added to cut the trim at an angle (Figure 13-1.26).

FIGURE 13-1.26 Example of void cutting solid

*TIP: Voids only cut solids created before they were; **otherwise you have to use the** Cut tool on the Modify tab.*

52. Using several of the techniques just covered, add the trim to the East side return as shown in Figure 13-1.27.

 a. Create a *Reference Plane* in your Floor Plan view.

 b. Name the *Reference Plane*.

 c. Create a section looking West.

 d. Create one extrusion and then copy it up.

53. In the front section view of the desk, use the *Join* tool and select the trim boards which meet at the outside corner.

The reception desk is almost done! The last step is to assign *Materials* for the wood (WD), plastic laminate (PLAM) and solid surface (SSURF).

FYI: Solid surface is a generic name for Corian or similar.

FIGURE 13-1.27 Wood trim returns

54. Select the desk and enter *Edit* mode if needed.

55. Select one of the wood trim boards.

56. Click in the *Material* cell in the *Properties Palette*. Click the button to open the **Materials** dialog.

This will assign the *Material* for just the selected element. You will create a *Material* specifically for the reception desk and then apply it to the other trim boards. All the reception desk materials will begin with "Recpt Desk – " to keep them together.

57. In the *Material Browser* search for **Maple**; double-click **Maple - Semigloss** in the *Autodesk Library*.

58. Right-click the name, just loaded into the project, and select *Duplicate*.

59. Enter the *Material* name: **Recpt Desk - WD1 - Maple Trim**.

60. On the *Identity* tab, enter **WD1** for the *Mark*.

61. On the *Graphics* tab, check **Use Render Appearance** for *Shading*.

62. On the *Appearance* tab, notice the Roughness map controls how shiny the surface looks (semigloss in this case); this map is light gray whereas the glossy version is a darker gray for the roughness map.

63. Click **OK**.

64. Select the remaining wood trim elements, all at once, and then set the *Material* to **Recpt Desk – WD1 – Maple Trim**.

Now you will repeat these steps for the remaining elements.

65. Create the following *Materials* (remember to duplicate the asset):

 a. **Recpt Desk – PLAM1 – Black**
 i. Search for **Plastic**, load **Plastic, Opaque Black** from *Autodesk Library*
 ii. Mark: **PLAM1**
 iii. *Check:* **Use Render Appearance** for *Shading*

 b. **Recpt Desk – SSURF1 – Gold**
 i. Copy from existing *Material*: "Recpt Desk – PLAM1 – Black"
 ii. *Mark:* **SSURF1**
 iii. *Check:* **Use Render Appearance** for *Shading*
 iv. *Appearance:*
 • **Duplicate** *and* **Rename**
 • Type: **Plastic (Solid)**
 • Color: **RGB 137, 107, 0**
 • Finish: **Matte**

66. Assign **PLAM1** to the two vertical elements behind the wood trim.

67. Assign **SSURF1** to the two horizontal surfaces, the transaction and work surfaces, and to the toe space; this will be the wall base color.

68. Finish the *In-Place* model and then **Save** your project.

In the elevation or section view of the front of the desk, set the *Visual Style* to **Consistent Colors** to see the *Material* colors.

Exercise 13-2:
Lobby Wall Sign

In this exercise you will create the sign for the law firm shown in the image below (Figure 13-2.1). The sign consists of a 1½″ thick glass panel held out from the wall a few inches by four corner buttons. You will use the *Model Text* tool to create the 3D letters.

FIGURE 13-2.1 Lobby wall sign to be created

The entire sign will be in its own *In-Place* family. First, you will create the glass panel utilizing the named *Reference Plane* previously created.

1. Switch to the section looking at the front of the reception desk.

2. Adjust the *Crop Region* if needed to see the entire wall East of the reception desk.

3. Select the **Model In-Place** tool.

4. Select **Specialty Equipment** to set the *Category*, and then enter the name **Lobby Wall Sign**.

5. Create the **Extrusion** for the glass panel:

 a. *Work Plane:* **Recpt Desk Sign Wall**

 b. *Start:* **0′ 0″**

 c. *End:* **0′ 1½″**

 d. See Figure 13-2.2.

FIGURE 13-2.2 Glass panel extrusion

Remember this extrusion is relative to the **Reference Plane** drawn earlier in this chapter. Recall that it is out from the wall 3″ (refer back to Figure 13-1.6). Before finishing the glass panel extrusion, you will assign the *Material* now.

6. Via the *Properties Palette*, open the **Materials** dialog.

7. Load **Glass, Frosted** from the *Autodesk Library*; select the one with an orange triangle in the preview.

8. *Duplicate* it, with the name **Recpt Desk – Art Glass**.

 a. Change the *Custom Color* to **RGB 58 200 172**

 b. Set the *Roughness* slider to about **0.75**, for a frosted glass appearance.

9. Check *Use Render Appearance* for *Shading* on the *Graphics* tab.

10. Click **OK** to close the *Materials* dialog.

11. Select the **green checkmark**, on the *Ribbon*, to finish the *Extrusion*.

Next you will create the four button-type connectors, one at each corner.

12. While still in the *In-Place* family *Edit* mode, **zoom in** to the lower left corner of the glass panel.

13. Create the ***Extrusion*** for the connectors:

FIGURE 13-2.3 Circle extrusion

 a. *Work Plane:* **Recpt Desk Sign Wall**

 b. *Start:* **-0′ 0 ½″ (note the minus sign)**

 c. *End:* **0′ 3″**

 d. Use ***Circle*** draw option; dia. of **2"**.

 e. See Figure 13-2.3.

 TIP: Draw circle at corner; move right 4″ and then move up 4″.

14. Set the *Material* to **Recpt Desk - PLAM1 – Black**.

 FYI: We will not be tracking the color and material for this connector so we will just map it to an adjacent color. You certainly could make a new material for this if you wanted to do so.

15. Finish the extrusion.

The minus ½″ for the start point causes the extrusion to begin ½″ behind the *Reference Plane*, i.e., out in front of the glass panel ½″. Keep in mind this is all based on how the *Reference Plane* was drawn in plan: left to right, or right to left.

Now you can copy/mirror the connector to the other three corners.

16. Use ***Copy*** and/or ***Mirror*** to create the other three connectors (Figure 13-2.4).

FIGURE 13-2.4 Four glass panel connectors properly positioned

The next step is to create the 3D letters. Most text added in Revit is 2D and does not show up in camera views, which also means it does not appear in renderings. Revit provides the **Model Text** tool to overcome this issue. This text entity is similar to the general *Text* tool in that you can create various heights and specify a specific font.

Unfortunately, the *Model Text* tool cannot be used in the *In-Place* family mode; notice the tool is grayed out on the *Architecture* tab.

17. Select **Finish Model** on the *Ribbon*.

18. Select **Architecture → Model Text** from the *Ribbon*.

Model Text

19. Select *Work Plane:* **Recpt Desk Sign Wall**.

20. Type **GREENE** (Figure 13-2.5).

21. Click OK.

22. Click somewhere on the glass panel to place the text.

Based on the default *Model Text* style, we see the text is way too large (Figure 13-2.6). We simply need to create a new *Type*.

23. Select the *Model Text* element in the model.

FIGURE 13-2.5 Enter text to appear 3D

24. Set the *Depth* to **1″** in *Properties Palette*.

25. Click **Edit Type**.

26. Pick **Duplicate**.

27. *Name:* **Lobby Sign – Large Letters**.

28. *Text Height:* **8″**

29. *Text Font:* **Felix Titling** (or similar).

FIGURE 13-2.6 Text placed but too large

30. Click **OK** to update the selected text.

31. **Move** the *Model Text* element up and to the right.

32. **Copy** the *Model Text* down **11″**.

33. **Copy** the second row of text down **1′-5″** to create a third row.

34. Select the third row and pick the **Edit** text option in the *Properties Palette* (Figure 13-2.7).

35. Change the text from GREENE to **VERT**.

36. Set the *Horizontal Alignment* to **Right** (Figure 13-2.7).

37. Click **OK**.

38. Create a second *Model Text* type:

 a. *Name:* **Lobby Sign – Small Letters**

 b. *Text Height:* **3½″**

 c. *Text font:* (same)

39. Add the word "**AND**" as shown in Figure 13-2.8.

FIGURE 13-2.7 Model text edit

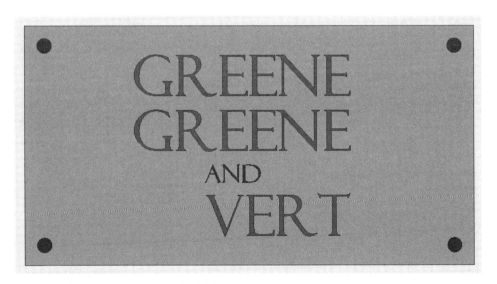

FIGURE 13-2.8 Final text layout for lobby sign

The last thing you will do is set the *Material* for the letters. The large letters will be gold and the small letters black.

40. Select the three large letter text elements and set their *Material* to **Recpt Desk - SSURF1 – Gold** via the *Properties Palette*.

41. Set the small text *Material* to **Recpt Desk - PLAM1 – Black**.

42. Create a *Camera* view looking from the main entry door into the lobby towards the reception desk and sign wall (Figure 13-2.1).

43. Adjust the **Crop Region** as needed.

44. Set the *Visual Style* to **Consistent Colors**.

45. Select one of the USG Geometric tiles, **right-click→Select all instances→In Entire Project**, and then set its *Material* to **Recpt Desk - SSURF1 – Gold**.

46. Type **VV**; uncheck *Specialty Equipment\Ceiling Graphics*.

47. **Save** your project.

As you can see in the three-dimensional *Camera* view, there are still a number of things which need to be cleaned up. Some soffits may still need to be added; the Level 2 *Railing* is missing and the material for the *Stairs* has not been set.

Exercise 13-3:
Controlling the Graphical Representation of Elements

This section will help Revit users understand the basics of how the graphical representation of various elements in Revit is controlled. The primary focus will be placed on controlling color, but line weight, line style and materials are controlled the same way. This discussion applies to all three "flavors" of Revit: Architecture, Structure and MEP. Be sure to look in the *Colored Images* folder, provided online, for several of the images using color in this section. **Nothing in this exercise needs to be applied to the *Law Office* model at this time.**

Project-wide Control

The project-wide settings of an element can be controlled in the ***Object Styles*** dialog, which is found on the *Manage* tab. In this dialog (Figure 13-3.1) you have access to the *Model* tab (i.e., things that actually get built) and the *Annotation* tab (i.e., notes, tags and dimensions). All the predefined categories can be found on these two tabs.

> *TIP: Check all options in the Filter list drop-down to see a complete list.*

FIGURE 13-3.1 Object Styles dialog; notice two door sub-categories changed to magenta.

Clicking the plus next to a category (e.g., *Doors*) reveals its subcategories. Notice how the two subcategories for the doors (*Panel* and *Plan Swing*) have been changed to **magenta** for this exercise; these subcategories represent the 2D line work displayed in a floor plan view for a door.

This exercise uses color to make the point more clear, as it is easier to discern than changing line weights. It is not recommended that categories be changed to color to match an AutoCAD layer standard. This will create printing issues for the entire design team.

Once a change has been made in the *Object Styles* dialog, all views in the project are instantly updated. Notice in Figure 13-3.2 all doors are magenta in color. The *Object Styles* dialog is the best place to go when you need to make a change to the entire project.

FIGURE 13-3.2 The project-wide graphical control of any element in Revit is controlled in the *Object Styles* dialog. Notice **all doors are magenta** due to the settings in Object Styles. This would be the case in any plan view in the project.

View Specific Control

On occasion, certain elements will need to be modified in a single view rather than at the global level just discussed. This type of change can be done in a view's **Visibility/ Graphic Overrides** (VG) dialog (Figure 13-3.3). Notice the door subcategories for the floor plan view have been changed to green (which will override the default magenta color). Any changes here will only affect the current view.

FIGURE 13-3.3 These settings are for the current view only and override the project-wide *Object Styles*. Notice the two door sub-categories are set to a green line color.

Once changes have been made in the view's *VG* the results will be seen immediately (Figure 13-3.4). This view now has green doors rather than magenta. Again, only this view is modified. Any other plan view will still have magenta doors (assuming they do not have *VG* modifications). This is true even for other views at the same Level (e.g., Level 1 Code Plan, Level 1 Finish Plan, Level 1 Area Plan, etc.).

FIGURE 13-3.4 The VG settings override the *Object Style* settings, making **all the doors green** rather than magenta in this view only.

View Specific Filters

Each view can have ***Filters*** applied to it. These are similar to *VG* settings in that they only apply to the current view. In fact, *Filters* are controlled in the *VG* dialog, the main difference being a more refined level of control of what elements are changed graphically. Rather than changing the color of an entire category (e.g., *Doors*), it is possible to make a change based on a specific characteristic about the doors (e.g., fire rating).

In Figure 13-3.5, notice the *Filter* dialog (a project-wide dialog) specifies the criteria for what is filtered and the *Filter* tab in *VG* specifies what to do when elements are found. In this example, any door which has a fire rating (i.e., a parameter set within the *Door* family) is changed to a color of red.

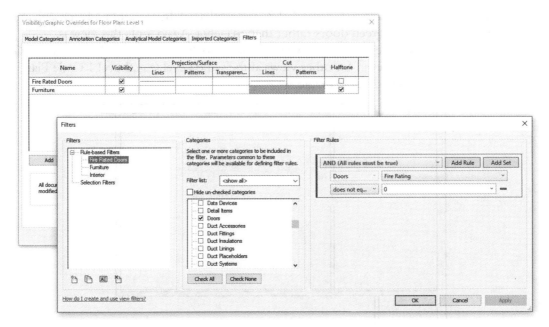

FIGURE 13-3.5 Similar to the *VG* settings, a *View Filter* changes the appearance of an element from the project-wide *Object Styles* settings for the current view only. Notice a filter is set up to change any door with a fire rating to be red.

The view shown in Figure 13-3.6 does not have any *VG* overrides other than the *Filters* shown in Figure 13-3.5. Notice that door number 1 changed to red because it has a one-hour fire rating. The other two doors do not have a fire rating so they remain magenta (based on the *Object Styles* settings). However, now that the filter is set up, any new or modified door that has a fire rating will automatically turn red.

FIGURE 13-3.6 Door number 1 has a one-hour fire rating set within its properties; the other two doors do not have a rating. Therefore, **door number 1 is red** and the **other two doors are magenta**, which is based on the *Object Style* settings. (Note this example does not have any VG overrides.)

Looking at Figure 13-3.7, one can see that a *Filter* cancels out any conflicting *VG* settings for a specific view. And, as we have already seen, the *VG* settings take precedence over the *Object Style* settings.

FIGURE 13-3.7 When both a *VG* and a *Filter* modification have been made to the same view, the *Filter* wins. Notice the *Object Styles* setting of magenta is not present at all. **Door number 1 is red** and the **other two doors are green**.

Phasing Related Overrides

When a project has phasing (e.g., *Existing, New, Future*, etc.), a view's **Phase Filter** setting can have a significant impact on how elements are represented in that view. Every [model] view has two phase related settings: *Phase* and *Phase Filter*. An example of these settings can be seen in Figure 13-3.8 below.

In our first phasing example, the plan view will contain all existing elements and no *VG* or *View Filter* modifications. The goal is to see what happens when a *Phase Filter* is imposed on a basic view and the project's *Object Style* settings.

Phase Filters are used to change the appearance of existing elements, so they are visually discernible from any new elements in the same view. Similarly, these filters can completely hide elements. For example, a proposed floor plan can hide all walls set to be demolished so the view is not so cluttered.

Notice the *Phase Filter* is set to **Show All** in Figure 13-3.8 for the plan view under consideration. What are the settings for this filter? These settings can be found in the *Phasing* dialog (via the *Manage* tab). This is a project-wide dialog, not view specific. Notice, again in Figure 13-3.8, that the *Phase Filter* uses two of the three possibilities: *By Category* and *Overridden* (*Not Displayed* being the third option). Anything existing will be overridden when the *Show All* filter is applied to a view. So, what are the overrides?

FIGURE 13-3.8 When a view has a *Phase Filter* applied to it, you have to pay attention to what the overrides are in the *Phasing* dialog. Notice, the view properties for the view we have been working with show a *Phase Filter* set to *Show All*; the *Phasing* dialog shows that anything existing in that view will be overridden.

The last tab in the *Phasing* dialog reveals exactly what the override is for a specific view. As can be seen in Figure 13-3.9, the overrides for *Existing* are gray lines.

FIGURE 13-3.9 This tab shows how anything in a specific view, which is *Existing*, will be overridden.

The impact of the *Show All* Phase Filter can be seen in Figure 13-3.10; remember – everything in this view is *Existing*. Comparing this plan to Figure 13-3.2 shows that everything is being controlled by the *Phase Filter* override. Both the black and magenta line work is now gray.

FIGURE 13-3.10 The phasing graphic overrides take precedence over everything in *Object Styles*. Note this view does not have any *VG* or *View Filter* overrides.

The final example is to see what happens when the prevision plan view has both *VG* and *View Filters* applied in addition to the *Phase Filter*. What will win – magenta, green, red or gray?

And the winners are... gray and red (Figure 13-3.11)! These are both *Filters*. Of course, the *View Filter* is the ultimate winner as it is overriding anything the *Phase Filter* would otherwise be doing to door number 1.

FIGURE 13-3.11 The phasing graphic overrides take precedence over everything in *Object Styles* and *VG*. However, *View Filters* still rule the view.

Conclusion

So the hierarchy for controlling the graphics of a view is as follows:

- View Filter *- View specific setting*
- Phase Filter *- Project-wide setting applied to a specific view*
- Visibility/Graphics Override (VG) *- View specific setting*
- Object Styles *- Project-wide setting*

There are a few other things that control how elements are displayed in a view, such as element specific overrides, linking, and variations on how content is created. Additionally, a view's *Discipline* setting affects how the architect's elements are displayed when set to *Mechanical* or *Electrical*. However, those will not be covered at this time.

Below is a more detailed list, including some features not covered in this discussion. This list was presented by an Autodesk employee, Ryan Duell, at Autodesk University 2011. The session was *Troubleshooting Autodesk Revit Architecture Using a Support Perspective*, AB6400.

- **Linework** tool *- View specific setting*
- Override Graphics in View → By Element → **Halftone** *- View specific setting*
- Graphics Display Options → **Silhouette** *- View specific setting*
- Override Graphics in View → **By Element** *- View specific setting*
- **View Filter** *- View specific setting*
- **Phase Filter** *- View specific setting*
- **Cut Line Styles** *- View specific setting*
- **VG Overrides** *- View specific setting*
- **Object Styles** *- Project-wide setting*

Self-Exam:

The following questions can be used as a way to check your knowledge of this lesson. The answers can be found at the bottom of this page.

1. It is important to name *Reference Planes* which are used to create extrusions from. (T/F)

2. Use the *Join* tool to make two or more extrusions appear monolithic. (T/F)

3. *Voids* only cut solids created after the void was created. (T/F)

4. Use the _____ tool to create 3D text in your model.

5. *Model Text* can be created within an *In-Place* family. (T/F)

Review Questions:

The following questions may be assigned by your instructor as a way to assess your knowledge of this section. Your instructor has the answers to the review questions.

1. An extrusion can be copied within an *In-Place* family. (T/F)

2. When creating an *In-Place* family you need to select a category such as *Casework* or *Specialty Equipment*. (T/F)

3. You used the _____ to make the wall conform to the underside of the stairs.

4. You can use a *View Filter* to highlight all fire rated doors. (T/F)

5. Materials cannot be added to solids within the *In-Place Edit* mode. (T/F)

6. What is the size of your Revit file? _____ MB

7. The entire reception desk was created in a single *In-Place* family. (T/F)

8. _____ is the highest level, project-wide control for an element's line weight, color and material.

9. The wall sign was created in a separate *In-Place* family. (T/F)

10. An *Object Style* setting overrides a *View Filter* setting. (T/F)

Notes:

Lesson 14
Furniture, Groups and Design Options:

This chapter will explore the process of accessing and placing furniture in your model using Autodesk® Revit®. We will discuss the huge benefits of manufacturer content and a few of the problems to overcome related to *Shared Parameters*.

Exercise 14-1:
Furniture Content

As you have already seen, there is a lot of content available with the Revit software and online in general. This is also true for furniture. There are a number of furniture options available for use in the BIM process. Autodesk provides several generic *Families* with the software to help get things started, especially for those just learning Revit. Finding all the content you require can take some time. Keep in mind that some content is not available at all – you have to make it yourself.

There are also a few user-to-user sites set up to share content. One example is www.RevitCity.com. This site allows anyone to upload content they have created. Anyone who has signed up for a free account on this website can then download any of the content for free.

Any of the above-mentioned resources have good and bad things about them. Some content is too simple, some is the wrong size, and others are too complex and can seriously degrade overall model performance. When a toilet partition family, for example, is the wrong size, there could be a code violation; not catching this until the room is built could cost a lot to fix. It takes experience to understand what is decent, good or bad. So users be warned.

In this chapter you will mainly be using *Steelcase*® furniture and their *Coalesse*® brand. However, another great resource for Revit content is *Haworth*®.

To help streamline things in this chapter, all the content you will use has been provided with the online files (with permission from Steelcase). This will save time downloading ZIP files for an entire series of a product and then extracting the correct file. Most of the files have not been changed in any way from what you would download with the following exceptions:

- The files have the *Type Mark* added to the beginning of the name.

- The desk assembly (i.e., the work surface, file storage, legs, etc.) has been saved as a group, and the main work surface has the *Type Mark* added to it.

When viewing one of the Steelcase websites, if a particular item has a Revit family, you will see a link as shown in the image below (Figure 14-1.1).

FIGURE 14-1.1 Furniture item has an associated Revit family available

We will now review a few of the common options one can expect to find when using the Steelcase content. The options can vary quite a bit from one family to another in order to control and select the various options and accessories available.

When looking at the *Instance Parameters* for a specific family you will see **Opaque** and **Draw in Foreground** toggles (Figure 14-1.2). These are pretty easy to understand. If you want to see the storage under a work surface or the legs under a conference room table, you uncheck *Opaque*. Or if you want a chair to look like it is tucked under a table partially, you would uncheck *Draw in Foreground*. The two examples in Figure 14-1.3 show the difference between *Draw in Foreground* toggled on versus off.

FIGURE 14-1.2 Instance parameters

One problem you may run into if you turn *Draw in Foreground off* is when you have solid elements above the floor, as we do with our ⅛″ floor finish (Figure 14-1.4). You have to weigh the pros and cons of which options you employ. In this tutorial you will have *Draw in Foreground* turned *off* for the chairs and not have the finish floor *on* in the same view as the chairs.

FIGURE 14-1.3 *Draw in Foreground*, for chairs, **on** (left) and **off** (right)

Looking at Figure 14-1.5 we see the *Family* comes with a lot of pre-populated information about the specific product. This is part of the "I" in BIM.

Looking at the *Instance* and *Type Parameters* you notice the location to set the *Materials* are not all in one location. But this is intentional. In this example, the legs would likely be the same for every chair of this *Type*, but the upholstery can vary for each chair (even at the same table), therefore making the upholstery an *Instance Parameter* that does not require a new *Family Type* to be created for each material option when the entire chair is otherwise the same.

FIGURE 14-1.4 *Draw in Foreground* problem (right)

Notice the *Lowback* and *Highback* options; Figures 14-1.6 and 14-1.7 show the results of changing this setting.

Some *Parameters* are built into the software and cannot be deleted. That appears to be the case for the *Width Parameter* as changing this does nothing visible to the model, which is good because this chair probably does not come in various widths.

FIGURE 14-1.5 Type Parameters for a chair

Also, notice the *Authoring Version* says Revit Architecture 2009. Content providers will often maintain content in an older version seeing as content cannot be saved backwards. This reduces the number of updates required, not having to update content for multiple versions.

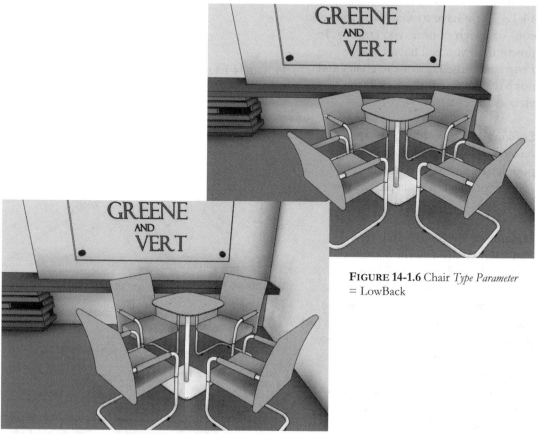

FIGURE 14-1.6 Chair *Type Parameter* = LowBack

FIGURE 14-1.7 Chair *Type Parameter* = HighBack

Some content is simply a single family like the chair we have been looking at. And some is made up of two or more elements. For example, some conference tables have one family for the top and another for the legs/support. Also, workstations have so many options you have to open a special file and piece together your exact configuration needed (Figure 14-1.8). This allows for maximum flexibility.

FIGURE 14-1.8 Workstation parts/resource files

Once the desired content is downloaded from the internet and extracted from the compressed ZIP file, you can easily browse the available content initially using *Windows Explorer* in many cases. Simply set the view setting to a *Small, Medium, Large* or *Extra Large* icon (Figure 14-1.9). You may also type something like "credenza" in the search box and quickly locate what you need.

Once the desired item is found, you can drag it into your open Revit floor plan to quickly load and place it in your model.

> ***FYI:*** *You can view the contents of a ZIP file via Windows Explorer, but you cannot drag a Family from a ZIP file into a Revit project.*

FIGURE 14-1.9 Previewing Revit content via *Windows Explorer*

You will be adding a *Type Mark* value to each family used in the project and then tagging that information in your furniture floor plans.

Typically, *Type Marks* have a common prefix for each type of furniture. For example, "CH" for chair and "TBL" for table. An example of this is shown in the image to the right (Figure 14-1.10).

FIGURE 14-1.10 Furniture tags added

Exercise 14-2:
Furniture Placement

Introduction

In this exercise you will place furniture throughout the building. You will lay out several types of rooms to give you exposure to various design techniques.

A client benefits from an experienced architect or interior designer. Many factors must be taken into consideration when designing a particular space. The following is just a few examples of design considerations related to space layout:

- Accessibility access and clearances
 - If furniture layouts are not done early enough, the room might need to grow in size to accommodate the program and accessibility codes. This type of revision can throw off estimates and create problems for adjacent spaces.

- Sightlines (Is tiered seating or privacy required?)

- Furniture required (desk, tables or simply chairs)
 - An experienced designer will take the time to understand the program and the intended use of the space to make sure all needs are met, sometimes suggesting staffing or procedural ideas that may improve the client's efficiency.

- Light switch locations (if room darkening is required often)

- Window treatment (again, if room darkening is required)
 - If unit is concealed in wall/ceiling, special details are required
 - Do exterior windows create excessive glare at certain times?

- General circulation within the space
 - Are people coming and going at the same time?
 - Does the space require two exits due to its size?

- LEED© and Sustainable Design
 - Daylighting requirements/opportunities.
 - Functional materials with recycled content and low/no VOCs.
 - LEED© credit tracking (must be thought about continuously)

The list above is, again, just a sample of the thought process that a designer employs when developing a space.

Loading Content

You will now load all of the content needed for this exercise. All the content used, and much more, can be found on the Steelcase website. However, because there is so much content and the specific item to be used may be found in a ZIP file amongst many other similar

items, we have placed the needed items online. A lot of the families will have multiple *Types* within them, which will require you to pick the correct options from those available.

1. **Open** your *Law Office* project.

2. Select ***Load Family*** from the *Insert* tab.

3. Browse to the **Furniture** folder saved from the provided online files for this book.

You can load multiple families at once. This is done by selecting the first file, and then while holding the *Shift* key you select the last file.

4. Select all the online files, saved to your computer, listed in the *Furniture* folder.

5. Click **Open** to load the content into the current project.

Your list of available items in the *Type Selector*, for the *Component* tool, just got a lot longer! As it turns out, due to the quantity and complexity of the content, the size of your Revit model just about doubled, or it will once you save.

The schedule below shows a list of all the content you will be using in this exercise (Figure 14-2.1). You will learn how to create a schedule like this later in the book. Information like the total quantity of items placed in the model can be listed. This is a live number and will automatically update with any changes made to the model.

		<Furniture Schedule - Type Mark>		
A	**B**	**C**	**D**	
Type Mark	Family	Type	Manufacturer	
CH1	CH1 Steelcase - Seating - Leap 462 Series - Stool	Arm	Steelcase	Seating
CH1a	CH1a Steelcase - Seating - Leap 462 Series - Work Chair	Arm and Headrest	Steelcase	Seating
CH2	CH2 Steelcase Coalesse - Kathryn - Seating - Guest	Fully Upholstered Back - Round Arm	Steelcase	Coalesse
CH3	CH3 Steelcase Coalesse - Switch - Seating - Sledbase Chair	Lowback - Polished Chrome Leg	Steelcase	Coalesse
CH4	CH4 Steelcase Coalesse - Bindu - Seating - Conference - Chair	Mid Back	Steelcase	Coalesse
CH5	CH5 Steelcase Coalesse - Coupe Grande - Seating	Chair	Steelcase	Coalesse
CH6	CH6 Steelcase Coalesse - Donovan - Seating	Two Seat Sofa	Steelcase	Coalesse
CH7	CH7 Steelcase Coalesse - Enea - Seating - Cafe Stool	Bar - Wood Seat	Steelcase	Coalesse
CH8	CH8 Steelcase Coalesse - Kart - Seating - Stack	Non Upholstered - Glide	Steelcase	Coalesse
DK1	Steelcase - Elective Elements 6 - Worksurface - Straight - 30" D	96" W	Steelcase	Elective
DK2	Steelcase Coalesse - Topo - Desktop - Straight with Extension	30" - 82"D - Right - Flat Edge	Steelcase	Coalesse
DK3	Steelcase - Elective Elements 6 - Worksurface - Desk - Straight	24" D x 90" W	Steelcase	Elective E
DK4	Steelcase Coalesse - Topo - Desktop - Straight with Extension1	24" - 70"D - Left - Hardwood Edge	Steelcase	Coalesse
TBL1	TBL1 Steelcase Coalesse - Denizen - Table - Round	Touch Latch Power Access - 27" H	Steelcase	Coalesse
TBL2	TBL2a Steelcase - Convene - Conference Table Top - Racetrack	Flat Profile - 60" D x 216" W	Steelcase	Convene
TBL3	TBL3 Coalesse_Host_Boat_8FT_FLARED	96" x 54" (2 TAPS)	COALESSE	
TBL4a	TBL4 Steelcase Coalesse - Enea Lottus - Table - Cafe	Counter Height - 24" W	Steelcase	Coalesse
TBL4b	TBL4 Steelcase Coalesse - Enea Lottus - Table - Cafe	Bar Height - 30" W	Steelcase	Coalesse
TBL5	TBL5 Steelcase Coalesse - Enea - Table - Single Base	Cafe - Bar Height - 24" Dia	Steelcase	Coalesse
TBL6	TBL6 Steelcase Coalesse - Calm - Table - Square	20" D x 20" W x 22" H	Steelcase	Coalesse
TBL7	TBL7 Steelcase Coalesse - Calm - Table - Rectangle	20" D x 38" W x 17" H	Steelcase	Coalesse
TBL8	Steelcase Coalesse - Andoo - Table - Dinning	Short Fixed Top	Steelcase	Coalesse
TBL9	TBL9 Steelcase Coalesse - Sidewalk - Table - Occasional	22" D x 22" W x 22" H	Steelcase	Coalesse
TBL10	TBL10 Coalesse_Akira_Rectangle_Fixed_20T	72" x 20" x 28 1/2"	COALESSE	

FIGURE 14-2.1 Project schedule showing furniture used in this exercise

Furniture Plans and View Templates

Next you will create a furniture plan for each level. This will be a simple, clean plan which focuses on the furniture and not column dimensions or section marks.

6. Right-click on the **Level 1 Floor Plan**.

7. Select **Duplicate View → Duplicate**.

By selecting this option, any view-specific notes, dimensions, tags, etc. will not be copied into the new view. All of these items are copied when "Duplicate with Detailing" is selected.

8. Rename the newly copied view to **LEVEL 1 FURNITURE PLAN**.

9. Repeat these steps to create **LEVEL 2 FURNITURE PLAN**.

> *TIP: Do not try to duplicate Level 1 and turn it into a Level 2 plan. That is not possible and will cause several visibility problems if you try. Rather, duplicate Level 2 Floor Plan.*

10. Do the following for the LEVEL 1 FURNITURE PLAN:

 a. *Detail Level:* **Coarse**

 b. Type **VV** and make these edits (Figure 14-2.2):

 i. Check *Halftone* for *Specialty Equipment*.

 ii. For the *Walls* category, set the *Cut Pattern* override to a dark gray color with a solid fill.

 iii. Turn off: *Grids, Elevations* and *Reference Planes* (*Annotation Categories* tab).

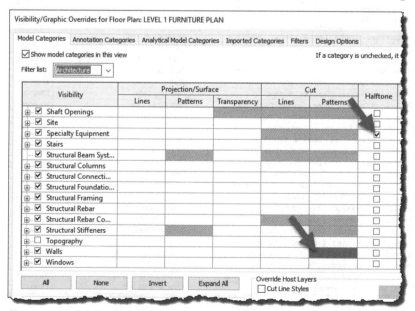

FIGURE 14-2.2 Adjusting the furniture plan view's graphics settings

Next you will create a **_View Template_** which can then be applied to the Level 2 Furniture Plan to quickly make its settings match Level 1.

11. In the _Project Browser_, right-click on the LEVEL 1 FURNITURE PLAN and select **_Create View Template from View_**.

12. Enter the name **Furniture Plans** and then click **OK**.

13. Notice all the settings which are saved for the new _View Template_ and then click **OK**.

Now you will apply the new _View Template_ to the LEVEL 2 FURNITURE PLAN. This is a onetime view settings update. Changing the _View Template_ later will not automatically update any views.

14. Right-click on the LEVEL 2 FURNITURE PLAN view and select **_Apply Template Properties_**.

15. Select **_Furniture Plans_** in the list on the right and then click **OK** (Figure 14-2.3).

Your grids, elevations and reference planes should now be gone. We made the _Specialty Equipment_ category _Halftone_ to make the furniture stand out when next to exercise equipment and other non-furniture items. We also set the walls to have a solid gray fill to force attention on the furniture rather than the wall construction. Using _View Templates_ is a great way to quickly make several views match. It can also be used to troubleshoot problem views.

FIGURE 14-2.3 Schedule listing furniture to be placed in this exercise

Notice, in the image above, that when applying a _View Template_ you can exclude certain things such as _View Scale_. This is good because you might have a few enlarged furniture plan views that require identical view settings except for the drawing scale.

Each view has an option to permanently assign a _View Template_ via the _Properties Palette_. When this is done, any adjustments made to the _View Template_ will automatically update the related views. Note in the image above, that no views are currently "assigned" to the _Furniture Plans_ template. In the previous steps you **applied** the properties, not **assigned** (or linked) them. When a _View Template_ is "assigned" to a view, the settings being controlled by the VT are greyed out; in this case you have to edit the VT via the view's properties.

Creating Groups

Next you place a table and four chairs in the lobby. Because this same arrangement will be used in multiple locations you will create a *Group*. This feature allows you to edit one *Group* and all other *Groups* in the model will be updated. When in *Edit Group* mode, you can add, remove and change elements. Not using a *Group* would require each location to be updated if the table needed to change from square to round, or if the type of chair changed. This requires you to remember where each configuration is located in the model. Not using *Groups* requires you to select five elements, in the case of our table and four chairs, rather than one group.

One thing that is really special about *Groups* is the ability to exclude elements without the need to **Ungroup**. When you select an element within a group, via tab and then selecting it, you can right-click and select **Exclude** from the pop-up menu. Doing this also excludes the element from schedules. Any changes, to a single *Group* instance, which require more than excluding elements need to be **Ungrouped** and possibly turned into a new *Group*.

16. In the *Lobby* on Level 1, place **CH2** and **TBL4**; refer back to Figure 14-2.1 for the specific *Type* to use for the selected family (Figure 14-2.4).

FIGURE 14-2.4 Place table and chairs in lobby

Now you will use the *Group* command to make the table, and chairs maintain the configuration you have chosen and allow you to easily place this configuration throughout your project.

17. Select the four chairs and the table.

18. Select the **Group** button from the *Ribbon*.

19. Name it **TBL4a** and click **OK**.

You now have a *Group* with five elements in it (Figure 14-2.5). The **Group Origin** icon in the center indicates the insertion point when placing future instances of the *Group*. The center will be fine for this example, but a typical office layout might want the origin to be in the corner of the office. You simply drag this icon to relocate the origin.

> **FYI:** *Moving this will not affect the position of any previously placed instances of that* Group.

This *Group* is listed in the *Project Browser* under *Groups/Model* near the bottom of the list.

FIGURE 14-2.5 Grouped furniture

Placing a Group

Now you will place two instances of this *Group* in the *Law Library*.

20. Zoom in on the *Law Library*.

21. From the *Architecture* tab, select **Model Group →** **Place Model Group**.

22. Select **TBL4a** from the *Type Selector*.

23. Place two instances of the *Group* in the law office approximately as shown in Figure 14-2.6.

FIGURE 14-2.6 Placing group in Law Library

Editing a Group

You now have three instances of the *Group* in your model. Next you will edit one of them and see how all three automatically update. You will change the chair to another type.

24. Select one of the three groups placed in your model.

25. Select **Edit Group** from the *Ribbon*.

You are now in *Edit Group* mode. While in this mode you will see this floating tool *Panel*. You can model new elements just as if you were not in *Group Edit Mode*. It is even possible to add walls and doors to a *Group* and they will clean up with walls not in the group. We will only be placing furniture in the *Groups* at this time.

While in *Group Edit* mode you will select the four chairs, swap them out via the *Type Selector* and then *Finish* the *Group*.

26. Select the four chairs.

27. Select **CH3**; see the table at the beginning of this exercise for the correct *Type*.

28. Click **Finish** to complete the *Group*.

Take a look at all three instances of the *Group*; they are all updated with the new chair. Using *Groups* can save a lot of time and reduce errors.

Entering the Type Mark

To set things up for tagging, we will add a *Type Mark* value to the chair and table in the *Group*. Because this is a *Type Mark* we only need to do this for one chair, and all the others will have this same value.

29. Edit the *Group* again.

30. Select one of the chairs.

31. Select **Edit Type** from the *Properties Palette*.

32. Set the *Type Mark* for the chair to **CH3** (Figure 14-2.7).

Parameter	Value	
Style Number(s)	1519PC	
Release Date	January 2011	
Product Line	Coalesse Seating / Occasional T	
Product	Switch	
Manufacturer	Steelcase	
Description		
Copyright	© 2010 Steelcase Inc.	
AutoCAD Tag	SWITCH	
Authoring Version	Revit Architecture 2009	
Assembly Code	E2020200	
Type Image		
Keynote		
Model		
Type Comments		
Cost		
Assembly Description	Furniture & Accessories	
Type Mark	CH3	
OmniClass Number		
OmniClass Title		
Code Name		

FIGURE 14-2.7 Entering type mark value

33. Set the *Type Mark* for the table to **TBL4a**.

34. **Finish** the *Group*.

Duplicating a Group

Anytime something needs to be different in a *Group*, you need a new *Group*. This configuration is very similar to one used in the *Lunch Room*. The table is the same model but taller and stool-type chairs are to be used. Next you will create a new group by copying this *Group* and editing it.

35. Right-click on the *Group* **TBL4a** in the *Project Browser* (near the bottom of the list under *Groups*).

36. Select **Duplicate**.

37. Right-click and **Rename** the new *Group* to **TBL4b**.

38. Place an instance of the *Group* in the *Lunch Room*.

39. Select the *Group*; click *Edit Group*.

40. Select the table and swap it out for the **Bar Height 30″ Wide** version of the same family.

41. Set the *Type Mark* for the table to **TBL4b**.

42. Select the four chairs and change them to **CH7** (Figure 14-2.8).

 a. *Move* the chairs and table back into position as needed.

FIGURE 14-2.8 TBL4b in *Lunch Room*

43. Set the *Type Mark* for the chairs to **CH7**.

44. Finish the *Group*.

Draw in Foreground Option

You will be instructed to create one more *Group*, but this time you will tuck the chair part way under the table. This involves adjusting the chair settings so it does not appear to be over the table.

45. Zoom in on one of the *Small Conference Rooms* at the center of the Level 1 Floor Plan.

46. Place **TBL1** and four **CH3** chairs as shown in Figure 14-2.9.

47. Select all four chairs and <u>uncheck</u> ***Draw in Foreground*** in the *Properties Palette.*

 TIP: Make sure your view is not set to wireframe.

 FYI: This is an Instance Parameter *so you need to do this for each chair as needed.*

48. Set the *Type Mark* values and create a *Group* called **TBL1**.

FIGURE 14-2.9 Small Conference Room

Keep in mind you can name the *Group* anything you like. It might be the department it is meant for, or the type of room it will be in rather than the type of table, especially if the chairs used vary.

Load as Group

To save time, the provided online files have four project files which contain complete desk configurations, well, nearly complete. This is a full-blown Revit project file and not a family file. Revit has a command that will load an entire project into your current project as a *Group*. You will do this for the desks.

49. Select **Load as Group** from the *Insert* tab on the *Ribbon*.

50. Browse to the *Furniture* folder from the provided online files for this book.

51. Load **DK1.RVT**.

FIGURE 14-2.10 Chairs now appear under table

FIGURE 14-2.11 Project file provided with online files

52. Repeat this for the other three desks.

53. Zoom into the *Visitor's Office* on Level 1.

54. Use the ***Place Model Group*** tool from the *Architecture* tab to place **DK2** as shown in Figure 14-2.12, along with the other furniture shown.

 a. *Rotate* and *Mirror* the group as needed.

 b. Do not edit the *Group*; the chairs and table can be by themselves.

 c. Tagging furniture will be covered in the next few steps.

 FYI: In the desk groups, only the main work surface has the Type Mark *set.*

FIGURE 14-2.12 Visitor's Office layout

Sometimes it is not appropriate to rotate *Groups* because the manufacturer you are working with might not make a reversed version of the item. In our case, we will assume this is not a problem.

Tagging Furniture

Next you will tag the furniture so the contractor can identify the various elements for pricing and placement within the building.

First you need to load the *Furniture Tag* from the library.

55. Select **Load Family** from the *Insert* tab.

56. Browse to **Annotations/Architectural**.

57. Select **Furniture Tag.rfa** and then click **Open**.

58. Select **Tag by Category** from the *Quick Access Toolbar*.

59. Set the *Type Selector* to **Standard**, and uncheck **Leader** from the *Options Bar*.

60. Zoom into the chairs and table in the *Lobby*.

61. Click on each element to add a tag.

As you can see, the *Type Mark* is the value being reported by this tag (Figure 14-2.13). The default tag location is the center of the item being tagged (when leader is turned off). When needed, you can select the tag and nudge it a bit with the arrow keys. Notice how Revit ignores the fact that the table and chairs are in a *Group* when using the *Tag by Category* command. This is exactly how you tag furniture not in a group.

FIGURE 14-2.13 Tags added to furniture

Adding the Rest of the Furniture

Now you will be presented several enlarged images showing the furniture locations throughout the law office, for both levels. You will add this content approximately as shown; edit the *Type Mark* and *Tag* each item.

62. Place the remaining furniture as shown in the remaining images in this exercise.

 a. The first two images are overall images for reference; use the images after them to read the item numbers.

 b. Rotate furniture as shown.

 TIP: Use the Space Bar *during placement.*

 c. Edit the *Type Mark* parameter for each family type.

 d. Add a **Tag** to each item; add only one for the main desk *Groups*.

e. The equipment used in the Exercise Room families, *Lifting Bench*, *Exercise Bike* and *Treadmill*, can be found with the download files for this book.

f. Remember, you can use the arrow keys to nudge the furniture into place.

g. Save often so you don't lose work.

h. Some of the images have been rotated to increase their size for more clarity.

FIGURE 14-2.14 Level 1 overall furniture plan – rotated image

FIGURE 14-2.15 Level 2 overall furniture plan – rotated image

FIGURE 14-2.16 Level 1 West side – partial view

FIGURE 14-2.17 Level 1 – partial view

FIGURE 14-2.18 Level 1 lunchroom

FIGURE 14-2.19 Level 1 lunchroom camera view with ambient shadows turned on

FIGURE 14-2.20 Level 1 lobby – rotated image

FIGURE 14-2.21 Level 2 West side – partial view

FIGURE 14-2.22 Level 2 – partial view

FIGURE 14-2.23 Level 2 – partial view East side – rotated image

Steelcase Coalesse Topo
Storage – Credenza and
Pedestal Tower

FIGURE 14-2.24
Level 2 – partial view

FIGURE 14-2.25 Level 1 overall 3D view

FIGURE 14-2.26 Level 2 overall 3D view

Exercise 14-3:
Design Options

Now you will explore a feature called *Design Options*. This feature allows you to present two or more options for a portion of your design without having to save a copy of your project and end up having to maintain more than one project until the preferred design option is selected.

This feature can be used during the early design phase of a project or to manage bidding alternates all the way through construction documents. *Design Options* are different from phasing, which manages changes on the model over time, in that it manages changes to the model within the same time frame.

Design Options Overview:

A Revit project can have several design option studies at any given time. You might have an (A) entry canopy options study, (B) an executive board room options study, and a (C) toilet room layout options study in a project. Each of these studies can have several design options associated with them. For example, the entry canopy study might have three options: 1) flat roof, 2) gable roof, and 3) sloped glass roof.

A particular study of an area within your project is called a *Design Option Set*, and the different designs associated with a *Design Options Set* are called *Options*. Both the *Design Options Set* and the *Options* can be named.

One of the *Options* in a *Design Option Set* is specified as the *Primary Option*; this is the option that is shown by default in all new and existing views. The other *Options* are called *Secondary Options*. However, you can adjust the *Visibility* of a *View* to show a different option. Typically, you would duplicate a *View*, adjust the *Visibility*, and *Rename* it to have each option at your fingertips.

When the preferred design is selected, by you or the client, you set that *Option* to *Primary*. Finally, you select a tool called *Accept Primary* which deletes the *Secondary Options* and the *Design Option Set*, leaving the *Primary Option* as part of the main building model.

Design Option Set

The following image (Figure 14-3.1) is an overview of the *Design Options* dialog:

FIGURE 14-3.1 Design Options dialog; "Modern Layout" Option selected

Notes about the *Design Options* dialog box:

Edit buttons:

You can edit a *Design Option* by selecting an *Option*, in the window area on the left, and then clicking the *Edit Selected* button. Next you add, move and delete elements in that *Design Option*.

When finished editing a *Design Option*, you reopen the *Design Options* dialog and click the *Finish Editing* button.

If you are currently in an *Option Editing* mode, the *Now Editing* area in the *Design Options* dialog displays the *Option* name being modified; otherwise it displays "Main Model."

Option Set buttons:

The *New* button is always available. You can quickly set up several *Option Sets*. Each time you create a new *Option Set*, Revit automatically creates a *Primary Option* named "Option 1."

The other buttons are only available when an *Option Set* label is highlighted, i.e., selected, in the list on the left.

The *Accept Primary* button causes the *Primary Option* of the selected *Set* to become a normal part of the building model and deletes the *Set* and *Secondary Options*. This is a way of "cleaning house" by getting rid of unnecessary information which helps to better manage the project and keeps the file size down.

Option buttons:

These buttons are only available when an *Option*, *Primary* or *Secondary*, is selected within an *Option Set*. You can quickly set up several *Options* without having to immediately add any content, i.e., walls, components, etc., to them.

The *Make Primary* button allows you to change the status of a *Secondary Option* to *Primary*. As previously mentioned, the *Primary Option* is the *Option* that is shown by default in previously created and new views.

The *Duplicate* button will copy all the elements in the selected *Option* into a new *Option*. This makes the file larger because you are technically adding additional content to the project. You can then use the copied elements, e.g., walls, furniture, etc., as a starting point for the next design option. This is handy if the various options are similar.

Now you will put this knowledge to use! In this exercise you will create one *Design Option Set*. This will be used to create an alternate furniture layout for the *Lobby* area.

Setting Up Design Options in Your Project:

First you will set up the *Option Sets* and *Options*.

1. Open your *Law Office* project.

2. Select the **Design Options** icon on the *Status Bar* (see image above).

You are now in the *Design Options* dialog box (Figure 14-3.2). Unless you have modified your file to have *Option Sets*, your dialog will look like this one.

3. In the *Option Set* area click **New**.

Notice an *Option Set* named <u>Option Set 1</u> has been created. Revit also automatically created the *Primary Option* named <u>Option 1</u> (Figure 14-3.3). Next you will rename the *Option Set* to something that is easier to recognize.

4. (See warning below.) Select the *Option Set* currently named <u>Option Set 1</u> and then click the **Rename** button in the *Option Set* area.

 WARNING! *Be sure you are not renaming the* **Option** *but, rather, the* **Option Set**.

5. In the *Rename* dialog type: **Lobby Furniture Options** (Figure 14-3.4).

6. Click **OK** to rename.

FIGURE 14-3.2 Design Options Dialog; initial view

FIGURE 14-3.3 Design Options dialog; new option set created

Giving the *Option Set* a name that is easy to recognize helps in managing the various options later, especially if you have several.

Next you will create a *Secondary Option* for the Lobby Furniture Options *Option Set*.

FIGURE 14-3.4 Rename Option Set Dialog

7. With the Lobby Furniture Options *Option Set* selected (or any option in that set), click **New** in the *Option* area.

Notice a secondary option was created and automatically named Option 2 (Figure 14-3.5). If you have descriptive names for the options in a set, you should apply them. In this example you can leave them as they are.

FIGURE 14-3.5 Option 2 created

The basic thinking with the *Design Options* feature is that you set up the *Option Set* and *Option*s and then start drawing the elements related to the current *Option*. So, you select the option you wish to edit via the *Design Options* drop-down list on the *Status Bar*, make the additions and modifications relative to that *Design Option*, and finally set the drop-down list back to "Main Model."

However, in your case, you want to move content already drawn to Option 1. Revit has a feature that allows you to move content to a *Design Option Set*, which means the content gets copied to each *Option* in the *Set* you select. This option will work for the *Lobby* because the second option will be similar to the first one.

8. Zoom in on the *Lobby* in the **LEVEL 1 FURNITURE PLAN** view.

9. Select all the furniture in the *Lobby* EXCEPT for the chair behind the reception desk. This includes 10 chairs, one of which is a couch, and three tables. Do not select the tags as they cannot be added to a *Design Option*.

 TIP: Be careful when dragging to select, so you don't select additional items or a Room element.

10. Select the **Add to Set** icon on the *Status Bar* (Figure 14-3.6).

FIGURE 14-3.6 Add to Set icon on status bar

11. Click **OK** (Figure 14-3.7).

The selected items are now in both *Option 1* and *Option 2* under the *Option Set: Lobby Furniture Options*.

From this point forward you can only modify the *Lobby* furniture when **in** *Option 1* or *Option 2* "edit mode" (in which case the tables are turned and you cannot edit the main building model; this is because "exclude options" is selected on the *Status Bar* as shown in Figure 14-3.6).

Do not add anything to the *Design Option* which does not need to change from one option to another. The reception desk chair will not change in either case, so we did not add.

FIGURE 14-3.7 Add to Set Dialog

12. Select *Lobby Furniture Options*: **Option 2** (i.e., *Option Set:* Lobby Furniture Options; *Option:* Option 2) from the *Design Options* list on the *Status Bar* (Figure 14-3.8).

FIGURE 14-3.8 Activating a design option edit mode

Now you should notice that the main building model is slightly grey and not editable. It is not editable because "Active Option Only" is selected on the *Status Bar*.

13. **Zoom** in on the furniture in the *Lobby*.

Now you will make changes to the furniture layout while in *Option 2* edit mode.

14. **Delete**, **Move**, **Copy** and **Rotate** the furniture as shown in Figure 14-3.9.

> *FYI: The tags disappeared because they are associated with the furniture in the* Primary Option, *which is* Option 1. *The furniture in* Option 2 *has not yet been tagged.*

FIGURE 14-3.9 Alternate lobby furniture composition

15. To finish editing the design option select **Main Model** from the *Design Options* list on the *Status Bar*.

It now appears like all your changes disappeared, right? Well, if you recall from the introduction to this exercise, the *Primary Option* is displayed by default for all new and existing views. So, when you finished editing *Lobby Furniture Options: Option 2* the LEVEL 1 FURNITURE FLOOR PLAN switched back to the *Primary Option*, which is currently set to *Option 1*.

Next you will create a new view and adjust its *Visibility* to display *Option 2* of the *Lobby Design Option Set*.

First you will create a duplicate copy of the view.

16. In the *Project Browser*, right-click on the **LEVEL 1 FURNITURE PLAN** label.

17. Select **Duplicate View → Duplicate** from the pop-up menu.

You now have a copy of the view named *Copy of LEVEL 1 FURNITURE PLAN*

18. Rename the new view to LEVEL 1 FURNITURE PLAN - LOBBY OPTION 2.

19. Switch to your new view (if required).

20. Type **VV** on the keyboard to open the *Visibility/Graphics Overrides* for the current view.

21. Click on the ***Design Options*** tab at the top of the dialog. Note that this tab did not exist until you created *Design Options* in the project.

22. Change the *Design Option* parameter to ***Option 2*** (Figure 14-3.10).

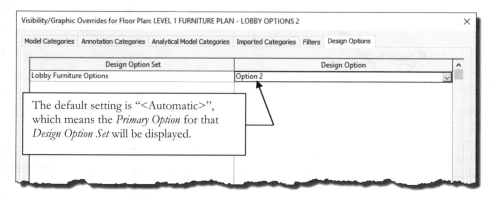

FIGURE 14-3.10 Visibility/Graphic Overrides dialog: modified design option visibility

23. Click **OK** to close the dialog.

24. Turn on **Cropping** and make the **Crop Region** visible via the *View Control Bar* at the bottom of the screen (Figure 14-3.11).

 FYI: Must not be in *Design Option* edit mode.

FIGURE 14-3.11 View Control Bar

25. Zoom out to see crop region (it will be outside of your building) and adjust it to crop the view as shown in Figure 14-3.12 below. You can turn off the crop region visibility so the line does not print.

26. Tag the furniture in this view.

FIGURE 14-3.12 Cropped floor plan view

Now, with the LEVEL 1 FURNITURE PLAN view and the <u>LEVEL 1 FURNITURE PLAN – LOBBY OPTION 2</u> view, you can quickly switch between design options. Both views could be placed on the same sheet and printed out for a design critique.

> *FYI: When a design option is "hard wired" in this way you cannot ever edit any other* Lobby Furniture Options *in this view as they are not visible in this view. This only works when the view is set to* Automatic *or to the specific option you want to edit.*

Remember, when you go to any other view in the project, or when new views are created, you will only see what is in <u>Option 1</u> because it is set to be *Primary*. If <u>Option 2</u> were set to be *Primary* in the *Design Options* dialog, the contents of <u>Option 2</u> will automatically appear in all views, assuming the *Design Option* is set to **Automatic** in the *VV* dialog.

When the camera view below was initially created, the <u>Option 1</u> furniture was showing. However, after typing *VV* and adjusting the *Design Options* tab to use <u>Option 2</u> in this view, the furniture changed.

FIGURE 14-3.13 Camera view with Design Option 2 selected via the VV dialog

This concludes the introduction to *Design Options*. Take a look at the *Help* system for more information on this topic.

Self-Exam:

The following questions can be used as a way to check your knowledge of this lesson. The answers can be found at the bottom of this page.

1. Most of the furniture content used in this chapter was provided by Steelcase. (T/F)

2. Some variations in a product can be accomplished within a single family; for example, high or low chair back. (T/F)

3. Use _____ to develop multiple furniture design ideas within the same space.

4. Use the _____ tool to insert another Revit file as a *Group*.

5. There is no limit to the number of *Options* in an *Option Set*. (T/F)

Review Questions:

The following questions may be assigned by your instructor as a way to assess your knowledge of this section. Your instructor has the answers to the review questions.

1. What is the size of your Revit project? _____MB.

2. When a furniture arrangement, e.g., chairs and tables, is used in several locations, you should use a _____.

3. Creating a _____ is an easy way to update multiple views at once.

4. Revit content cannot be saved backwards to an older version. (T/F)

5. A group cannot be mirrored. (T/F)

6. You can drag a family from a zip file into a Revit project. (T/F)

7. The furniture was tagged using the *Tag by Category* tool. (T/F)

8. The *Primary Option* appears by default in all views. (T/F)

9. Moving the *Group Origin* will cause all instances of that *Group* to move. (T/F)

10. It is not possible to duplicate a Level 1 view and change it to a Level 2 view. (T/F)

Lesson 15
Detailing:

This chapter will cover the ways details are created in Autodesk® Revit®.

Exercise 15-1:
Introduction to Detailing

In this lesson you will draw details, which are larger scale drawings of smaller areas used to convey more intricate information to the contractor.

Many design firms have detail libraries containing many of the types of details drawn in this section. This allows the designer to spend more time designing the "big picture" portions of the project and less time on the "typical" details. A designer can flip through the three-ring binder or (better yet) an intranet for details that apply to the project currently being worked on. The marked details can then be loaded into the Revit model by the designer or a technician.

Not every detail is standard. In fact, many items that simply come out of a box do not usually need to be detailed unless you need to show how it connects or relates to the building. For example, windows are manufactured in a factory, but they are still detailed to show how they are to be attached and aligned within the wall opening.

Some conditions within a building are very much custom, unique to one project only, and must be thoroughly detailed so the contractor knows how to build it. Sometimes these details can be started from a standard detail that is similar to the custom design and other times it will be started from scratch.

The key to detailing is to think smart: draw what needs to be drawn, make sure all necessary dimensions, grid lines and floor level references are given. You should also try to reuse as much accurate and previously drawn content as possible, whether from a detail library or a previous project.

Exercise 15-2:
Exercises

This exercise shows you how to develop 2D details which are not tied to the 3D model in any way. Why would you want to do this, you might ask? Many design firms have developed detail libraries over the years for typical conditions. These details contain a significant amount of embodied knowledge of the firm as a whole. Many notes and dimensions have been added which cover certain situations that have come up and created problems or cost the firm money. For example, a window head detail might show flashing, which directs any moisture in the wall out – rather than into – the window or inside the building. Well, a note might have been added to instruct the contractor to turn the flashing up at each side of the window to ensure the moisture does not just run off the end of the flashing and stay in the wall.

If every detail were a live cut through the model, the designer would have to spend the time adding all these notes and dimensions, and more importantly try not to forget any, even if typing them from a printed reference page – similar to what you are doing with this book. (Have you missed anything yet, and had to go back and make a correction?) Furthermore, if the part of the model changes, the detail could be messed up. Or, an item being detailed from the live model might change and not be the typical condition anymore.

So, as you can see, there are a number of reasons a design firm maintains and utilizes static 2D details. *NOTE: Sometimes these details are used as starting points for similar details. This saves time not having to start from scratch.*

It should also be pointed out that standard details should always be reviewed before "dumping" them into a project. If a note says, "Apply fireproofing to underside of metal roof deck" and your building has precast concrete plank, you need to change the note and the drawing. All other parts of the detail may perfectly match the project design you are working on.

Linking an AutoCAD Drawing

This first exercise will explore linking AutoCAD drawings into Revit when the need to use legacy details arises.

It is better to recreate these details in native Revit format rather than linking an AutoCAD drawing. Any external files linked in have the potential to slow your BIM experience and introduce corruption. This is especially true with site plans created in AutoCAD or AutoCAD Civil 3D. Site plans are often a great distance from the origin (i.e., 0,0,0 coordinate in an AutoCAD drawing) and this creates several issues.

In general, it is best to avoid AutoCAD DWG files within Revit. However, when it is required, they should always be *Linked* in and not *Imported*, and never *Exploded*. Importing DWG files makes them difficult to manage and exploding them creates lots of extra text styles, fill patterns and other items that clutter the BIM database.

1. Open your Law Office project.

AutoCAD DWG files can be linked directly into a plan view and be used as an underlay to sketch walls and place doors and windows, when modeling an existing building in Revit that has been drawn in a traditional CAD program.

In our example, we have a DWG file which contains a detail we want to reference and place on a sheet for our office building project. To do this, you create a *Drafting View* and link the CAD file into the drafting view. A *Drafting View* is a 2D drawing within the BIM project that has no direct relationship to the 3D model.

2. Click **View → Create → Drafting View** from the *Ribbon*.

Drafting View

Now you are prompted for a name and scale for the new drafting view; this can be changed later if needed.

3. Enter the following (Figure 15-2.1):

 a. *Name:* **Typical Roof Drain Detail**

 b. *Scale:* **1½" = 1'-0"**

4. Click **OK** to create the new *Drafting View*.

FIGURE 15-2.1 Creating a drafting view

You now have a new section, under *Views* in the *Project Browser*, called *Drafting Views* (Detail). Within this section is your new *Drafting View* – Typical Roof Drain Detail.

Within this new drafting view you could begin sketching a detail from scratch using the various tools on the *Annotate* and *Modify* tabs. Or, in this example, you may link in a DWG file.

This roof drain detail is a good example of why 2D details are still useful in Revit, either DWG or native Revit. As previously mentioned, many firms spend years developing standard details. These details have notes that have been added to and edited as building materials change and problems occur. It would take a lot of time to cut a section at a 3D roof drain in the model and then add all the notes, if one can even remember what all the notes are. Now, repeat this for 20 to 50 other items throughout the building project.

Now you will link in the DWG file using the provided online file.

5. While in the newly created drafting view, select **Insert → Link → Link CAD** from the *Ribbon*.

Link CAD

6. Browse to the **DWG Files** folder within the files downloaded from the publisher's website.

7. Select the file **Typical Roof Drain Detail.DWG**.

8. Set the *Colors* options to **Black and White** (Figure 15-2.2).

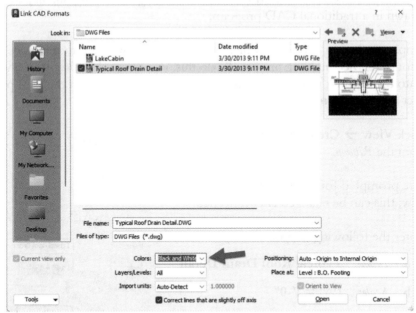

FIGURE 15-2.2 Linking an AutoCAD detail file

9. Click **Open** to place the linked AutoCAD DWG file.

10. Type **ZF** (for *Zoom Fit*) on the keyboard; do **not** press **Enter**.

You should now see the roof drain detail, with line weights.

FIGURE 15-2.3 DWG file linked into drafting view

The drawing can be selected and moved around within the drafting view, but it cannot be edited. If you need to make changes to this drawing, you would have to do it in AutoCAD. Revit will automatically update any linked files when the project file is opened. It can also be done manually using the *Manage Links* tool.

When DWG files are linked into Revit, a specific set of line weights is used. These settings can be seen by clicking the small arrow (i.e., the dialog launcher) in the lower right corner of the *Import* panel on the *Insert* tab.

FIGURE 15-2.4 DWG color to Revit line weight conversion

Applying line weights is a onetime conversion process when the DWG file is linked in. Changing the line weight setting after a DWG has been linked in has no effect on it (only on new DWG files to be linked in).

> **TIP:** *Go to* **Manage** → **Settings** → **Additional Settings** → **Line Weights** *to see what each line weight number is equal to.*

If you type **VV** in the drafting view and then select the *Imported Categories* tab, you can see the AutoCAD *Layers* that exist in the imported DWG file (Figure 15-2.5). Unchecking a *Layer* will hide that information within the drafting view. You can also control the color and line weight of the lines on each *Layer*.

FIGURE 15-2.5 Controlling DWG layer visibility in Revit

Creating 2D Details

Autodesk Revit has a large array of 2D detail components that can be used to create 2D details. These components allow for efficient detail drafting and design. Not every detail in Revit is generated from the 3D model; the amount of modeling required to make this happen is restricted by time, file size and computing power. The following is an outline of the overall process; this will be followed by a few exercises for practice:

A. To create a 2D detail one would create a **drafting view** via *View > Create > Drafting View*, providing a name and selecting a scale.

B. Once the *drafting view* has been created, **Detail Lines** and **Filled Regions** (via the *Annotate* tab) can be added.

C. In addition to *Detail Lines* and *Filled Regions*, one can insert pre-drawn items from the <u>Detail Library</u>.

 i. Select **Load Autodesk Family** from the *Insert* tab, on the Ribbon.

 ii. **Browse** to the specific "CSI organized" folder; for example, *Detail Items* → *Div 5-Metals* → *052100- Steel Joists Framing* → *K-Series Bar Joist-Side*.

 iii. Click **Load** to place the component.

D. Use the **Detail Component** tool on the *Annotate* tab to place the bar joist.

E. Add notes and dimensions to complete the 2D detail.

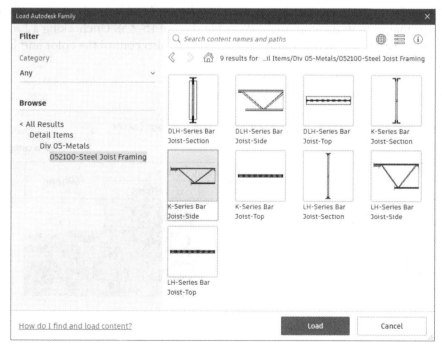

FIGURE 15-2.6
Loading detail components

Flooring Details

The first two details you will draw are simple details consisting of *Detail Lines*, *Filled Regions*, *Text* and *Dimensions*.

You will draw a high-end floor and wall base detail known as terrazzo. This finish is poured in a liquid state, allowed to dry and then polished to a smooth finish. The colors and aggregate options are virtually unlimited (for example, you could use a clear epoxy resin and place leaves within the flooring).

1. Per the steps previously covered in this section, create a new drafting view.

 a. *Name*: **Terrazzo Base Detail**
 b. *Scale*: **3″=1′-0″**

You will now draw the detail shown below. See the next page for specific steps.

Terrazzo floor example with brass inlay

FIGURE 15-2.7 Terrazzo floor and wall base detail

2. Using the **Detail Lines** tool from the *Annotate* tab, draw the floor line **8″** long using **Wide Lines**.

> *FYI:* *The eight inch dimension is random and does not represent anything other than a portion of the floor surface.*

3. Draw a vertical line, also **8″** long, using the **Medium Lines** style. This line should be about 2″ from the right edge of the floor line (just so your detail is generally proportional to the one presented in the book).

Next, you will offset the two lines just drawn to quickly create the terrazzo floor and base.

4. Select **Modify → Offset** on the *Ribbon*.

5. On the *Options Bar*, enter an offset value of **3/8″**.

6. Pick the floor line when the preview line appears above the horizontal line.

7. Now **Offset** the vertical line **3/8″** to the left.

Your drawing should now look like the image to the right (less the arrows). Next you will change the top horizontal line to a lighter line weight, offset it up 6″ to create the top of the wall base and then use the *Fillet Arc* feature to round off the corners.

8. Select the top horizontal line and change the *Line Style* from *Wide Lines* to **Medium Lines** via the *Ribbon*.

9. **Offset** the top horizontal line upward **6"** inches.

10. Select the new horizontal line and then drag its right end grip over to the vertical line.

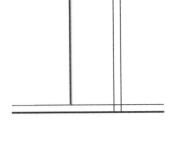

Next you will trim and round the corners in one step.

11. Select the **Detail Line** tool from the *Annotate* tab.

 a. Select **Fillet Arc** from the *Draw* panel.
 b. Set the *Line Style* to **Medium Lines**.
 c. Check and set the *Radius* to **¼"**.

12. Click on the portion of the two lines you want to remain (see the two numbered clicks in the image to the right).

The line is now trimmed and an arc has been added.

13. Repeat these steps to round off the top edge of the wall base.

Next you will add a *Filled Region* to represent the terrazzo material with a pattern when viewed in a section. When creating the perimeter of a *Filled Region* you also specify a *Line Style* (similar to the *Detail Line* tool). In this case you will use *Thin Lines* for all but the left edge of the floor thickness. There you will change the *Line Style* to be an invisible line so as not to suggest a joint or the end of the flooring but rather that the flooring material continues.

14. Select **Annotate → Detail → Region → Filled Region** from the *Ribbon*.

 a. Select the **Pick Lines** option from the *Draw* panel.

 b. Select **Thin Lines** from the *Lines Style* panel.

Selecting *Pick Lines* will allow you to quickly create the boundary of your *Filled Region* based on lines already drawn. So you simply click on a line rather than pick two points to define the start and endpoints of each edge. However, with the *Line* and *Arc* options you could also snap to the endpoints of the previously drawn linework. The one drawback to using the *Pick Lines* option is you will have to trim a few corners because the *Filled Region* tool requires that a clean perimeter be defined (similar to the floor and roof tools).

15. Pick the five lines and two arcs which define the edges of the floor and wall base.

16. Switch to the **Line** option in the *Draw* panel and then set the *Line Style* to **<Invisible Lines>**.

While in *Sketch Mode* (i.e., the green check mark and red X are visible) for the *Filled Region* tool, the invisible lines are not actually invisible. This allows you to select and modify them as needed.

17. Draw a line to close off the open edge of the flooring on the left hand side.

18. Use the **Trim** tool to clean up the two corners where the lines run past.

19. Click the **green check mark** to finish the *Filled Region*.

Your drawing should look like the one shown to the right. The default pattern is a cross hatch. You will change this next.

20. Select the *Filled Region*; you must click on one of the edges (and you may need to use *Tab*).

21. Expand the **Type Selector** to see the options currently available.

Looking at the list, we notice an option for concrete is not listed (which is what we decided we want). Next you will learn how to add this.

22. Press **Esc** to close the *Type Selector*.

23. Click **Edit Type**.

24. Click **Duplicate**.

25. Enter **Concrete** for the name.

26. Click in the *Foreground Fill Pattern* field and then **click the icon** that appears to the right.

27. Select **Concrete** from the list and click **OK**.

Notice the option to hatch the fill background opaque or transparent and the line weight setting.

28. Click **OK** to close the *Type Properties*.

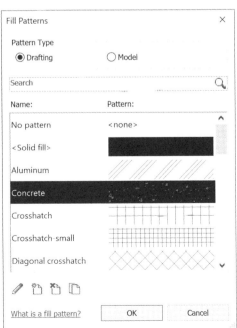

Your terrazzo now has a concrete pattern. You can import additional fill patterns (using any AutoCAD hatch pattern file) or create custom ones with specific line spacing. This *Filled Region* tool can be used in floor plans as well; maybe you want to highlight the corridors or private office areas. When a *Filled Region* is selected the square foot area is listed in the *Properties Palette*. Thus, you could quickly create a *Filled Region* just to list the area and then delete the *Filled Region*.

The last thing you will do is add the notes and dimensions. These will be the correct scale based on the *View Scale* setting (which should be 3″ = 1′-0″). Once you place the dimension you will learn how to adjust the dimension style so the 0′ does not show up.

29. Add the dimension and two notes as shown.

30. Select the dimension.

31. Select **Edit Type**.

32. Click the button to the right of **Units Format**.

33. Uncheck **Use Project Settings**.

34. Check **Suppress 0 feet**.

35. Click **OK** twice to close both open dialog boxes.

Your dimension should now only say 6″ rather than 0′-6″. Because you changed this in the *Type Properties*, all dimensions will have this change applied (both previously saved and new). If you want to have both options, you would first need to *Duplicate* the dimension type, which is similar to creating new wall, door and window types!

The last thing you will look at is adjusting the arrow style for the notes.

36. Select one of the notes.

37. Click **Edit Type**.

38. Change the *Leader Arrowhead* to **Arrow Filled 30 Degree**.

39. Click **OK**.

Your detail should now look similar to the one presented at the beginning of this exercise. This detail can now be placed on a sheet (covered in chapter 12). You can also export this detail and save it in a detail library that you and others in your firm can utilize. To export the detail, simply right-click on the view name in the *Project Browser* and select *Save to new File*. Place the file on a server so everyone can get at it. To load this file into another project select **Insert** → **Insert from File** → **Insert Views from File**.

Now you will draw another detail using the same tools and techniques. Keep in mind, every detail needs to go in its own *drafting view*. This is required for Revit to manage the reference bubbles on the sheets.

40. Create a new **drafting view**:
 a. *Name:* **Floor Transition Detail**
 b. *Scale:* **3″ = 1′-0″**

41. Draw the detail per the following guidelines:
 a. The detail will be plotted at 3″=1′-0″ (this determines the text and leader size).
 b. Tile pattern is **Diagonal Up–Small**; this requires a new *Fill Region* style (see the previous exercise for more information).
 c. The grout (i.e., area under tile) is to be hatched with **Sand – Dense**; this also requires a new *Fill Region* style.
 d. Draw the tile ¼″ thick and 4″ wide.
 e. The grout is ¼″ thick.
 f. The resilient flooring is shown ⅛″ thick.
 g. The solid surface (e.g., Corian) threshold is 1⅞″ wide; draw an arc between the two floor thicknesses.
 h. Hatch the threshold with the solid hatch:
 i. **Duplicate** the **Solid Black** *Filled Region* style.
 ii. Name the new style **Solid Gray**.
 iii. Set the **foreground pattern color** to a light gray (RGB color 192).
 i. The bottom concrete floor line is to be the heaviest line.

RESILIENT FLOORING

SOLID SURFACE THRESHOLD

THIN SET CERAMIC TILE

CONCRETE FLOOR

FIGURE 15-2.8 Floor transition detail: ceramic tile to resilient flooring

The previous drawing would typically occur in a door opening and the location of the door would also be shown in the detail. This lets the contractor know that the threshold is to occur directly below the door slab.

Base Cabinet with Drawers

This section will dive right into drawing cabinet details. These are often based on industry standard dimensions so many of the dimensions and material thicknesses can be omitted (assuming the project manual/specification covers this).

42. Create a **drafting view**:

 a. *Name:* **Base Cabinet Detail – Drawers**
 b. *Scale:* **1″ = 1′-0″**

Revit provides many *Detail Components* which aid in creating 2D details. Things such as side views of bar joists, section views of steel beams and angles, and more are available in the *Detail Component* library. The detail below takes advantage of three *Detail Components* which ship with Revit: particle board, lumber and the countertop. The only things drawn with the *Detail Line* tool are the tops of the drawers, the drawer pulls (i.e., handles) and the heavy wall/floor lines.

FIGURE 15-2.9 Cabinet section (with drawers)

43. Using the *Detail Line* tool, draw the floor and wall line, shown in the image to the right, using the **Wide Lines** style. (Do not add the dimensions.)

Next, you will load the countertop *Family* from the *Detail Component* library.

44. Select **Annotate → Detail → Component → Detail Component** from the *Ribbon*.

45. Click **Load Family** from the *Ribbon*.

46. Open the **Detail Items** folder.

47. Now, browse to *Div 12-Furnishings → 123000-Casework →123600-Countertops*.

48. Select **Countertop-Section.rfa** and then click **Open**.

Countertop-Section with type *24" Depth* is current, in the *Type Selector*, and ready to be placed.

49. Place the countertop as shown:
 a. Aligned with wall
 b. 3'-0" above the floor (dimension to the line second from the top; the open heavy line is an exaggeration to highlight the added plastic laminate surface).
 c. With the countertop selected, adjust the values in the *Properties Palette*:
 i. *Backsplash Depth:* **0' 1"**
 ii. *Counter Depth:* **2'-1"**
 iii. *Thickness:* **0' 1⁵⁄₈"**
 d. Do not add dimensions yet.

Be careful not to click on the grips when the countertop is selected as this will adjust its dimensions; this is because the values are associated with an *instance parameter* rather than a *type parameter*.

Next, you will load and place the 2x lumber. The two on the floor are 4" high, which are cut down from a 2x6. So you will load a 2x4 *Family* and then create a duplicate and adjust the height from 3½" to 4".

50. Per the step just covered, load **Nominal Cut Lumber-Section.RFA** from the following location: *Detail Item → Div 06-Wood and Plastic → 061100-Wood Framing*.

To minimize the number of *Types* for the lumber *Family*, you are presented with the *Specify Types* dialog. This lets you pick just the sizes you want – more can be added later.

51. Hold the **Ctrl** key and select **1x4** and **2x4** and then pick **OK**.

52. Select **2x4** in the *Type Selector*.

53. Click **Edit Type**.

54. Click **Duplicate**.

55. Enter **2x4 Base Cabinets** for the name.

56. Change the *Height* from 3½″ to **4″**.

57. Click **OK** to close the *Type Properties*.

58. Place the two **2x4 Base Cabinet** detail components on the floor as shown in the image to the right. (Do not add the dimension.)

59. Place the two **1x4** components as shown.

Next, you will load the detail component used to draw particle board. This *Family* is somewhat like the *Wall* tool. You pick two points and two lines and a fill pattern is generated. The *Type Selector* also has a number of standard thicknesses ready to use.

60. Per the step just covered, load **Particle board-Section.rfa** from the following location: *Detail Items → Div 06-Wood and Plastic → 061600-Sheathing*.

61. Select the **3/4″** type from the *Type Selector.*

62. Draw the two pieces of particle board shown in the image to the right; these are the top and bottom of the base cabinet. ***TIP:*** *Use the space bar to flip the thickness while drawing, if needed.*

63. Draw the cabinet back; use the **3/8″** thickness option and extend it **1/4″** into the top and bottom boards. See image below. ***TIP:*** *Draw temporary detail lines so you have a place to pick if needed. Delete them when done.*

The newest drawn particle board automatically shows up on top of any previously drawn particle board. This is how the notch is created. There was no trimming or erasing required. If you need to change the order of the overlap, simply select the component and use the *Arrange* options on the *Ribbon.*

64. Draw the rest of the base cabinet using the techniques previously covered and the following information:

 a. The "EQ" dimensions are 6″.

 b. The drawer bottoms and backs are all ½″ particle board; everything else is ¾″.

 c. Use **Thin Lines** for the pulls and the top edge of the side drawer panel (seen in elevation in **Figure 15-2.9**). Use **Medium Lines** for the rubber base.

 d. Add dimensions not provided which can be approximated; make it look like the image in the book as much as possible.

 e. Add all the notes and dimensions shown on the first image only.

Some things you should know about detailing

In the previous steps you drew a typical detail showing a standard base cabinet with drawers. Interior designers occasionally draw these details, but more often, they simply review them for finishes.

The countertop material needs to match that which is specified in the **Project Manual** and intended for the project. For example, a PLAM (i.e., plastic laminate) countertop would not be appropriate in a laboratory where chemicals would be used.

The base of the cabinet typically has the same wall base as the adjacent walls. For example, if the walls have a rubber base (also referred to as a resilient base), the toe-kick area of the base cabinet would also receive a rubber base; this is the type of base shown in the previous detail. If the project only had ceramic tile wall base, you would show that.

The **notes for details** (or any drawing) should be simple, generic and to the point. Notice in previously drawn detail that the note for the base does not indicate whether the base is rubber, tile or wood. This helps avoid contradictions with the room finish schedule. The note simply says the cabinet and floor are to receive a finish and instructs the contractor to go to the *Room Finish Schedule* to see what the finish is. This is particularly important in buildings that have several variations of floor and base finishes.

Notes should not have any **proprietary or manufacturers' names** in them either. For example, you should not say "Sheetrock" in a note because this is the brand name; rather, you should use the generic term "gypsum board." Similarly, you would use the term "solid plastic" rather than "Corian" when referring to countertops or toilet partitions. In any event, whatever term you use on the drawings should be the same term used in the Project Manual!

One last comment: the *Construction Documents* set should never have **abbreviations** within the drawings that are not covered in the *Abbreviations* list, usually located on the title sheet. *Construction Documents* are legal, binding documents, which the contractor must follow to a "T." They should not have to guess as to what the designers meant in various notes all over the set of drawings. It is better to spell out every word, if possible, only abbreviating when space does not permit. You would not want a bunch of abbreviations in your bank loan or mortgage papers you were about to sign! Plus, non-documented abbreviations would probably not have much merit before a judge or arbitrator in the case of a legal dispute!

65. Using the same steps just covered, draw the base cabinet shown below. Name the drafting view **Base Cabinet Detail**. The *Scale* is also **1″ = 1′-0″**.

TIP: *Duplicate the Base Cabinet Detail – Drawer view and modify it to be this detail; much of this detail is exactly the same.*

FIGURE 15-2.10 Cabinet section (door + drawer)

Cabinet details do not need to have every nook-and-cranny dimensioned because they are very much a standard item in the construction industry. Furthermore, the Project Manual usually references an industry standard that the contractor can refer to for typical dimensions, thicknesses and grades of wood.

The vertical dimension shown in the cabinet detail above says "VARIES - SEE ELEV." This notation, rather than an actual number, allows the detail to represent more than one condition. The interior elevations are required to have these dimensions, which may be the standard 36″ or the lower handicap-accessible height.

Ceiling Detail

Next, you will draw a typical recessed light trough detail at the ceiling. This is used often in commercial/public toilet rooms.

66. Create a new drafting view:

 a. *Name:* Toilet Room Ceiling Detail

 b. *Scale:* 1½" = 1'-0"

 c. *See the next page for additional notes.*

67. Load the following *Detail Components* into your project:

 a. Div 09-Finishes → 092000-Plaster and Gypsum Board → 092200-Supports → 092216-Non-Structural Metal Framing

 i. **Interior Metal Runner Channels-Section.rfa**

 ii. **Interior Metal Studs-Side.rfa**

 b. Div 09-Finishes → 092000-Plaster and Gypsum Board → 092900-Gypsum Board → **Gypsum Wallboard-Section.rfa**

 c. Div 09-Finishes → 095000-Ceilings → 095100-Acoustical Ceilings

 i. **Suspension Wall Angle-Section.rfa**

 ii. **Suspended Acoustic Ceiling-Square Edge-Section.rfa**

FIGURE 15-2.11 Ceiling detail

 TIP: The notes in details should align on one edge as shown. Additionally, the leaders should not cross dimensions or other leaders unless it is totally unavoidable.

When using the *Detail Components* you will occasionally run into a few challenges getting things to look exactly the way you want them. For example, many firms have traditionally shown gypsum board as if it were continuous – not indicating the joints. Most contractors understand this and the designer usually does not want to imply the gypsum board be installed in a specific way or order. However, the *Detail Component* feature forces these edge lines to appear. Unfortunately, we **cannot** use the *Linework* tool to set some of the lines in a *Detail Component* to be invisible.

Another problem you will run into, when using *Detail Components*, is the fact that everything is drawn true to life size. You may think, "Isn't that a good thing?" Usually it is, but some things often need to be exaggerated so they are legible on the printed page. Take the metal stud runner for example; when placed next to the *Gypsum Board* detail component it is totally hidden because the gypsum line weight is heavier than the runner stud. The best thing to do is to edit the *Family* so the runner stud has a heavier line weight and its thickness is exaggerated inward (you don't want to change the overall width). However, this type of change is outside the scope of this exercise, so you will do the following:

68. Erase the runner stud *Detail Components* and draw **Detail Lines** using **Medium Lines** – in bound from the gypsum board.

69. Make the light fixture **4½″ x 1¾″** with a **1″** circle for the light. Use the **Solid Gray** *Fill Region* previously created.

70. Sketch the egg-crate lens using **Detail Lines**. Create a new *Fill Region* using the **Vertical-Small** fill pattern.

71. Add the notes and dimensions shown.

Fixed Student Desk at Raised Seating Classroom

This is a continuous student desk in a classroom setting.

72. Create a new **Drafting View**:
 a. *Name:* **Fixed Student Desk**
 b. *Scale:* **1½" = 1'-0"**
 c. *See the next page for additional comments*

PLAM over 1" PARTICLE BOARD

NOTE: ALL EXPOSED SURFACES TO RECEIVE PLAM FINISH

1x2 TRIM

TAPERED WOOD BLOCKING

1x4 TRIM

PLAM over 1" PARTICLE BOARD SUPPORT at 36" O.C.

PLAM over 1/2" PARTICLE BOARD - BOTH SIDES OF STUD WALL

3 5/8" METAL STUDS AT 16" O.C.

BASE, SEE RM FIN SCHED

FLOOR

FIGURE 15-2.12 Fixed desk detail

73. Load the following detail component:
 a. Div 06-Wood and Plastic\062200-Millwork\Standard Millwork-Section.rfa
 i. Load the **1x2** and **1x4** sizes.

74. Develop the fixed student desk following these guidelines:
 a. All particle board and trim to be Detail Components.
 b. Tapered wood blocking to have **Wood 2** *Fill Pattern*.
 c. The curvy line pointing to the 1'-0" dimension should be drawn like this:
 i. First add the "PLAM" note using a regular leader;
 ii. Use the **Detail Line** tool;
 iii. Select the **Spline** *Draw* option;
 iv. Set the *Line Style* to **Thin**;
 v. Sketch the curvy line starting at the corner of the leader.
 d. Add a horizontal Detail Line at the top of the sloped work surface so you have something to pick when adding the angle dimension.

As you can see, some of the *Detail Components* have odd line weights when placed side-by-side. Both the wood trim and particle board are in section so they should be the same line weight. You would have to edit the family to make this change, which will not be covered at this time.

Using the Keynotes Feature

The content that ships with Revit, both 2D and 3D, has a default keynote value assigned to it. Keynotes are used to save room and make details look neater; it is a reference number rather than a full note, and then an adjacent legend lists what each number means. This legend is for all the details on a sheet. You will learn how this works next. You will make a copy of the *Fixed Student Desk*, add keynotes and then create a keynote legend.

75. Right-click on the **Fixed Student Desk** item in the *Project Browser*.

76. Select **Duplicate → Duplicated with Detailing**.

77. Rename the new view **Fixed Student Desk – Keynotes**.

78. **Erase all the notes**, but leave the dimensions.

79. Select **Annotate → Tag → Keynote** from the *Ribbon*.

FIGURE 15-2.13 Keynote added

80. With the *Keynote* tool active, click the ½″ particle board shown in Figure 15-2.13; you will see it highlight just before selecting it.

81. Click two additional points to define a leader and text location, just like placing text with a leader.

You now have a keyed note placed in your drawing. This only works on *Detail Components* (in drafting views) and not *Detail Lines* as they are too generic. Next, you will see where this keynote notation is coming from.

82. Select the ½″ particle board; go to its *Type Properties*.

Notice the *Keynote* value listed. This was defined in the *Family* you loaded.

Now you will view the *keynote text file* so you can see how a family could be changed to "mean" something else. You might change a family directly or create a duplicate *Type* first.

83. Click in the cell listing the *Keynote*.

84. Click the **small icon** that appears to the right (not in the "equals" column).

You now see a rather extensive listing of keyed notes. Take a minute to explore the various sections and descriptions for the keynote references.

Notice the path listed at the top. This is the location of the text file being used or the keynotes. The path to this file is set via the *Keynoting Settings* icon located in the *Tag* panel expanded area on the *Annotate* tab.

85. After reviewing the keynote text file click **Cancel**.

86. Click **OK** to close the *Type Properties* dialog.

The last thing to learn is how to create the **Keynote Legend**. This legend can be added to any sheet with keynotes. If the "By sheet" option is selected in the dialog box shown to the right, only keynotes actually found on that specific sheet will be listed.

87. From the *View* tab, select **Legend → Keynote Legend**.

88. Click **OK** to accept the name **Keynote Legend**.

89. Click **OK** to accept the default properties and to create the legend.

You should now see the *Keynote Legend* shown below.

FYI: *Make sure the* numbering method *is still set to* By keynote *in order to see all the keynotes at this time.*

Later, in Chapter 12, you will learn how to place views on sheets.

Keynote Legend	
Key Value	Keynote Text
06 16 00.	1/2" Particleboard
06 16 00.	1" Particleboard
06 22 00.	1x2 Wood Trim
06 22 00.	1x4 Wood Trim
09 22 16.	3 5/8" Metal Stud Framing
09 22 16.	3 5/8" Metal Runner
09 29 00.	5/8" Gypsum Wallboard

90. Add the remaining keynotes and then **Save**.

Finally, you can create elevations and sections that reference a drafting view rather than a true view of the model. This is another one of those places where you are breaking the intelligence of Revit's drawing sheet and number coordination.

To do this, you select the *Elevation* or *Section* tool, and rather than picking points in the drawing right away, you select a view that already exists in the project from the *Options Bar*. Even though you have not placed any roof drains, you could switch to the **Roof** plan view and add a section mark that references the roof drain detail.

The following steps do not need to be performed at this time:

91. Switch to the **Roof** floor plan view.

92. Select the **Section** tool on the *View* tab.

93. On the *Ribbon/Options Bar* settings (Figure 15-2.14):

 a. *Type Selector:* **Detail View: Detail**

 b. *Reference other view:* **check**

 c. *Reference other view drop-down list:* **Typical Roof Drain Detail**

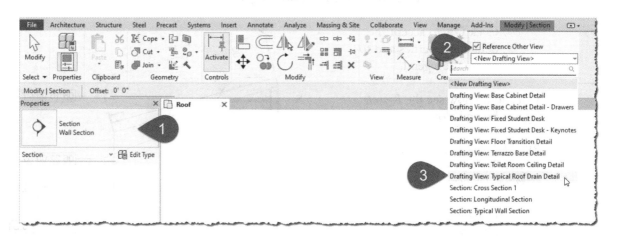

FIGURE 15-2.14 Placing a section that references the roof drain detail

94. Pick two points, roughly as shown in the image to the right.

Try double-clicking on the blue bubble head; it should bring you to the roof drain detail. Once the detail is placed on a sheet, the bubble head will be automatically filled out!

95. Save as **15-2.rvt**.

Exercise 15-3:
Live Detail – Reception Desk

This last exercise on detailing will look at creating a live detail section through something in your model. This is often done with custom construction which is unique to the project, and there are no 2D details readily available. In this case we will cut a live section through the element(s) in the model and then add notes and dimensions in the live section view. Because this section is live, you have to be careful not to move the item in the model as this could mess up the notes and dimensions in the section view. Sometimes this is necessary and the section just needs to be updated. This would never happen with a static 2D drafting view, but in that case the detail might get overlooked and not properly updated for the change. Even if the live section/detail did not get properly updated, the view would likely look messed up and sooner or later someone would notice the problem.

In this exercise you will cut a section through the reception desk, and then add notes and dimensions to it.

1. Zoom in on the reception desk in the Level 1 Floor Plan (not the furniture plan).

2. With the *Type Selector* set to **Detail**, create a **Section** as shown in Figure 15-3.1 below.

FIGURE 15-3.1 Adding a section at the reception desk

3. Set up the new view (Figure 15-3.2):

 a. Hide the *Levels* and ***Reference Planes***.

 b. Apply the **Floor Finish** *Filter* to hide them.

 c. Adjust the ***Crop Region***.

 d. Hide the ***Crop Region***.

 e. Detail Level: **Fine**

 f. View Scale: **1½″ = 1′-0″**

FIGURE 15-3.2 Initial detail view at reception desk

Keep in mind additional elements can appear in this view if added to the model, e.g., file cabinets below the countertop. Now you can add notes, dimensions and detail components as needed to properly convey the construction intent to the contractor.

4. Add the *Text*, *Dimensions*, *Detail Components*, and *Fill Regions* shown in Figure 15-3.3.

 a. *Filled Regions* used: **Wood 2** and **Sand** with **Wide Lines**.

 b. *Detail Components* used:

 i. Particleboard=Section: ¾″

 ii. Nominal Cut Lumber-Section: 2x4

FIGURE 15-3.3 Live reception desk detail

If the reception desk moves or is changed in the model, this view will highlight that change because things will be out of alignment.

You can also create a 3D view, lock it so it does not move, crop it with the *Section Box* tool and then add notes, tags and dimensions to it. This is a great way to note things which are hard to show with two-dimensional drawings. Notice the example below has the door tagged (Figure 15-3.4).

FIGURE 15-3.4 3D detail of reception desk

Self-Exam:

The following questions can be used as a way to check your knowledge of this lesson. The answers can be found at the bottom of the page.

1. The plan is updated automatically when an elevation is modified, but not the other way around. (T/F)

2. You can use the *Elevation* tool to place both interior and exterior elevations. (T/F)

3. You can rename elevation views to better manage them. (T/F)

4. You have to resize the *Level* datum and annotations after changing a view's scale. (T/F)

5. The controls for the section mark (when selected) are similar to the controls for the elevation tag. (T/F)

Review Questions:

The following questions may be assigned by your instructor as a way to assess your knowledge of this section. Your instructor has the answers to the review questions.

1. The visibility of the crop window can be controlled. (T/F)

2. You have to manually adjust the line weights in the elevations. (T/F)

3. As you move the cursor around the building, during placement, the elevation tag turns to point at the building. (T/F)

4. When a section mark is added to a view, all the other related views automatically get a section mark added to them. (T/F)

5. You cannot adjust the "extent of view" (width) using the crop window. (T/F)

6. What is the first thing you should do after placing an elevation tag?

7. Although they make the drawing look very interesting, using the _____ feature can cause Revit to run extremely slow.

8. It is possible to modify objects (like doors, windows and ceilings) in section views. (T/F)

9. You need to adjust the _____ to see objects, in elevation, that are a distance back from the main elevation.

10. AutoCAD DWG files should never be exploded. (T/F)

Lesson 16
Schedules:

Now that your model has several elements in it, using Autodesk® Revit® you will leverage some of the scheduling features you learned about earlier in this book (see Chapter 5).

Exercise 16-1:
Room Finish Schedule

In this exercise you will create a *Room Finish Schedule*. This schedule defines the finishes for the walls, floor and ceiling in each room. The first thing you need to do is edit the room numbers. Most room numbering systems have 100s for level 1, 200s for level 2 and so on. Additionally, the numbering starts at the main entry and then works counterclockwise around the building. This can vary quite a bit, but this is the methodology we will employ.

1. **Open** your *Law Office* project.

2. Open your **Level 1 Floor Plan** (does not need to be the finish plan).

3. **Zoom in** on the main entrance, near the *Lobby*.

A few rooms have yet to receive a *Room* and *Room Tag*. You will add most of the missing *Rooms* in the next few steps.

4. Add a **Room** to the vestibule.

 a. Make sure **Tag on Placement** is selected on the *Ribbon*.

5. Select the **Room Tag** (Figure 16-1.1):

 a. Click on the room name and then change it to **Vestibule**.

 b. Click on the room number and then change it to **100**.

FIGURE 16-1.1 Room added to vestibule

6. Add and/or edit the remaining rooms as shown in Figures 16-1.2 and 16-1.3.

 FYI: The tags have been enlarged for the book in these images. Yours will be smaller (the View Scale *should be* ¼″ = 1′ 0″*).*

FIGURE 16-1.2 Level 1 room numbers – rotated image

FIGURE 16-1.3 Level 2 room numbers – rotated images

Create a Room Finish Schedule:

Remember to refer back to Chapter 5 for more information on how schedules work.

7. Select **View** → **Create** → **Schedule** → **Schedule/Quantities** button from the *Ribbon*.

8. Select ***Rooms*** under *Category*.

9. Type **Law Office Room Schedule** for the *Name*.

10. Set the *Phase* to ***New Construction***.

11. Click **OK** to create the schedule (Figure 16-1.4).

12. In the ***Fields*** tab of the *Schedule Properties* dialog, add the following fields to be scheduled (Figure 16-1.5):

 a. Number
 b. Name
 c. Base Finish
 d. Floor Finish
 e. Wall Finish
 f. Ceiling Finish
 g. Area

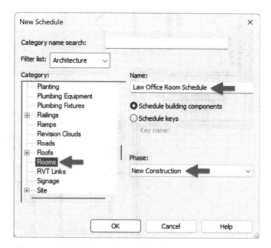

FIGURE 16-1.4 New Schedule dialog

Area is not typically listed on a room finish schedule. However, you will add it to your schedule to see the various options Revit allows. It is possible to have multiple "room" schedules that list various things about the room. If you recall, the ***Program Statement*** is a *Room Schedule* as well.

13. On the ***Sorting/Grouping*** tab, set the schedule to be sorted by the ***Number*** field.

14. On the ***Appearance*** tab, set the *Header* and *Body Text* to **3/32" Arial** (Figure 16-1.6).

15. Select **OK** to generate the room schedule.

FIGURE 16-1.5
Schedule Properties - Fields

FIGURE 16-1.6
Schedule Properties - Appearance

Your schedule should look similar to Figure 16-1.7.

16. Resize the **Name** column so all the room names are visible. Place the cursor between *Name* and *Base Finish* and drag to the right until all the names are visible (Figure 16-1.7).

The formatting (i.e., *Bold* header text) will not show up until the schedule is placed on a sheet.

It is easy to add custom parameters that track something about the element being scheduled (rooms in this case). Simply click the *Add Parameter* button seen in Figure 16-1.5.

Place cursor here to resize the column

<Law Office Room Schedule>

A	B	C	D	E	F	G
Number	Name	Base Finish	Floor Finish	Wall Finish	Ceiling Finish	Area
100	Vestibule					297 SF
101	Waiting / l					304 SF
102	Reception					70 SF
103	Workroo					342 SF
104	lunchroo					321 SF
105	Mechanic					431 SF
106	Public Re					143 SF
107	Public Re					143 SF
108	Janitor cl					46 SF
109	File room					248 SF
110	Fitness c					215 SF
111	Visitor off					118 SF
112	Mock cou					389 SF
113	library					251 SF
114	Medium C					240 SF
115	Large Co					426 SF
116	Small Con					123 SF
117	Imprompt					106 SF
118	Small Con					102 SF
200	Retired p					252 SF
201	Secretary					67 SF
202	Office ma					80 SF
203	Intern					64 SF
204	Paralegal					80 SF
205	Paralegal					80 SF
206	Paralegal					80 SF
207	General S					172 SF

FIGURE 16-1.7 Room Schedule initial view

Modifying and Populating a Room Schedule:

The room schedule is a tabular view of the building model. So, you can change the room name or number on the schedule or in the plans. It is all pointing to the same information within the database.

17. In the **Law Office Room Schedule** view, change the name for room *105* (currently called *Room*) to ***Mechanical Room***.

 > ***TIP:*** *Clicking on the current room name and then clicking on the down-arrow that appears gives you a list of all the names in the current project; otherwise you can type a new name (which you need to do here).*

18. Switch to the **Level 1** view to see the updated *Room Tag*.

In a schedule view you can only enter information one cell at a time. However, in a floor plan view you can quickly enter *Finish* information for several *Rooms* at one time. You will do this next.

19. In the **Level 2 Plan** view, select the *Rooms* (not the *Room Tags*) for the offices shown – 4 total (Figure 16-1.8).

 > ***REMEMBER:*** *Hold the Ctrl key down to select multiple objects.*

 > ***TIP:*** *Move the cursor near the* Room Tag *but not over it to select the room; the large "X" will appear when the* Room *is selectable (see image below).*

FIGURE 16-1.8 Level 2 rooms selected

20. Notice the parameters listed in the *Properties Palette* (aka, *Instance Properties*).

The *Parameters* listed here are the same as the *Fields* available for display in the *Room Schedule*. When more than one tag is displayed and a *Parameter* is not the same, that value field is left blank. Otherwise, the values are displayed for the selected *Room*. Next you will enter values for the finishes.

21. Enter the following for the finishes (Figure 16-1.9):

 a. *Base Finish:* **WD.1**

 b. *Ceiling Finish:* **ACT.1**

 c. *Wall Finish:* **PT.1**

 d. *Floor Finish:* **CPT.1**

22. Click **Apply**.

> *FYI: Just moving your cursor back into the drawing window also applies your edits.*

23. Switch back to the **Law Office Room Schedule** view to see the automatic updates (Figure 16-1.10).

You can also enter data directly into the **Room Schedule** view.

FIGURE 16-1.9 Instance Properties

24. Enter the following data for the *Men's* and *Women's Toilet Rooms (both levels)*:

 a. *Base:* **COVED CT.2**

 b. *Ceiling:* **GYP BD**

 c. *Wall:* **CT.1, CT.2, PT.2**

 d. *Floor:* **CT.3**

> *TIP: Remember, once you entered the information once, it can be selected from a drop-down list for the remaining rooms.*

211	first year social paralegal					150 SF
212	Associate office					150 SF
213	Associate office					150 SF
214	Associate office					150 SF
215	Full time graphic design office	WD.1	CPT.1	PT.1	ACT.1	95 SF
216	Associate office	WD.1	CPT.1	PT.1	ACT.1	109 SF
217	Associate office	WD.1	CPT.1	PT.1	ACT.1	133 SF
218	Partner / principal office					247 SF
219	Partner / principal office					242 SF
220	Partner / principal office					196 SF
221	Associate office	WD.1	CPT.1	PT.1	ACT.1	108 SF
222	Partner / principal office					227 SF
223	Impromptu meeting space					73 SF

FIGURE 16-1.10 New data added to schedule

Hopefully, in the near future, Revit will be able to enter the finishes based on the wall, floor and ceiling types previously created! The alternative is to use the *Material Tag* in the finish floor plans and try to identify all *Materials*, and not use a schedule.

Setting Up a Color-Coded Floor Plan:

With the *Rooms* in place you can quickly set up color-coded floor plans. These are plans that indicate, with color, which rooms are *Offices*, *Circulation*, *Public*, etc., based on the room name or department, etc. This does not directly relate to schedules, but ties into the concept of organizing *Rooms* and room information.

25. Right-click on the **Level 1 Floor Plan** view in the *Project Browser*.

26. Select **Duplicate View → Duplicate** (without detail) from the pop-up menu.

27. Rename the new view *Copy of Level 1* to **Level 1 – Color**.

28. Select **Annotate → Tag → Tag All**.

29. Select *Room Tags:* **Room Tag** from the list.

Property Line Segment Tags	Property Line Tag-Radius
Room Tags	Room Tag
Room Tags	Room Tag : Room Tag With Are
Room Tags	Room Tag : Room Tag With Vol
Structural Framing Tags	Structural Framing Tag : Boxed

30. Click **OK**.

Now, all the *Rooms* have a *Room Tag* in them!

31. In the *New* view, select **Annotate → Color Fill → Legend**.

32. Click just to the right side of the plan to place the *Legend Key*.

Color Fill Legend

33. Select **Department** and then **OK** to the prompt (Figure 16-1.11).

You now have a color-coded plan where the colors are assigned by department name; e.g., all the rooms in the "*Accessory Spaces*" department have the same color (Figure 16-1.12).

It is possible to sort by other parameters other than *Department*. For example, you could sort by *Room Name*.

FIGURE 16-1.11 Color fill prompt

FIGURE 16-1.12 Color filled floor plan

> **TIP:** *Notice how the color coincides with the perimeter of the* Room *element previously placed. The room* Separation Lines *also control the room/color.*

Some furniture and equipment is white and other furniture is filled with the "fill" color of the room. The colored elements have the 3D geometry turned off in the plan view for increased performance. You will adjust this next.

34. Go to the **Level 1 – Color** view's *Properties* (*Properties Palette*).

35. Set the *Color Scheme Location* to **Foreground** and click **Apply**.

The colors are now consistent; the only drawback is the colors go to the center of the walls now.

36. Select the *Room Legend* shown in **Figure 16-1.12**.

37. Click *Edit Scheme* on the *Ribbon* (Figure 16-1.13).

Each unique room name will get a different color. Before you finish, you will change one *Color*.

Edit
Scheme
Scheme

FIGURE 16-1.13
Edit Scheme button

38. Click on the *Color* for the **Accessory Spaces**, or the darkest color, which makes things within those rooms hard to see (Figure 16-1.14).

39. Click the **Color Books** button to select a standard *Pantone®* color (Figure 16-1.15).

FIGURE 16-1.14
Edit Color Scheme dialog

FIGURE 16-1.15 Color selector

40. Pick any color you like (Figure 16-1.16).

41. Click **OK** to accept.

FIGURE 16-1.16 PANTONE Color Picker

Your plan should now have the new color you selected for the *Accessory Spaces*. Interior designers can create a color-filled plan based on the floor finishes. For example, all the rooms with *CPT-1* get one color and rooms with *CT-1* flooring get another color.

42. Use the **VV** shortcut to turn off the *Grids*, *Sections* and *Elevations* in the **Level 1 – Color** view.

43. Follow the steps just covered to set up a **Level 2 – Color** view.

44. **Save** your project.

With the new "color" views set up, you can leave them in the project and refer back to them as needed. If you set up the colors in your main design views, you would likely need to remove the colors (*View Properties* → *Color Scheme* → *None*), making extra work when you needed the color view again.

Exercise 16-2:
Furniture Schedule

In this chapter you will create a *Furniture Schedule*. This schedule will list specific product information about the furniture, most of which is already entered into the content. You will also add cost information. Please note that the cost values provided are meant to be close to reality; they should not be considered accurate. This goes without saying, but you should consult your local rep for real-world numbers.

Creating the Schedule:

1. Select the **Schedules → Schedule/Quantities** tool from the *View* tab.

2. Select *Furniture* for the *Category*, and ***New Construction*** for the *Phase*.

3. Click **OK** to create the schedule.

4. Add the following *Fields* (Figure 16-2.1):

 FYI: Some of these fields did not exist until the furniture was loaded.

FIGURE 16-2.1 Fields for furniture schedule

5. Set the **Filter** as follows (Figure 16-2.2):

FIGURE 16-2.2 Filter for furniture schedule

6. Set **Sorting/Grouping** as follows (Figure 16-2.3):

FIGURE 16-2.3 Sorting/Grouping for furniture schedule

7. Click **OK** to finish the schedule creation process.

Your schedule should now look similar to Figure 16-2.4; adjust the column widths as necessary to see all the information. Next you will add the cost information.

8. Enter the *Cost* values shown in Figure 16-2.4, or if you have researched more accurate costing information you may enter that.

 a. TBL3 should be 0.00 for a future step.

 FYI: Cost *is a* Type Parameter, *but because this schedule is not set to itemize every instance Revit does not prompt you about changing a Type Parameter which will affect multiple instances.*

<Furniture Schedule - Type Mark>						
A	B	C	D	E	F	G
Type Mark	Manufacturer	Product Line	Style Number(s)	Product	Cost	Count
CH1	Steelcase	Seating	46267179	Work Chairs	879.00	14
CH1a	Steelcase	Seating	46216179	Work Chairs	1329.00	22
CH2	Steelcase	Coalesse Seating / Occasional Tables	2547R-1	Kathryn	792.00	41
CH3	Steelcase	Coalesse Seating / Occasional Tables	1519PC	Switch	1860.00	61
CH4	Steelcase	Coalesse Seating / Occasional Tables	CO300	Bindu	1235.00	14
CH5	Steelcase	Coalesse Seating / Occasional Tables	476	Coupe Grande	1178.00	2
CH6	Steelcase	Coalesse Seating / Occasional Tables	356-2	Donovan	1673.00	4
CH7	Steelcase	Coalesse Seating / Occasional Tables	H350-S31	Enea	527.00	12
CH8	Steelcase	Coalesse Seating / Occasional Tables	126111	Kart	526.00	8
DK1	Steelcase	Elective Elements 6	E6WS3096	Worksurface	22050.00	4
DK2	Steelcase	Coalesse Office	PDE8230R	Topo	13854.00	9
DK3	Steelcase	Elective Elements 6	E6WD2490	Worksurface	7784.00	12
DK4	Steelcase	Coalesse Office	PDEW7024	Topo	9840.00	1
TBL1	Steelcase	Coalesse Office	CODRNDP4827	Denizen	2419.00	7
TBL2	Steelcase	Convene	CWFK60216	Conference Table Top	16000.00	1
TBL3	COALESSE				0.00	1
TBL4a	Steelcase	Coalesse Seating / Occasional Tables	COEL3624	Enea Lottus	1946.00	3
TBL4b	Steelcase	Coalesse Seating / Occasional Tables	COEL4230	Enea Lottus	1746.00	3
TBL5	Steelcase	Coalesse Seating / Occasional Tables	H35-T2442W	Enea	846.00	2
TBL6	Steelcase	Coalesse Seating / Occasional Tables	72-2020H	Calm	1306.00	1
TBL7	Steelcase	Coalesse Seating / Occasional Tables	72-2038LR	Calm	1006.00	1
TBL8	Steelcase	Coalesse Seating / Occasional Tables	COANDOTBL67	Andoo	6110.00	3
TBL9	Steelcase	Coalesse Seating / Occasional Tables	43-T222222	Sidewalk	1316.00	7
TBL10	COALESSE				878.00	10
Grand total: 243						

FIGURE 16-2.4 Furniture schedule

Notice a few things about the schedule above before moving on.

- **TBL3** and **TBL10** are missing information. Given the massive amount of information Steelcase has to enter, this is to be expected while they continue to develop content and their process for delivering this information.

- The **Count** column automatically changes to show total count when not *itemizing every instance*.

- The **Cost** value currently is for one instance.

- Furniture without a **Type Mark** does not appear in this schedule. Most of the furniture within the desk groups does not have *Type Marks* – just the main work surface does so we can just schedule the *Group*. (Unfortunately, you cannot schedule the *Group* itself here.)

Next you will adjust a few things related to the cost.

9. Click **Edit** next to *Formatting* in the *Properties Palette*.

10. Select *Cost* and check **Calculate totals** (Figure 16-2.5).

11. Click **OK**.

FIGURE 16-2.5 Furniture schedule formatting options

The totals are now calculated based on the *Count* parameter (Figure 16-2.6). If you also wanted a unit cost you could create a *Calculated Value* (via the **Fields** tab) and have it equal to *Cost*. You will do this next.

12. Click **Edit** next to *Fields* in the *Properties Palette*.

13. Click the **Calculated Value...** button in the center.

14. Enter the following (Figure 16-2.7).

E	F	G
Product	Cost	Count
s	12306.00	14
s	29238.00	22
	32472.00	41
	113460.00	61
	17290.00	14
ande	2356.00	2
	6692.00	4
	6324.00	12
	4208.00	8
ce	88200.00	4
	124686.00	9

FIGURE 16-2.6 Calculated totals

FIGURE 16-2.7 Calculated value

15. Click **OK**.

16. Move the new parameter just before the *Cost* column via the *Fields* tab.

17. Click **OK** to update the schedule.

Your schedule should now look like Figure 16-2.8.

Next you will adjust the formatting so the numbers are right aligned and have a dollar sign.

18. Click **Edit** next to *Formatting*.

19. Select *Cost*.

20. Select **Right** for *Alignment*.

21. Click the **Field Format...** button.

22. Make the adjustments shown in Figure 16-2.9.

23. Make these same changes for *Unit Cost*.

Your schedule should now look like Figure 16-2.10. The last thing you will do is apply *Conditional Formatting* to the schedule so any cost below a certain value is highlighted.

24. Click **Edit** next to *Formatting*.

25. Click *Unit Cost*.

26. Select *Conditional Formatting*.

27. Set the *Conditional Formatting* as shown in Figure 16-2.11; set the background color to **Yellow**.

28. Click **OK** twice.

E	F	G	H
Product	Unit Cost	Cost	Count
hairs	879.00	12306.00	14
hairs	1329.00	29238.00	22
n	792.00	32472.00	41
	1860.00	113460.00	61
	1235.00	17290.00	14
Grande	1178.00	2356.00	2
an	1673.00	6692.00	4
	527.00	6324.00	12
	526.00	4208.00	8
urface	22050.00	88200.00	4
	13854.00	124686.00	9
urface	7784.00	93408.00	12
	9840.00	9840.00	1

FIGURE 16-2.8 Calculated value in schedule

FIGURE 16-2.9 Cost format dialog

F	G	H
Unit Cost	Cost	Count
$879.00	$12306.00	14
$1329.00	$29238.00	22
$792.00	$32472.00	41
$1860.00	$113460.00	61
$1235.00	$17290.00	14
$1178.00	$2356.00	2
$1673.00	$6692.00	4
$527.00	$6324.00	12
$526.00	$4208.00	8
$22050.00	$88200.00	4

FIGURE 16-2.10 Added formatting

The schedule now shows a yellow cell for the *Unit Cost* of item TBL3 (Figure 16-2.12). This alerts you to the fact that the value is below the threshold of what would normally be possible. Your schedule, a cost estimate in this case, could be way off and you might not notice without the aid of the colored cells.

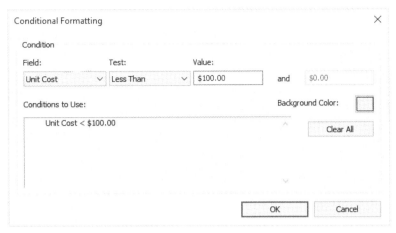

FIGURE 16-2.11 Conditional Formatting

FYI: This color fill does not print unless "Show conditional formatting on sheets" is checked on the Formatting tab (See Figure 16-2.5).

DK4	Steelcase	Coalesse Office	PDEW7024	Topo	$9840.00	$9840.00	1
TBL1	Steelcase	Coalesse Office	CODRNDP4827	Denizen	$2419.00	$16933.00	7
TBL2	Steelcase	Convene	CWFK60216	Conference Table Top	$16000.00	$16000.00	1
TBL3	COALESS				$0.00	$0.00	1
TBL4a	Steelcase	Coalesse Seating / Occasional Tables	COEL3624	Enea Lottus	$1946.00	$5838.00	3
TBL4b	Steelcase	Coalesse Seating / Occasional Tables	COEL4230	Enea Lottus	$1746.00	$5238.00	3
TBL5	Steelcase	Coalesse Seating / Occasional Tables	H35-T2442W	Enea	$846.00	$1692.00	2
TBL6	Steelcase	Coalesse Seating / Occasional Tables	72-2020H	Calm	$1306.00	$1306.00	1
TBL7	Steelcase	Coalesse Seating / Occasional Tables	72-2038LR	Calm	$1006.00	$1006.00	1
TBL8	Steelcase	Coalesse Seating / Occasional Tables	COANDOTBL67	Andoo	$6110.00	$18330.00	3
TBL9	Steelcase	Coalesse Seating / Occasional Tables	43-T222222	Sidewalk	$1316.00	$9212.00	7
TBL10	COALESS				$878.00	$8780.00	10
Grand total: 243						$624815.00	

FIGURE 16-2.12 Conditional Formatting shown in schedule

You will fix the schedule to have an accurate cost, but the way we set it up, it can no longer be changed here. The *Cost*, with **Calculate Total** enabled, cannot be edited (you would have to undo that temporarily), and the *Unit Cost* is based on another value (i.e., *Cost*) so it cannot be changed either.

29 Switch to the **Level 1 Furniture Plan**.

30. Select **TBL3** in the *Medium Conference Room*.

31. Click *Edit Type*; set the *Cost* to **4545.00**. Click **OK**.

Your schedule now has all the *Cost* parameters provided in the schedule and they are all over $100. Thus, there are no more colored cells. Take note of how quickly you were able to calculate over a half-million dollars' worth of furniture! There are many more variations on this, and you can even create multiple *Furniture Schedules* – the one that goes on the drawings would not have the cost information, for example.

Exercise 16-3:
Floor Finish Schedule

Because you created separate floor types for each floor finish, except for the sealed concrete, you can easily schedule the floor finishes. You will do so in this chapter, listing the floor finish, square footage and which level the finishes are on.

Creating a Floor Finish Schedule:

1. Open your *Law Office* file if needed.

2. Create a new schedule:

 a. *Category:* **Floors**

 b. *Phase:* **New Construction**

 c. *Name:* **Floor Finish Schedule**

3. Add the following *Fields* (in order):

 a. Type Mark

 b. Area

 c. Type

 d. Cost

4. Click **OK** to create the schedule.

In this schedule we want to filter it so we only see the ⅛″ finish floors. However, for some reason Revit does not let us *Filter* by *Type*. So we will have to use the *Type Mark* parameter which is currently empty.

<FloorFinish Schedule>			
A	**B**	**C**	**D**
Type Mark	Area	Type	Cost
	5814 SF	4" Concrete Slab	
	4735 SF	LW Concrete on Metal Deck	
	332 SF	LW Concrete on Metal Deck	
	91 SF	Floor Finish – CPT1	
	383 SF	Floor Finish – CPT1	
	251 SF	Floor Finish – CPT1	
	241 SF	Floor Finish – CPT1	
	111 SF	Floor Finish – CPT1	
	409 SF	Floor Finish – CPT2	
	1070 SF	Floor Finish – CPT2	
	144 SF	Floor Finish – CPT1	
	144 SF	Floor Finish – CPT1	
	144 SF	Floor Finish – CPT1	

FIGURE 16-3.1 Initial flooring schedule

You will sort the schedule by *Type*, enter the *Type Mark* information and then *Sort* and *Filter* by *Type Mark*. This is not perfect but is what we need to do in this case. The good thing is this will give you a better grasp on how schedules work.

5. Select **Edit** for *Sorting/Grouping* in the *Properties Palette*.

6. Sort by **Type** and **uncheck** *Itemize every instance*.

7. Click **OK**.

Your schedule is now sorted by *Type*; that is, *Family Type* (Figure 16-3.2). Now you can easily enter the *Type Mark* information.

> *FYI:* Notice some rows have an area listed and others don't. The ones that are blank mean there is more than one instance and the areas vary. In the image below, it is possible that there are two instances of EM1 and they are both exactly 299SF. We would need the Count column added to know for sure. Of course, that is not the case in this project.

<FloorFinish Schedule>			
A	**B**	**C**	**D**
Type Mark	Area	Type	Cost
	5814 SF	4" Concrete Slab	
		Floor Finish - CT3	
	299 SF	Floor Finish - EM1	
	203 SF	Floor Finish – RF1	
		Floor Finish – RF2	
		Floor Finish – CPT1	
		Floor Finish – CPT2	
		Floor Finish – CPT3	
		Floor Finish – CPT4	
		Floor Finish – CPT5	
		LW Concrete on Metal Deck	

FIGURE 16-3.2 Flooring schedule sorted by [Family] Type

8. For the "Floor Finish" rows, add a *Type Mark* value:

 a. For *Floor Finish – CT3*, enter *Type Mark* value **CT3**.

 b. For *Floor Finish – CPT1*, enter *Type Mark* value **CPT1**.

 c. Etc. (Figure 16-3.3)

You can now *Filter* and *Sort* the schedule by *Type Mark*. Because the *Type Mark* is a *Type* parameter, you only have to enter it once and it automatically applies to all instances of that floor *Type*, even future instances.

9. Set the schedule to **Filter** by *Type Mark* **Contains** [blank] as shown in Figure 16-3.4.

10. Set the ***Sorting/Grouping*** to sort by *Type Mark*, and check **Grand Totals:** *Totals only* (Figure 16-3.5).

11. Click **OK**.

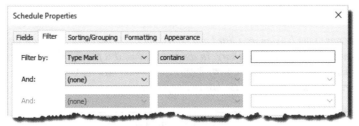

FIGURE 16-3.3 Flooring Schedule: *Type Mark* values added

FIGURE 16-3.4 Filter by blank *Type Mark*

FIGURE 16-3.5 Sort by *Type Mark* and display totals

Now your schedule is only showing the floor finish *Floors* (Figure 16-3.6). The next adjustment you need to make is on the ***Formatting*** tab, in the *Schedule Properties* dialog box. We will hide the *Type* column and set the *Area* column to add up the totals.

	<FloorFinish Schedule>		
A	**B**	**C**	**D**
Type Mark	Area	Type	Cost
CPT1		Floor Finish – CPT1	
CPT2		Floor Finish – CPT2	
CPT3		Floor Finish – CPT3	
CPT4		Floor Finish – CPT4	
CPT5		Floor Finish – CPT5	
CT3		Floor Finish – CT3	
EM1	299 SF	Floor Finish – EM1	
PR1	203 SF	Floor Finish – RF1	
RF2		Floor Finish – RF2	

FIGURE 16-3.6 Floor schedule sorted by *Type Mark*

12. Click *Edit* next to *Formatting* in the *Properties Palette*.

13. Select *Area* and check the **Calculate totals** option (Figure 16-3.7).

14. Select *Type* and check the **Hidden Field** option.

15. Click **OK**.

FIGURE 16-3.7 Adjusting formatting

Your schedule now reports the total areas for each finish (Figure 16-3.8). You could add *Cost* data and also create a **Calculated Value**, <u>Cost x Area</u>, to list the total cost per floor finish.

\<FloorFinish Schedule\>		
A	**B**	**C**
Type Mark	Area	Cost
CPT1	3894 SF	
CPT2	1479 SF	
CPT3	664 SF	
CPT4	148 SF	
CPT5	132 SF	
CT3	1724 SF	
EM1	299 SF	
PR1	203 SF	
RF2	668 SF	
	9211 SF	

FIGURE 16-3.8 Total areas shown

The last steps you will perform are to duplicate the current floor schedule and make the new one sort the floor finishes by *Level*.

16. Right click on the *Floor Finish Schedule* in the *Properties Browser* and select **Duplicate View → Duplicate**.

17. Rename the new schedule view to **Floor Finish Schedule by Level**.

18. Open the **Schedule Properties** dialog and make the following changes:

 a. *Fields* tab: Add *Level* to the schedule.

 b. *Sorting/Grouping* tab: Make adjustments shown in Figure 16-3.9.

 c. *Formatting* tab: Set *Level* to be a **Hidden field**.

FIGURE 16-3.9 Sorting and grouping by *Level*

When finished your schedule should look like Figure 16-3.10. You now have a total for each floor. This may not be necessary for the floor finish schedule but will most certainly be applicable to several schedule types.

<Floor Finish Schedule by Level>		
A	B	C
Type Mark	Area	Cost
Level 1		
CPT1	1059 SF	
CPT2	409 SF	
CPT3	135 SF	
CPT4	148 SF	
CPT5	132 SF	
CT3	1724 SF	
EM1	299 SF	
PR1	203 SF	
RF2	668 SF	
	4778 SF	
Level 2		
CPT1	2835 SF	
CPT2	1070 SF	
CPT3	529 SF	
	4434 SF	
	9211 SF	

FIGURE 16-3.10 Final floor finish schedule

Facilities Management:

It is easy to add additional parameters, associated with floors, to track installation date, material care instructions and more to aid in facilities management. This information could be utilized by the design firm as additional services or by the client once the project is complete, or both parties!

Reusing Schedules:

Schedules can be created in your template and thus be ready to use at the very beginning of a project. In the case of flooring, as soon as you place the first ⅛″ floor finish, the schedule would automatically begin to report square footage.

If a new schedule is created in a project and you would like to use it in other projects, you can right-click on the schedule title in the *Project Browser* and select **Save to New File**. The file created is essentially an empty project file which only contains the schedule, which is also empty because there is nothing in the file to schedule. To load this schedule in another project or template file, select **Insert → Load from File → Insert Views from File**.

Conclusion:

It is easy to see how powerful schedules are in Revit. You are encouraged to read more about schedules and learn to leverage this feature in your work.

Self-Exam:

The following questions can be used as a way to check your knowledge of this lesson. The answers can be found at the bottom of this page.

1. The area for a room is derived from the *Room* element. (T/F)

2. A *Room Tag* can have a leader. (T/F)

3. Use the _____ tool to add color to the rooms in a plan view.

4. When "Itemize every Instance" is checked, there is a row for every instance. (T/F)

5. The *Count* field can be used in formulas. (T/F)

Review Questions:

The following questions may be assigned by your instructor as a way to assess your knowledge of this section. Your instructor has the answers to the review questions.

1. It is not possible to add finish information, e.g., base finish, wall finish, etc., to multiple rooms at once. (T/F)

2. What is the size of your Revit project file? _____MB

3. _____ can be used to change the color of a cell in a schedule.

4. The schedule formatting only shows up when you place the schedule on a sheet. (T/F)

5. Schedules can be exported for use in other projects. (T/F)

6. The colors in a *Legend Color Scheme* can be changed. (T/F)

7. Schedules cannot be used for facility management. (T/F)

8. You cannot export a schedule to a file. (T/F)

9. A schedule can be filtered to show only furniture which costs more than $100. (T/F)

10. Schedules can sort elements by which Level they are on in the model. (T/F)

Lesson 17
Lighting and Rendering:

This chapter will cover the development of artificial lighting and photorealistic renderings using Autodesk® Revit®.

The image below has pendant lights hanging below the wavy ceiling above the *Lobby*. You should also be able to spot the lights in the private offices next to the stairs. In this rendering the sun is shining brightly into the space, so the artificial lighting is not as pronounced.

You will learn how to create images like this!

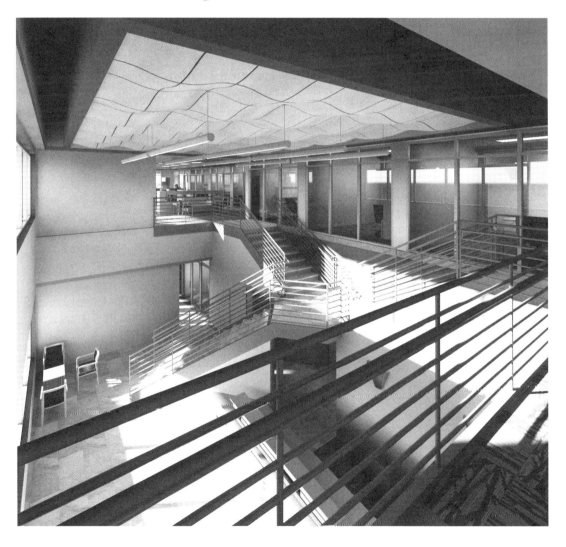

Exercise 17-1:
Lighting Design

To give lighting design a proper introduction, a separate book would be required. Here we will simply cover a few basic concepts to set things up for our renderings. You will learn a little about what content is available, the settings and how to place them into your model.

Loading Content:

You will only be using content which comes with Revit in this chapter. However, there are a few lighting manufacturers who are now providing lighting fixtures in Revit format (RFA).

1. **Open** your law office project.

2. **Load** the *Families* listed below from the **Lighting Fixtures** folder:

Family	Type	Count
Downlight - Recessed Can	6" Incandescent - 277V	105
Downlight - Spot	75 Degree Tilt - 277V	3
Pendant Light - Linear - 1 Lamp		23
Sconce Light - Uplight	100W - 277V	10
Troffer Light - 2x2 Parabolic	2'x2'(4 Lamp) - 277V	38
Troffer Light - 2x4 Parabolic	2'x4'(4 Lamp) - 277V	61

Note that most lighting fixtures today at LED, which these sample Autodesk families are not. However, they will be used for this tutorial for convenience.

You will refer back to this list to determine the *Types* to use when placing the content within the model. The count listed is not particularly important as you may decide to change the lighting design if desired, if your instructor allows it.

With Revit architectural content, the *Lighting Fixtures* are *ceiling hosted*. This content does not work when the ceilings are coming from a linked model. Face based content is needed when the ceilings are linked. All the MEP content that comes with Revit is face based because the MEP engineers usually work within the context of a linked architectural model. The one problem with face-based content, when placed on a linked ceiling, is that it cannot cut a hole in the ceiling.

> **FYI:** *A link is one Revit model referenced into another.*

Lighting fixtures can be placed by the **Competent** tool or the **Lighting Fixture** tool on the Systems tab.

Placing Light Fixtures:

3. Using the *Component* tool, place the **Troffer Light - 2x4 Parabolic** *family* as shown in the image below (Figure 17-1.1).

 a. See the list on page 17-2 for the specific *Type* to use.

 b. All fixtures shall be centered or equally spaced, as shown.

 c. Adjust ceiling grid positions if needed.

 d. Place the light on a ceiling, *Rotate* if required, and then use *Align* to move it into place.

 e. Once you have a fixture placed and aligned, it is sometimes easier to use the *Copy* tool, snapping to ceiling grid intersections to reference.

 f. See Step 4 for instructions on placing the pendant lights in the *Large Conference Room*.

FIGURE 17-1.1 Level 1 Reflected Ceiling Plan with light fixtures added – West portion of building

The wavy feature ceiling in the *Large Conference Room* is not in the *Ceiling* category so the pendant lights cannot be directly placed on them. Even if they were, the irregular shape would cause problems. If you recall, the wavy ceiling, in the *Specialty Equipment* category, was placed on a regular 2'x2' ACT ceiling system. If we temporarily delete the wavy ceiling, we can then place the pendant lights. Once the lights are placed, we can then use the ***Paste Aligned to Same Place*** tool to put the wavy ceiling back. The sections will not look perfect, as the pendant stems will be poking up above the ceiling. However, this is the best we can do without getting into more involved family editing.

4. Place the pendant lights in the *Large Conference Room* per the following steps:

 a. **Select** all the wavy ceiling tiles.

 b. Delete them using **Ctrl+X**. (This copies them to the *Clipboard* first.)

 c. Place the five pendant lights (Figures 17-1.1 and 17-1.2).

 d. **Modify** *(tab)* → **Paste** *(drop-down)* → **Aligned to Same Place** to replace the wavy ceiling tiles.

FIGURE 17-1.2 Adding pendant lights in large conference room

When placing lighting fixtures, you often realize the ceiling tiles and grids need to shift. Initially the goal is to have the ceilings centered on the room. However, it is more important to have an equal light distribution, rather than having lights offset to one side or not equally spaced. Lights which are too close to the wall also create noticeable hotspots.

5. Follow the same steps to place the light fixtures shown in Figure 17-1.3.

 a. The spot lights are hosted by the soffit and should point towards the wall sign – these three fixtures are highlighted in the image by arrows.

FIGURE 17-1.3 Adding more lighting fixtures in the level 1 ceiling plan

You are almost done with the Level 1 lights. The last fixture to add is a wall sconce. These will be added in the floor plan rather than the ceiling plan.

6. Switch to the **Level 1 Floor Plan** view.

7. Add the **Sconce Light – Uplight** family to the walls (Figure 17-1.4a).

 a. Family is wall hosted.

 b. Default height above floor is 6′-4″; do not change.

 FYI: This family has a special vertical line, set to be invisible, which extends down to the floor. This forces the family to pass through the view's cut plane and thus appear in the plan view.

The wall sconces are used to light the *Entry Vestibule* and a portion of the *Lobby*. Because we have elected to use irregular ceiling tiles, placing light fixtures in them is not easy. Additional lighting would likely be required, but we will not add anything else at this time.

The camera view below shows a few of the lights on the wall. Wall sconces come in all sorts of shapes and sizes.

FIGURE 17-1.4A Wall sconces added to Level 1 Floor Plan

FIGURE 17-1.4B Camera view showing wall sconces

8. Using the same steps previously outlined, place the fixtures shown in Figures 17-1.5, 6, 7 and 8.

 a. You may wish to add fixtures to areas not currently considered.

 b. For the pendant fixtures above the *Lobby*, use the **96″** version and adjust the *Type Properties* **Drop From Ceiling** value to **2′-0″**.

FIGURE 17-1.5 Level 2 Reflected Ceiling Plan with light fixtures added – West portion of building

FIGURE 17-1.6 Level 2 Reflected Ceiling Plan with light fixtures added – Northeast portion of building

FIGURE 17-1.7 Level 2 floor plan with wall sconces added – Northeast portion of building

The image below shows eight pendant light fixtures within the wavy ceiling above the *Lobby*.

FIGURE 17-1.8 Level 2 reflected ceiling plan with light fixtures added – Southeast portion of building

Content Properties:

When you select a light fixture and select *Edit Type*, you can see several properties which directly relate to how this family will render. For example, *Initial Intensity* can be adjusted for a higher wattage which will result in more light output. This can help when doing interior renderings, as they can be too dark at times. However, it is best to leave these at the intended real-world values, if known, and use the *Adjust Exposure* settings in the render dialog box, which will be covered in the next exercise.

FIGURE 17-1.9 Lighting fixture type properties

IES Photometric Files:

Revit has the ability to attach an *IES photometric file* to a lighting fixture family. This is a text file which defines the actual light distribution for a real-world fixture. This data is available from most major lighting manufacturers. This format has been around for some time and is used by many other programs. When these are used, the light cast in a rendering is more accurate. When not used, the light can still look very realistic and differences may not even be noticeable to most people. None of the content used in this chapter has an IES file attached to it.

Lighting Analysis in Revit:

It is possible to do accurate point-by-point artificial lighting analysis in Revit using a third-party add-in. One such add-in is ElumTools. This add-in is developed by the makers of AGI32, which is widely used throughout the industry by professional lighting designers. The ElumTools add-in can provide the designer with accurate data on how well a space is lit based on type and quantity of fixtures as well as the surface materials in the room. For example, a high-gloss wall or floor will reflect more light which will increase the ambient light in a space. ElumTools can calculate this!

When this software is installed on your computer, you will have a new tab on the *Ribbon* called ElumTools (Figure 17-1.10). On this tab you have several tools to define how to analyze the lighting and what the materials are. Initially the *Material Mapping* is based on the Revit *Materials* but can be overridden if needed.

FIGURE 17-1.10 ElumTools add-in for Revit adds a new tab to the *Ribbon* when installed

Once everything is set up in ElumTools, you can select one of the "calculate" options. This opens a separate window in which the calculations are done (Figure 17-1.11). However, when done the results are automatically pushed back into the *Room* element. This allows you to schedule the results using a schedule based on *Rooms*. Using tools such as this, plus energy analysis and structural analysis, is what really makes the world of BIM shine! Making decisions early on, before it is too difficult to change, is huge in the world of design.

Be sure to check out the ElumTools tutorial provided with the online files for this book. Students can use the software for free; visit www.elumtools.com for details!

Exercise 17-2:
Creating a Rendering of the Lobby

Autodesk Revit has a high-quality rendering engine which can create photo-realistic images. You will explore a number of rendering techniques in this section. The first thing you will do is set your geographic location to ensure the sunlight coming through the exterior glass is shown accurately.

Project Location:

A first step in setting up a rendering is to specify the location of the project on the Earth. This is important for accurate daylight studies.

1. Select **Manage → Project Location → Location**.

Location

2. In the *Project Address* field, enter **Minneapolis, MN**.

 a. You may enter your location if you wish.

 b. It is even possible to enter the actual street address.

3. Press **Enter**.

You should now see an internet-based map of your project location, complete with longitude and latitude (Figure 17-2.1).

4. Click **OK**.

FIGURE 17-2.1 Location Weather and Site dialog

Sun Settings (day and time):

The next step in preparing a rendering is to define the sun settings. This includes the month, day and time. Revit allows you to save specific settings, which enables you to easily switch between various lighting situations within a single view. You will explore the various options available. These settings affect renderings and regular views when shadows are turned on.

5. Select **Manage → Settings → Additional Settings** (*drop-down list icon*) **→ Sun Settings**.

6. Make the following changes (Figure 17-2.2):

 a. Solar Study: **Still**

 b. **Uncheck** *Ground Plane at Level*.

 c. Select **Summer Solstice** on the left.

 d. Click the ***Duplicate*** icon in the lower left.

 e. Enter the following name: **Law Office 6-30 at 3pm**.

 f. Adjust the *Month*, *Day* and *Time* to match the name just entered.

 g. Change the year to *2025*.

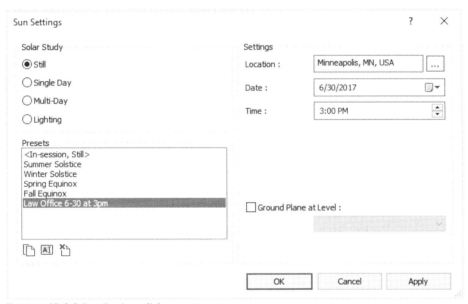

FIGURE 17-2.2 Sun Settings dialog

7. Click **OK** to close the dialog.

The previous sun settings apply to the entire project, not just the current view, which is the intent of the *Manage* tab. However, each view can be set to different sun settings. You could have one view setup for daytime and another for nighttime.

Specifying True North:

Another important step in rendering sunlight is telling Revit what *True North* is. All architectural drawings assume a project north being straight up for convenience of design and drafting. To set *True North*, you would do the following (but do not make any changes now):

- Switch to the **Site Plan** view.
- Set the *Orientation* to *True North* in the *View Properties*.
- Select **Position → Rotate True North** on the *Manage* tab.
- Rotate the model accordingly.

Properly setting *True North* will make the renderings more realistic and daylighting studies valid. This applies to both interior and exterior renderings, and any view with shadows on.

All views set to *Project North* will not be rotated. Do not modify *True North* at this time.

Custom Materials:

Some of the furniture you downloaded uses custom render appearance image files. For example, the Steelcase content uses a number of different images to represent the various types of fabrics available with their products. You will download the Steelcase image files and tell Revit where they are.

8. Browse to **Revit.Steelcase.com** using your web browser.

9. Click the **Steelcase Material Library** link to download a ZIP file.

 Tip: This file can also be found with the provided online files for this book.

10. Once downloaded, open the file and *Copy/Paste* the image files to a folder on your computer's hard drive, or a network location. *TIP:* The folder name cannot have a period in it. You must move the files out of the original ZIP file.

11. In Revit, select **File Tab → Options**.

12. Select the *Rendering* label on the left, and then click to add a new render appearance search path (Figure 17-2.3).

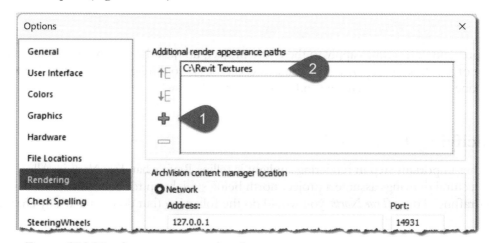

FIGURE 17-2.3 Render appearance search path to your custom textures

Rendering the View:

You have limited options for setting up the building's environment. If you need more control than what is provided directly in Revit, you will need to use another program such as Autodesk *3ds Max Design 2025* which is designed to work with Revit and can create extremely high-quality renderings and animations; it even has day lighting functionality that helps to validate LEED® (Leadership in Environmental and Energy Design) requirements.

13. Create a *Camera* view in the *Lobby*.

14. Select the ***Show Rendering Dialog*** icon on the *View Control Bar*; it looks like a teapot (Figure 17-2.4).

> ***FYI:*** *This icon is only visible when you are in a 3D view; the same is true for the* ViewCube.

FIGURE 17-2.4 Rendering dialog icon

The *Rendering* dialog box is now open (Figure 17-2.5). This dialog box allows you to control the environmental and quality settings and actually create the rendering.

15. In the *Lighting* section, click the down arrow next to *Scheme* to see the options.

16. Select **Interior: Sun and Artificial** (Figure 17-2.6).

The lighting options are very simple choices: Is your rendering an interior or exterior rendering, and is the light source *Sun*, *Artificial* or both? You may have artificial lights, light fixtures like the ones you placed in the office, but still only desire a rendering solely based on the light provided by the sun. *Sun only* will be faster as artificial takes much longer to calculate.

FIGURE 17-2.5 Rendering dialog

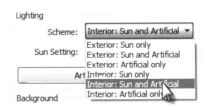

FIGURE 17-2.6 Lighting options

17. In the *Lighting* section, set the *Sun Setting* to <Still> **Law Office 6-30 at 3pm**. (This relates to the Figure 17-2.2 settings.)

> *FYI: In the Sun and Shadow Settings dialog box, Revit lets you set up various "scenes" which control time of day. Two examples would be:*
> - *Law Office Daytime, summer*
> - *Law Office Nighttime, window*
>
> *Looking back at Figure 17-2.2, you would click Duplicate again and provide additional names. The names would then be available from the Sun drop-down list in the Render dialog box.*

18. Click on the **Artificial Lighting** button (Figure 17-2.6).

You will now see a dialog similar to the one shown to the right (Figure 17-2.7).

You will see several light fixtures. The light fixtures relate to the fixtures you inserted in the Reflected Ceiling Plans. It is very convenient that you can place lights in the ceiling plan and have them ready to render whenever you need to (i.e., render and cast light into the scene!) Here you can group lights together so you can control which ones are on (e.g., exterior and interior lights).

FIGURE 17-2.7 Scene Lighting dialog

19. Click **Cancel** to close the *Artificial Lighting* dialog.

20. Click the down-arrow next to *Style* in the *Background* area (Figure 17-2.8).

21. Select **Sky** for Background style.

FIGURE 17-2.8 Background Style

Notice that for the background, one option is *Image*. This allows you to specify a photograph of the site, or one similar. It can prove difficult getting the perspective just right, but the end results can look as if you took a picture of the completed building.

22. Make sure the *Quality* is set to **Medium** and Resolution to **Printer/150 DPI**.

> *FYI: The time to process the rendering increases significantly as the quality level is raised.*

23. Click **Render** from the *Rendering* dialog box.

You will see a progress bar while the rendering is processing (Figure 17-2.9).

FIGURE 17-2.9 Rendering progress

You may see a number of warnings about missing render appearance images. These are image files which were assigned to some of the content you downloaded from the internet. These files were not readily available when you originally downloaded the content. You would have to go back and look for it, but it is more likely that you would have to request these image files or substitute different ones.

After a few minutes, depending on the speed of your computer, you should have a medium quality rendered image similar to Figure 17-2.10 below. However, this is a high-quality rendering, not a draft. You can increase the quality of the image by adjusting the quality and resolution settings in the *Render* dialog. However, these higher settings require substantially more time to generate the rendering.

The last step before saving the Revit project file is to save the rendered image to a file.

> *FYI: Each time you make changes to the model, which are visible from that view, you will have to re-render the view to get an updated image.*

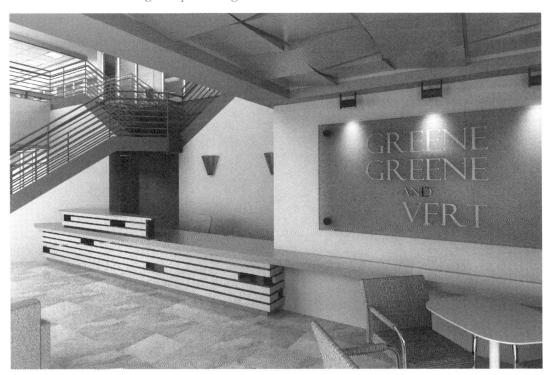

FIGURE 17-2.10 Rendered view

Exporting to a File

Once the rendering is done, it has not been saved. If you close the project or your computer loses power, you will need to re-render the view. Once the rendering is saved to a file you can use it in other programs such as *Microsoft Word* or *Microsoft PowerPoint*.

24. From the *Rendering* dialog select **Export**.

> *FYI: The* 'Save to Project' *button saves the image within the* Revit Project *for placement on* Sheets. *This is convenient but does make the project size larger, so you should delete old ones from the* Project Browser!

25. Select a *location* and provide a *file name*.

26. Set the **Save As** type to **JPEG** or **PNG**.

27. Click **Save**.

Realistic Display Setting:

In addition to *Shade* and *Hidden Line* display settings you can also select **Realistic**. This will show the *Render Appearance* materials in your building without the need to do a full rendering. Although the shades and shadows are not nearly as realistic as a full rendering, turning on **Ambient Shadows** in the *Properties Palette > Graphic Display Options* comes closer. This is a great way to get an idea of how the materials look but you will find it best to minimize the use of this setting as Revit will run much slower.

Exercise 17-3:
Rendering an Isometric in Section

This exercise will revisit the *Section Box* tool, but in the context of renderings. The two features together can produce some interesting results.

Setting Up the 3D View:

1. Open your *Law Office* file.

2. Switch to the **Default 3D View** via the **3D** icon on the *QAT*.

3. Set the *Visual Style* to **Consistent Colors** or *Realistic* if your computer can handle it.

4. Make sure nothing is selected so you see the **View Properties** in the *Properties Palette*.

5. Activate the **Section Box** parameter and then click **Apply**.

You should see a box appear around your building, similar to Figure 17-3.1. When selected, you can adjust the size of the box with its grips. Anything outside the box is not visible. This is a great way to study a particular area of your building while in an isometric view. You will experiment with this feature next.

Section Box (shown selected)

6 grips to manipulate the Section Box – only visible while the *Section Box* is selected

FIGURE 17-3.1 3D view with Section Box activated

6. To practice using the *Section Box*, drag the grips around until your view looks similar to **Figure 17-3.2**.

FYI: Grips only show when Section Box *is selected.*

TIP: This will require use of the ViewCube *tool as well.*

FIGURE 17-3.2 3D view with adjusted Section Box

This creates a very interesting view of the project. What client would have trouble understanding this drawing?

7. Select the *Render* icon.

 a. Select the *Scheme*: **Exterior: Sun only**.

 b. Set the *Sun Setting* to **Sunlight from Top Left**.

 c. Background *Style* to **Color** (set the color to white).

8. Leave all other settings as they are and **Render** the view.

After a few minutes, the view is rendered and can then be saved (Figure 17-3.3). Notice how the ground is rendered in section. Some of the model elements are gray because the various elements have the default material selected. If you take the time to specify a material for everything, you can generate another rendered image that is much better looking.

FIGURE 17-3.3
Rendered 3D view with Section Box

Two more examples are shown in Figure 17-3.5 and 17-3.6. Here we see an isolated area of the building enlarged to show more detail. It is possible to duplicate these 3D views so they can be easily accessed again. This view can even be placed on a sheet next to the plan and elevation views of the same area. This is a great way to create a presentation board for a specific area you are meeting with the client about.

Adjusting the **_Exposure Control_** before exporting the image can make a huge difference in how realistic the image looks. The previous image on this page has no adjustments, whereas the next two have the adjustments shown in Figure 17-3.4 to the right. For interior renderings, setting the _White Point_ to about 8600, or warmer, for artificial light makes the materials look more realistic. See the colored images provided with the online files for this book.

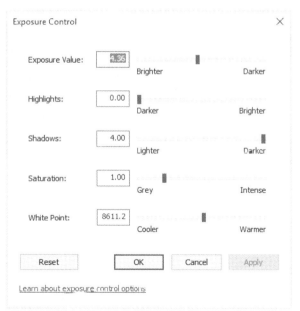

FIGURE 17-3.4 Exposure Control dialog

FIGURE 17-3.5 Rendered 3D view with Section Box – Level 1

FIGURE 17-3.6 Rendered 3D view with Section Box – Level 2

Exercise 17-4:
Creating another Interior Rendering

In this exercise you will create another interior rendering. This one will be in the *Lunch Room*.

Setting up the Camera View:

1. Open your *law office* project.

2. Open **Level 1** view.

3. From the *View* tab, select **3D View → Camera**.

4. Place the *Camera* as shown in **Figure 17-4.1**.

FIGURE 17-4.1 Camera placed in Lunch Room

Revit uses default heights for the camera and the target. These heights are based on the current level's floor elevation. The two points you picked are 5'-6" above the floor, which is the average eye height of a person. These reference points can be edited via the camera properties.

Revit will automatically open the newly generated camera view. Your view should look similar to Figure 17-4.2.

> *FYI: Make sure you created the camera on Level 1 and picked the points in the correct order.*

FIGURE 17-4.2 Initial interior camera view

We want to see more of the room, so we will adjust the *Crop Region*. This change is similar to adding a wide-angle lens to your camera.

5. Set the *Detail Level* to **Fine** and then adjust the **Crop Region** rectangle. Modify the view to look like Figure 17-4.3.

FIGURE 17-4.3 Modified interior camera; Consistent Colors view with Ambient Shadows turned on

6. Switch back to **Level 1** to see the revised camera view settings.

 a. While in plan view, right-click on the camera view name in the *Project Browser*, and then select **Show Camera**.

Notice the field of view triangle is wider based on the changes to the *Crop Region* (Figure 17-4.4).

FIGURE 17-4.4 Modified camera – Level 1

Creating the Rendering:

Next you will render the view.

7. Switch back to your interior camera view for the *Lunch Room*.

8. Select **Show Render Dialog** from the *View Control Bar*.

9. Set the *Quality* setting to **Medium**.

10. Set the lighting *Scheme* to **Interior: Sun and Artificial**.

 TIP: Be sure to select the "Interior" *option.*

11. Set the *Sun* to **Law Office 6-30 at 3pm**.

12. Click **Render** to begin the rendering process.

This will take several minutes depending on the speed of your computer. When finished, the view should look similar to Figure 17-4.5. Before exporting the image to a file, you will adjust the exposure settings to improve the quality of the image. The second image (Figure 17-4.6) below has the exposure adjusted.

FIGURE 17-4.5 Rendered interior view

FIGURE 17-4.6 Rendered interior view with exposure adjustments (see next page)

Adjusting Exposure:

13. In the *Rendering* dialog, click the *Adjust Exposure* button.

14. Adjust the dialog similar to Figure 17-4.7.

15. Click **Apply**.

> *TIP: Increase Exposure Value if your image is too dark.*

You should play with the settings and apply/render several times to see how the image changes with the various adjustments made.

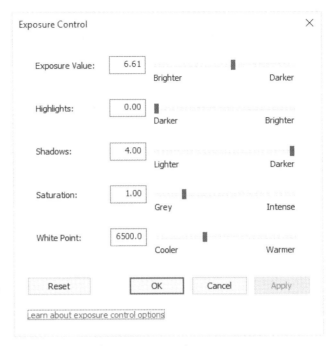

FIGURE 17-4.7 Adjust exposure settings

Save the Image:

16. Click ***Export*** from the *Rendering* dialog box to save the image to a file on your hard drive. Name the file **Lunch Room.jpg**.

You can now open the *Lunch Room.jpg* file in Adobe Photoshop or insert into a word processing program to manipulate or print.

> *TIP: A quick way to print is to right-click on the file in Windows Explorer and select **Print**.*

To toggle back to the normal hidden view, click ***Show the Model*** from the *Rendering* dialog box or simply close the *Rendering* dialog. To save time rendering, turn on the *Section Box* for the view. This will crop portions of the building not seen. This cuts out elements and lighting fixtures from the rendering process.

There are many things you can do to make the rendering look even better. You can add interior props (e.g., pictures on the wall and items on the countertop). You can adjust the *Sun* setting to nighttime and then render a night scene.

> *TIP: Setting the output to Printer rather than Screen allows you to generate a higher resolution image. Thus, between the Quality setting and the Output setting, you can create an extremely high-quality rendering, but it might take hours, if not days, to process!*

Revit also gives you the ability to set a material to be self illuminating. This will allow you to make something look like it is lit up. You can also set a lamp shade to glow when a light source has been defined under it, so it looks more realistic.

17. **Save** your project.

Exercise 17-5:
Adding People to the Rendering

Revit provides a few RPC people to add to your renderings. These are files from a company which provides 3D photographic content for use in renderings (www.archvision.com). You can buy this content in groupings, such as college students, or per item. In addition to people, they also offer cars, plants, trees, office equipment, etc.

Loading Content into the Current Project

1. Switch to **Level 1 Furniture Plan** view.

2. Using *Load Family*, select the **RPC Notebook Computer**, **RPC Male** and **RPC Female** files, using the *Ctrl* key to select all three from the *Entourage* folder.

3. Using the *Component* tool, place the **laptop**, three **Males** and three **Females** as shown in Figure 17-5.1. Set laptop offset to 29″; *it will still be hidden by the table in plan.*

FIGURE 17-5.1 Level 1 Large Conference Room– RPC people added

The RPC content often has additional visibility settings, as seen for the laptop in the image to the right.

The line and point in the symbols (Figure 17-5.1) represent the direction a person is looking. You simply rotate the object to make adjustments.

4. Create a **camera** view in the *Large Conference Room* as shown in Figure 17-5.1.

5. Adjust the new view's ***Crop Region*** to look like Figure 17-5.2.

Notice how the camera view now has abstract versions of people in the scene (Figure 17-5.2). These are placeholders which are replaced with photo-quality people. The people are in the correct perspective and create shadows.

You should switch back to plan view and adjust the people, if needed, so they are not hidden behind a chair or curtain wall mullion.

FIGURE 17-5.2 Camera view in Large Conference Room with RPC people added

6. **Render** the *Large Conference Room* view with the settings previously used.

Your rendering should now have people in it and look similar to Figure 17-5.3.

FIGURE 17-5.3 Large Conference Room rendering with people added

Adding people and other "props" gives your model a sense of scale and makes it look more realistic. After all, architecture is for people. These objects can be viewed from any angle. Try a new camera view from a different angle to see how the people adjust to match the view and perspective.

7. **Save** your project.

> *FYI: As with other families and components, the more you add to your project, the bigger your project file becomes. It is a good idea to load only the items you need and delete the unused items via the* Project Browser. *Your project should be about 45MB at this point in the tutorial.*

Sun Path

It is possible to turn on the feature called ***Sun Path***.

This shows the path of the sun over a day and a year. This is accessed from the *View Control Bar* while in a 3D view.

Once on, you can click and drag the sun along its daily or yearly path! Below are additional views rendered from the Law *Office* model. You are encouraged to create a few more camera views and see how they render. Try increasing the settings and see how much longer it takes to complete. Be sure to compare the quality. Often, a rendering set to *Medium* at **300dpi** is sufficient, especially if you are going to apply a filter, e.g., watercolor, in a graphics program such as Photoshop.

> *FYI: The images below were rendered with the* **Interior: Artificial Lights Only** *option. This eliminates the horizon and other possible distractions from the interior scene. Plus it renders faster without the sunlight.*

FIGURE 17-5.4 Level 2 open office in west wing, rendered with only artificial lighting

FIGURE 17-5.5 Level 1 hallway near Large Conference Room

FIGURE 17-5.6 Level 1 hallway, looking West at Small Conference Room

FIGURE 17-5.7 Level 1 Lobby view

Exercise 17-6:
Rendering Tips and Tricks

This section will present a number of tips and tricks you can use to improve your renderings in Revit.

Basic Tips

- **Lock 3D/Camera View**

 The ability to lock a 3D view is nice when you get your view set up the way you want it. Simply click the ***Lock 3D View*** icon on the *View Control Bar* while in the specific view.

 > ***WARNING:*** *There are still a few ways this view can be changed (unfortunately). A number of the* Navigation Wheel *tools can change a locked view. For example,* Walk, Up/Down, Pan, *etc. Also, using the* Show Camera *option (mentioned later) in a plan view lets you change the camera graphics but does not affect the view (it remains locked).*

- **Show Camera option**

 The camera graphics you see in plan view, when initially creating a camera view, can easily be made visible at a later time. While in a plan view, simply right-click on the camera view name in the *Project Browser* and select ***Show Camera***. Sometimes it is easier to reposition the camera in a plan view rather than using the *Navigation Wheel* while in the camera view.

 > ***TIP:*** *A similar process works to view a* Section Box *applied to 3D view.*

- **Use Navigation Wheel to Adjust View**

 The *Navigation Wheel* is a great way to make quick adjustments to a camera view. Once the view is set up and you want to adjust the view to look a little to the left, the *Navigation Wheel* will let you do this without "moving your feet" in the building. Moving forward or backwards a little can sometimes be tricky as it may move too much and place you behind a wall (behind you, for example).

- **Far Clip Active Setting**

 Each camera view has a ***Far Clip Active*** setting in its *Properties Palette*. The initial ***Far Clip Offset*** is based on your second pick (i.e., direction the camera is looking) when creating the camera. This can sometimes inadvertently crop portions of the model. You can either turn this off all together or adjust the offset value.

- **Crop Region Size**

 Crop Region Size is an often overlooked function of the final rendering product. In a camera view, when the *Crop Region* is selected, you can select the *Size Crop* button on the *Ribbon*. You are now in the *Crop Region Size* dialog (Figure 17-6.1). When changes are made here, the default is to adjust how much is seen in the 3D view (i.e., more or less of a given space). This can also be done in the view using the grips on the *Crop Region*.

 However, the other option is called ***Scale (locked proportions)***. This feature changes the overall size of the image rather than how much is seen within the same sized "window." Revit can actually render billboard sized images!

 The dimensions shown in the dialog in Figure 17-6.1 would not be good for a presentation board. When the image is enlarged to fit on the board, it becomes pixelated.

 FIGURE 17-6.1 Crop Region Size dialog

 When the size of the image is increased, the render time is also increased. Notice the change in pixels when the original dimensions (image on left, below) are doubled (image on right). The images below are from the *Render* dialog, before and after adjusting the *Crop Region*.

 Initial image size

 Crop region adjusted

- **Orient to View**

 One way to view your model is to look at it in a 3D view with a *Section Box* applied based on another view. To do this, follow these steps:

 1. Switch to your **default 3D** view (make a copy if you wish).

 2. Right-click on the ***ViewCube*** (Figure 17-6.2).

 a. Select ***Orient to View*** from the pop-up menu.

 b. Expand ***Sections***.

c. Select the named camera view (i.e., the view you are rendering)

3. Drag on the *ViewCube* to adjust the view (17-6.3).

FYI: The Section Box in this view is not in any way tied to the Section Box in your camera or section view. They will not update together.

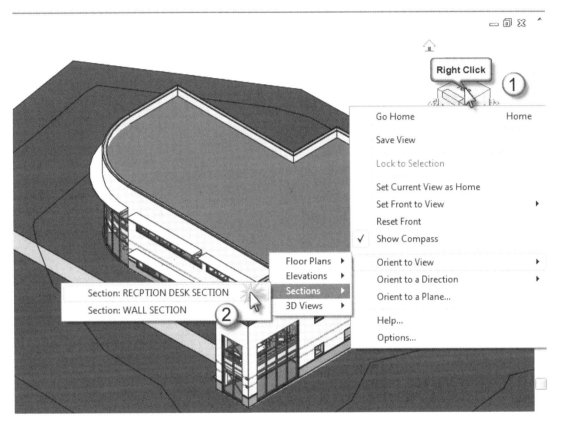

FIGURE 17-6.2 Using ViewCube's Orient To View feature to crop model

The result of *Orient to View* is the image below. The *Section Box* extents relate to the view depth and crop region size of the section view. The *Section Box* in this view can be turned off via **VV**.

FIGURE 17-6.3 Orient to view results

Rendering Dialog

- **Quality Settings (custom options)**
 The quality setting relates to how "hard" Revit has to think about the rendering while it is being created. For example, how many times should light bounce around the room. The more times, the more realistic; this takes more time (Figure 17-6.4).

FIGURE 17-6.4 Render quality options

Draft and *Low* should be used for early stages so you are not wasting time; you might let one render overnight at *Medium*. Artificial lighting really needs to be at least at a *Medium* setting for the final rendering to look decent. The *High* and *Best* settings are only needed for the very last rendering, once you know everything is set up right. And even then, most of the time you can get by with *Medium* quality.

Clicking the *Edit* option opens the **Render Quality Settings** dialog (Figure 17-6.5). You don't typically need to change things, but it is interesting/good to know what can be tweaked here. Revit provides one custom setting which allows these controls to be adjusted. For large complex models the Render Duration needs to be increased so the image is not grainy looking.

FIGURE 17-6.5 Render quality settings dialog

TIP: If your walls appear blotchy, use the custom option and check the three Daylight Portal *options at the bottom of the list. Start with another option, e.g.,* Medium, *and then click the* Copy to Custom *button.*

- **Artificial Lights (group)**
In the *Artificial Lights* dialog (Figure 17-6.6) the *Dimming* value can be changed for individual fixtures or for the entire group. To change the dimming value for the entire group, simply change the value to the right of the group name. Figures 17-6.9 thru 17-6.11 show how adjusting the dimming value changes the rendering; notice the hotspots on the back wall. This could be done for the three spot lights in the *Lobby*, at the *Sign Wall*.

FIGURE 17-6.6 Artificial lighting dialog

Placing lights in *Groups* allows you to control which lights emit light in a rendering. To create or add lights to a *Group*, simply select them in the model and pick from the *Light Group* list on the *Options Bar* (Figure 17-6.7).

FIGURE 17-6.7 Assigning light fixtures to a group in the model

TIP: Turn on the **Light Source** *sub-category, under* Lighting Fixtures, *to see the mesh representing the artificial light source in a view (Figure 17-6.8).*

FIGURE 17-6.8 Light source mesh made visible

The light source cannot be behind a solid object, unless that object is transparent. This can be a problem when the light fixtures are coming from a link. Linked light fixtures do not cut a hole in the ceiling. Thus, the ceiling obstructs the light. The light source in the fixture can be moved down just below the ceiling via the family editor.

FIGURE 17-6.9 Rendering with one light set to 75% dimming factor

 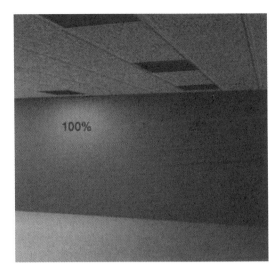

FIGURE 17-6.10 Renderings with additional dimming value adjustments (100%, 50% and 25%)

It is important to select render materials from the appropriate category (i.e. appearance asset). This will help to ensure they have the proper reflectance values, which contribute to the overall lighting of a space. The images above were rendered with *Interior: Artificial lighting only*, so the only light source is from the fixtures placed in the model.

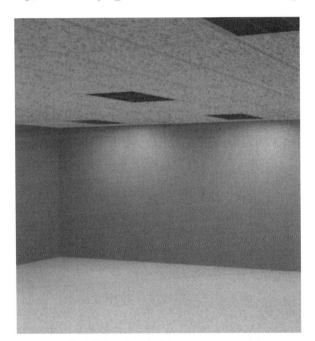

FIGURE 17-6.11 Rendered with concrete material

WALL PAINT SETTINGS

CONCRETE SETTINGS

IES files (photometric data)

An IES file is a definition of photometric data for a specific light fixture. This information is used by Revit to create more realistic artificial lighting. The IES file is available from most lighting manufacturers. A new *Add-In* for Revit called ElumTools, from the AGI32 folks, uses this photometric data to do point-by-point lighting analysis right in Revit! This means "professional grade" lighting design and analysis!

Take a look at this short article on IES files:
http://www.cgarena.com/freestuff/tutorials/max/ieslights/
Note the discussion on the IES *Viewer*.

Render Material Tips

* **Chipboard Rendering (using Phases)**
 Creating a chipboard (or paper/cardboard) rendering is easy and great for massing studies. The process can be done in the actual project without messing anything up. The trick is to create a "future" phase, set the 3D view to that future phase and assign a chipboard type material to the "existing" phase override in the *Phasing* dialog.

 Here are the steps:

 o Create a future phase; call this something like ***Paper Model*** (Figure 17-6.12).

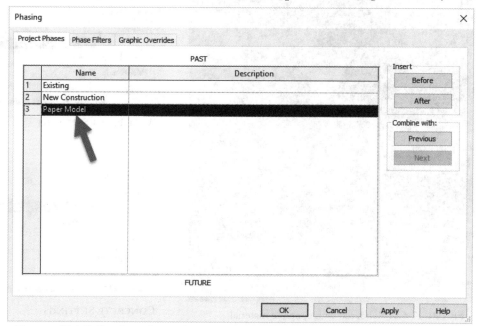

FIGURE 17-6.12 Creating a future phase

- o Set the **Graphic Overrides** for *Existing* to a *Material* (Figure 17-6.13).
 - This will be a material which represents the chipboard.
 - If your project has existing conditions, don't change the line and pattern overrides as they do not affect the rendering and will mess up the main project.

FIGURE 17-6.13 Setting the material override for existing elements in the Phasing dialog

- o In your 3D view you wish to render, adjust the view settings:
 - *Phase* = Paper Model
 - *Phase Filter* = Show All

Now, when you render, the elements will be overridden to use the "existing" material specified in the *Phasing* dialog. The key here is that the *Phase Filter* has the existing elements overridden. So, by setting the view to a fake future phase, everything in the view is considered existing. Thus, a universal material is applied for rendering!

FIGURE 17-6.14 Changing the 3D view's phase

In the 3D view you can tweak a few things to adjust the results. You can turn off the sub-categories for glass, under *Doors* and under *Windows*. You can also turn off the *Curtain Panels* category. This will make the glass hidden and allow the interior to be visible; otherwise the glass would be the chipboard material, not transparent. Another option would be to move certain elements (e.g., curtain wall or site) into the "future" phase so you

can control their materials separately. This will cause problems with the project in terms of documentation, so you might want to make a copy of the model for this final adjustment just before rendering.

The sample rendering shown below is the result (Figure 17-6.15). The renderings seem to process faster given the singular material and lack of reflectance and transparency. Thus, it is possible to increase the quality and print resolution higher than would otherwise normally be needed. This is nice in that the shadows are softer. The sample image also employs a *Section Box* which nicely crops the site.

FIGURE 17-6.15
Chipboard model rendering

Another variation on this process, or any rendering, is to set the time to sometime before sunrise. The initial result is the black blob seen to the right. Your initial thought when seeing this is – what a waste of time! But wait; there is actually a lot of hidden data here. BEFORE closing the file or doing another rendering, use the **Exposure Control** dialog to adjust the settings as shown in the image on the next page. The results are a nice lightly shadowed, realistic looking cardboard model. Other than adjusting the lightings, NOTHING else was changed.

FIGURE 17-6.16 Initial rendering with time changed to 4am

High quality chipboard rendering

FIGURE 17-6.17 Cardboard style rendering after exposure control adjustments

Making minor adjustments to exposure has a major impact on the image. So start out by making small adjustments and then click *Apply* to see the image change without closing the dialog. If things get too messed up, simply click the *Reset to Default* button and start over. Another option is to click OK, and then *Export* a good-looking image, and then try more changes.

> *TIP: Hover your cursor over the sliders to see a tooltip describing each setting.*

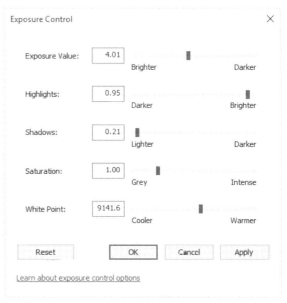

FIGURE 17-6.18 Exposure control adjustments

The following is a quote from Autodesk's Wiki Help document:

> "When rendering an image, exposure control (or tone mapping) is just as important as the lighting and materials used. Exposure control helps to convert real-world luminance values into a realistic image. It mimics the human eye response to luminance values with regard to color, saturation, contrast, and glare."

- *Material Options (using Transfer Project Standards)*

 Often, during the design phase there is a need to experiment with various color palettes. Revit does not have a way to do this. *Design Options* is certainly not something appropriate for this task. The easiest workflow, given the available functionality in Revit, is to use the **Transfer Project Standards** tool. Below are the steps:

1. Create a dummy Revit file as a "container" for the materials you want to use; the first file will be **Option A**.

2. Delete all the materials in the dummy project except the ones needed (Figure 17-6.19).

3. [Optional Step] Create an element for each material and add *Model Text* next to it as a reference (Fig. 17-6.21).

4. Copy the **Option A** file to create additional options. Adjust the materials as needed in each model. Name the files **Option B**, **Option C**, etc.

FIGURE 17-6.19 Materials list

5. Open the project file; make sure you have the same material names set up in this file.

6. In the project file, use *Transfer Project Standards* to import the material options.

 a. Make sure one of the "Option" files is open; select it from the drop-down list at the top.

 b. Click the *Check None* button.

 WARNING: *The project file can get really messed up if you miss this step.*

FIGURE 17-6.20 Transfer Project Settings dialog

 c. Check only **Materials** (Figure 17-6.20).

 d. Click **OK**.

 e. Click **Overwrite** when prompted.

The materials are now updated for the selected option. You can now render the model or print shaded 3D views. Once done, repeat these steps using another "Option" file to import those material settings. If the "drafting" settings (i.e., Materials → Graphics tab) are kept the same in all files, this method can be done in the main model – even during construction documents – without causing problems.

FIGURE 17-6.21A Option A "container" model **FIGURE 17-6.21B** Option B "container" model

Rendering Props Tips

- **Managing "props" Visibility (using design options)**
 This is a simple, yet powerful tip. Create a ***Design Option Set***, make the *Primary* option called "Empty" and a second (or more) option called "Entourage" (Figure 17-6.22). Only make the "Entourage" *Design Option* visible in your rendering views. All other views will automatically be set to the "Empty" view so your people, trees, cars, etc., will not appear in construction document views.

FIGURE 17-6.22 Design options

Enscape

Students are encouraged to check out Enscape, by Chaos Group, the real-time rendering engine. This is an add-in for Revit, and also works with SketchUp, Rhino, ArchiCAD and is free for students. This author has used this software to create the cover images for all of his Revit books for several years in a row. Visit www.Enscape3d.com for more information.

Self-Exam:

The following questions can be used as a way to check your knowledge of this lesson. The answers can be found at the bottom of this page.

1. You cannot point the RPC people in a specific direction. (T/F)

2. Revit's architectural light fixtures are ceiling hosted. (T/F)

3. *True North* must be set properly to get accurate sunlight. (T/F)

4. The *Adjust Exposure* settings make a huge difference on the quality of the final rendering. (T/F)

5. A *Section Box* crops a view which can then be rendered, producing compelling graphics. (T/F)

Review Questions:

The following questions may be assigned by your instructor as a way to assess your knowledge of this section. Your instructor has the answers to the review questions.

1. Several artificial lights can take hours, if not days, to render. (T/F)

2. Light fixtures cut a hole in the ceiling. (T/F)

3. Lights cannot be dimmed in a rendering. (T/F)

4. After inserting a light fixture, you need to adjust several settings before rendering. (T/F)

5. RPC content can only be viewed from one angle. (T/F)

6. Setting the render output to *Printer* rather than *Screen* results in better images. (T/F)

7. If the quality settings are too high, a rendering can take days to complete. (T/F)

8. Adjust the _____ to make more of a camera view visible.

9. IES files can be used to make artificial lighting more accurate. (T/F)

10. What is the file size of your model? _____ MB

Lesson 18
Sheets and Revisions:

This lesson will look at bringing everything you have drawn thus far together onto sheets using Autodesk® Revit®. The sheets, once set up, are ready for plotting. Basically, you place the various views you have created on sheets. The scale for each view is based on the scale you set while drawing that view, which is important to have set correctly because it affects the text and symbol sizes. When finished setting up the sheets, you will have a set of drawings ready to print, individually or all at once.

Exercise 18-1:
Creating Sheets

Creating a Sheet View:

1. Open your *Law Office* project with Revit.

The project you started with did not have any sheets created in it. You will learn how to create new sheets and place views on them to create a set of *Construction Documents*.

2. Select **View → Sheet Composition → Sheet**.

Next, Revit will prompt you for a *Titleblock* to use. The template file you started with has two (Figure 18-1.1).

3. Select the **E 34x44 Horizontal** titleblock and click **OK**.

That's it; you have created a new sheet that is ready to have views and/or schedules placed on it!

> **NOTE:** *A new item shows up in the* Project Browser *under the heading:* Sheets. *Once you get an entire set ready, this list can be very long.*

Revit also lets you create *placeholder* sheets which allow you to add sheets to the sheet index without the need to have a sheet show up in the *Project Browser*. This is helpful if a consultant, such as a food service designer, is not using Revit. This feature allows you to add those sheets to the sheet index. *Placeholder* sheets can be

FIGURE 18-1.1 Select a Titleblock

added in the sheet list schedule by using the *New Row* tool on the *Ribbon*. Once a *placeholder* sheet exists, it can be turned into a real sheet by selecting from the lower list in the dialog (Figure 18-1.1).

FIGURE 18-1.2 Initial Titleblock view

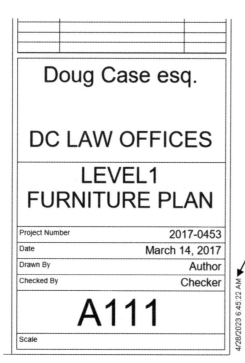

FIGURE 18-1.3 Revised Titleblock data

4. **Zoom** in to the sheet number area (lower right corner).

5. Adjust the text to look similar to **Figure 18-1.3**.

 TIP: Select the titleblock, hover the cursor over the text you wish to edit, and then click.

 NOTE: *The time and date stamp helps to remember when a sheet was plotted, especially if you forget to update the date in the titleblock before printing.*

6. With the sheet fully visible, click and drag the **Level 1 Furniture Floor Plan** view from the *Project Browser* onto the sheet view (Figure 18-1.4). Do not click on the sheet yet.

You will see a box that represents the extents of the view you are placing on the current sheet. In this example, we are using a larger sheet size, so place the view in the lower right corner of the sheet.

> ***FYI:*** *The most common sheet size is 22"x34". A half-size print for 22"x34" is an 11"x17" sheet, which prints quickly on most office copier/printers.*

7. Move the cursor around until the box is positioned as shown below.

Your view should look similar to Figure 18-1.4.

FIGURE 18-1.4 Sheet view with Level 1 Furniture Plan added

8. Click the mouse in a white area, not on any lines, to deselect the Level 1 view. Notice the box goes away.

9. **Zoom in** on the lower left corner to view the drawing title symbol, which Revit automatically added (Figure 18-1.5).

1)	LEVEL 1 FURNITURE PLAN
	1/4" = 1'-0"

FIGURE 18-1.5 Drawing title

NOTE: The drawing number for this sheet is added. The next drawing you add to this sheet will be number 2.

The view name is listed. This is another reason to rename the views as you create them. Also, notice that the drawing scale is listed. Again, this comes from the scale setting for the view.

10. To **Zoom out** to see the entire sheet again, type **ZF**.

Setting Up the Interior Elevations:

Next, you will create a sheet and place your interior elevations on it. Because you are using a larger sheet, the elevation will appear very small on it. However, on a larger project, this means the design team can get more drawings on a single sheet.

11. Create sheet **A500**.

12. Name the sheet **INTERIOR ELEVATIONS**.

13. Place the three interior elevation views on the sheet:

 a. Lunch Room

 b. Men's Public Restroom

 c. Workroom

Your sheet should look like Figure 18-1.6. Notice that each drawing has its own number. They are numbered based on the order they have been placed on the sheet. This can be changed, but you cannot have two drawings with the same number. So you might have to renumber one drawing to some random number, such as 100, and then change it back once the proper number is available.

It is good practice to align the floor line of interior elevations which are side-by-side.

TIP: On a sheet, if you right-click on a view, you can select Activate View. This will allow you to edit the model/drawing as if you were in the view. If you right-click again, you can select Deactivate View to get back to the sheet. Also, double-clicking on a schedule will open the schedule view.

FIGURE 18-1.6 Views placed on sheet

Anything that is visible in a view will print. This is the case with *Crop Regions* as well. You do not usually have this problem with interior elevations because the *Crop Region* coincides with the perimeter of the drawing: the walls, ceilings and floor. However, when you want to turn the *Crop Region* off, simply click the *Hide Crop Region* icon on the *View Control Bar* (Figure 18-1.7). You have to be in the view to make this change, not the sheet.

FIGURE 18-1.7 Crop controls on the View Control Bar

Now you will stop for a moment and notice that Revit is automatically referencing the drawings as you place them on sheets.

14. Switch to **Level 1** (see Figure 18-1.8).

Notice in Figure 18-1.8 that the number **A500** represents the sheet number that the drawing can be found on. The number 1 (one) is the drawing number to look for on sheet A500.

FIGURE 18-1.8 Elevation tags automatically filled out

Set Up the Remaining Sheets:

Next you set up sheets for the remaining views that have yet to be placed on a sheet.

> **TIP:** *Two sheets cannot have the same number.*
>
> *Sheets with views placed on them will have a plus sign next to the sheet name in the* Project Browser.

15. Create the following sheets **and** place the appropriate views on them; the view name will generally match the sheet name:

• A101	Level 1 Floor Plan
• A102	Level 2 Floor Plan
• A103	Level 1 Reflected Ceiling Plan
• A104	Level 2 Reflected Ceiling Plan
• A111	Level 1 Furniture Plan
• A112	Level 2 Furniture Plans
• A121	Level 1 Finish Plan
• A122	Level 2 Finish Plan
• A500	Interior Elevations
• A600	Details*
• A800	Schedules

Question: On a large project with hundreds of views, how do I know for sure if I have placed every view on a sheet?

Answer: Revit has a feature called *Browser Organization* which can hide all the views that have been placed on a sheet. You will try this next.

16. Take a general look at the example *Project Browser* to see how many views are listed (Figure 18-1.10).

17. Select **Views (All)** at the top of the *Project Browser* (Figure 18-1.9).

18. Select **Not on sheets** from the *Type Selector*.

19. Click **Apply**.

20. Notice the list in the *Project Browser* is now smaller (Figure 18-1.11).

FIGURE 18-1.9 Project Browser and Properties Palette

The *Project Browser* now only shows drawing views which have not been placed onto a sheet. Of course, you could have a few views that do not need to be placed on a sheet, but this feature will help eliminate errors.

Next you will reset the *Project Browser*.

21. Switch the *Browser Organization* back to **All** and click **Apply**.

FIGURE 18-1.10
Project Browser; Views (Discipline)

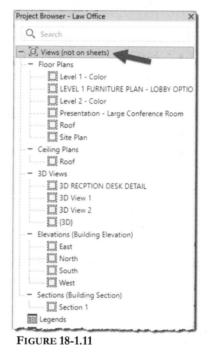

FIGURE 18-1.11
Project Browser; Views (not on sheets)

FIGURE 18-1.12
Project Browser; Sheets (All)

It is possible to customize the *Browser Organization* to sort views and sheets as needed. See page 31 in this chapter for more details on this.

22. **Save** your project.

Large Buildings – Matchlines and Dependent Views:

When the building floor plan is too large to fit on a single sheet you need to use the *Matchline* tool and the *dependent views* feature. Each can be used exclusive of the other, but they generally are used together. You will not do this for the book project, but the information is good to know.

The *Matchline* tool can be found on the *View* tab. It is similar to the *Grid* tool in that you add it in plan view and it will show up in other plan views because it has a height. When the *Matchline* tool is selected, you enter *Sketch* mode. Here you can draw multiple lines which do not have to form a loop or even touch each other.

Matchline

A *dependent view* is a view which is controlled by (i.e., dependent on) a main overall view, but has its own *Crop Region*. A main view can have several *dependent views*. Any change made in *Visibility/Graphics Overrides* is made to the main view and each dependent view. The dependent view's *Crop Regions* generally coincide with the *Matchlines*. However, the *Crop Regions* can overlap.

Additionally, if a dependent view's *Crop Region* is rotated, the model actually rotates so the *Crop Window* stays square with the computer screen. However, the other dependent views and the main view are unaffected. This is great when a wing of the building is at an angle. Having that wing orthogonal with the computer screen makes it easier to work in that portion of the building and it fits better on the sheet in most cases.

To create a dependent view, you right-click on a floor plan view and select **Duplicate View > Duplicate as Dependent**. The image to the right shows the result in the *Project Browser*, the **dependent views** are indented.

Once the *Matchline* is added and the dependent views are set up, you can add *View References* (from the *Annotate* tab) which indicate the sheet to flip to in order to see the drawing beyond the *Matchline*.

Finally, on large projects with multiple levels and floor plan views, you only have to set up the dependent views once (i.e., number and crop regions). With one plan set up, you right-click on the main view and select *Apply Dependent Views*. You will be prompted to select all the floor plan views for which you would like to have the dependent views created.

> *TIP:* View References can also be used to reference a detail in a note. This is done by overlapping the View Reference and the Note on the drawing. You could also create a View Reference family that looks like an interior elevation tag and use it to reference 3D views or photographs which have been placed on sheets. You would need one family per view direction as the pointer part of the fake elevation tag cannot rotate without rotating the text.

Exercise 18-2:
Creating a Sheet Index

Revit has the ability to create a sheet index automatically. You will study this now.

Creating a Sheet List Schedule

1. Open your project.

2. Select **View → Create → Schedule → Sheet List**. Sheet List

You are now in the *Sheet List* dialog box. This process is identical to creating schedules. Here you specify which fields you want in the sheet index and how to sort the list (Figure 18-2.1).

3. Add *Sheet Number* and *Sheet Name* to the right:

 a. Select the available field on the left.

 b. Click the **Add →** button in the center.

4. Click **OK**.

Now you should notice that the *Sheet Names* are cut off because the column is not wide enough. You will adjust this next.

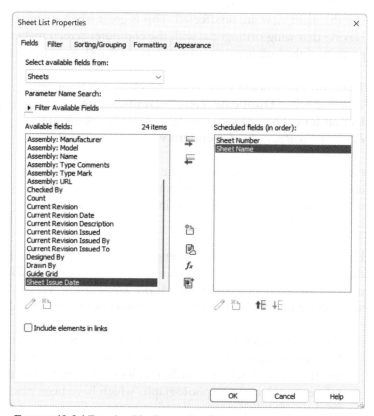

FIGURE 18-2.1 Drawing List Properties dialog; sheet number and name "added"

5. Move your cursor over the right edge of the *Sheet List* schedule and click-and-drag to the right until you can see the entire name (Figure 18-2.2 & 18-2.3).

6. Edit ***Sorting/Grouping*** via the *Properties Palette* and sort the list by ***Sheet Number***.

FIGURE 18-2.2 Sheet List view; notice sheet names are cut off in the right column

FIGURE 18-2.3 Sheet List view; sheet names are now visible

Setting Up a Title Sheet

Now you will create a title sheet to place your sheet index on.

7. Create a new Sheet:
 a. Number: **T001**
 b. Name: **TITLE SHEET**

8. From the *Schedules/Quantities* category of the *Project Browser*, place the schedule named **Sheet List** on the *Title Sheet*. Once on the sheet, drag the "triangle" grip to adjust the column width.

Next you will place one of your rendered images that you saved to file (raster image). If you have not created a raster image, you should refer back to Lesson 17 and create one now (otherwise you can use any BMP or JPG file on your hard drive or the provided online files if necessary).

9. Select **Insert → Import → Import Image.**

10. Browse to your raster image file (one of your exported renderings), select it and click **Open** to place the image.

11. Click on your *Title Sheet* to locate the image. Drag the corner grips to scale image.

Your sheet should look similar to Figure 18-2.4. *FYI: The text is added in the next step.*

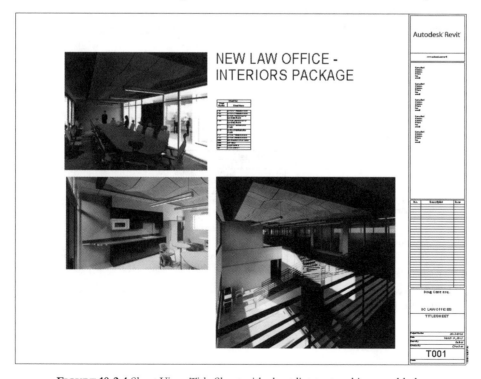

FIGURE 18-2.4 Sheet View: Title Sheet with sheet list, text and image added

12. Use the *Text* command to add the title shown in Figure 18-2.4.

 a. You will need to create a new text style and set the height to **1″**. Select the text tool, *Edit Type*, *Duplicate*, etc.

When you have raster images in your project, you can manage them via the *Manage Links* dialog.

13. Select **Insert → Links → Manage Links**.

You are now in the *Manage Links* dialog which gives you information about the image and allows you to delete it from the project (Figure 18-2.5).

FIGURE 18-2.5 Manage Links dialog, Image tab

 TIP: You can create a custom parameter (via the Add Parameter *button; see Figure 18-2.1) and then sort the schedule by that parameter. This allows you to force the "T" sheet to the top.*

14. Click **OK** to close the *Raster Images* dialog.

15. **Save** your project.

Exercise 18-3:
Presentation Boards

When orchestrating the interior design of a building, it is often necessary to present the finishes and furniture selections to the client. Traditionally this has been done by affixing samples and product photos to a board, similar to the one shown below (Figure 18-3.1). You will take a look at how this can be done in Revit and then presented with a projector.

FIGURE 18-3.1 Law Office material selection board

The process involves collecting images and samples, ideally as digital images such as JPG file format. When necessary you can scan samples or take a digital picture. Once your images are collected you can then arrange them on a sheet in Revit. The sheet can have a different, less formal, titleblock. You will try this with a handful of images provided and any you may choose to use from other sources.

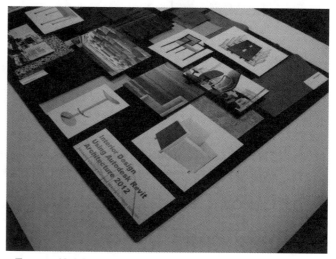

FIGURE 18-3.2 Law office material selection board enlarged

Loading another Titleblock

Revit allows you to have multiple titleblocks loaded into your project. It is relatively easy to customize a titleblock in the *Family Editor*. To save time, one has been provided with the online files that are available with this book.

1. Use the *Load Family* tool to import the **Presentation Titleblock** from the *Families* folder from the provided online files for this book.

 a. If prompted to place an instance of the titleblock, because you are in a sheet view, press the **Esc** key to cancel.

2. Create a new sheet, **View** (tab) → **Sheet**.

3. Select the **Presentation Titleblock** option from the list.

4. Edit the sheet name and number as shown in Figure 18-3.3.

March 14, 2019

Large Conference Room
Sheet: 1.0

FIGURE 18-3.3 Presentation Titleblock – sheet name and number

Next you will adjust a setting that will prevent this sheet from showing up in the sheet index. The sheet index, or sheet list, is primarily meant for construction documents, at least it is in our case.

5. With the *Presentation Sheet* current on the screen, in the *Properties Palette*, uncheck **Appears in Sheet List**.

You now have a 22″x34″ sheet with a simple border ready for images and views to highlight the *Large Conference Room* design concept. The final result can be printed to paper, printed to a PDF, or exported to an image file via *Application Menu* → *Export* → *Images and Animations* → *Image*.

The first thing you will do is create a 3D cutaway view of the *Large Conference Room* and place it on the sheet.

6. Switch to the **Default 3D view**, via the *Quick Access Toolbar*.

7. Right-click on the bold view name in the *Project Browser*, select **Duplicate View →
 Duplicate**.

8. Rename the new view to **Presentation - Large Conference Room**.

9. In the *Properties Palette* for the new view, check to turn on *Section Box*.

10. Adjust the *Section Box* to look similar to Figure 18-3.4.

11. Type **VV** and turn off the following categories:

 a. Entourage

 b. Toposurface

 c. Section Box

FIGURE 18-3.4 3D view of Large Conference Room; highlighted wall selected in next step

12. Select the wall pointed out in the image above.

13. Right-click and select **Override Graphics in View → By Element**.

14. Adjust *Surface Transparency to* **30** (Figure 18-3.5).

15. Click **OK**.

16. With nothing selected in the model, in the *Properties Palette*, select *Edit* next to **Graphic Display Options**.

17. Check *Show Ambient Shadows* (Figure 18-3.6).

> *FYI: If this option is grayed out you need to enable* **Hardware Acceleration** *in* Application Menu → Options → Graphics.

> *TIP: Notice the option to adjust the transparency for all surfaces in the dialog. This is nice if you want the table supports and all parts of the furniture to be visible. Give it a try.*

18. Click **OK**.

19. Set the *View Scale* to **½″ = 1′-0″**.

20. Place the 3D view on the new sheet. (See Figure 18-3.7 on the next page.)

21. Place an enlarged plan view of the *Conference Room* on the sheet (Figure 18-3.7).

 a. **Duplicate** the LEVEL 1 FURNITURE PLAN view; name: **Presentation - Large Conference Room**.

 b. **Crop** the view down to just the *Conference Room*.

 c. Turn on **Shadows**.

 d. In *Graphic Display Options*, set the *Shadow Intensity* to **25**.

 e. Turn off *Stairs*, *Railings* and *Entourage*.

 f. View Scale: **½″ = 1′-0″**

FIGURE 18-3.5 Making selected wall somewhat transparent.

FIGURE 18-3.6 Graphic Display Options dialog with all sections expanded – show ambient shadows has been checked.

FIGURE 18-3.7 Presentation sheet with 3D view and enlarged plan view

Next you will place raster image files on the sheet, scaling and arranging them to help the client understand the overall design concept of the space.

22. While in the sheet view, place the images provided in the *Presentation Board* folder saved from the provided online files: **Insert** (tab) → **Image**.

 a. Arrange approximately as shown in Figure 18-3.8.

 b. Select image and use **Bring to Front** or **Back** options listed on the *Ribbon*.

23. Add text as shown.

FIGURE 18-3.8 Final presentation board

One could create a presentation sheet like this for each room or for the overall project.

Be sure to explore the **Sketchy Lines** graphic display style. This works best in conjunction with the anti-aliasing option. Both can be enabled in the *Graphic Display Options* dialog shown to the right. Here you can adjust the line's jitter and extension length at its ends.

Exercise 18-4:
Printing a Set of Drawings

Revit has the ability to print an entire set of drawings, in addition to printing individual sheets.

Printing a Set of Drawings

1. **Open** your project.

2. Select **Application Menu → Print**.

3. In the *Print Range* area, click the option ***Selected views/sheets*** (Figure 18-4.1).

4. Click the **Select...** button within the *Print Range* area.

FIGURE 18-4.1 Print dialog box

FIGURE 18-4.2 Selecting sheets for printing

You should now see a listing of all views and sheets (Figure 18-4.2).

Notice at the bottom you can **Show** both **Sheets** and **Views**, or each separately. Because you are printing a set of drawings, you will want to see only the sheets.

5. **Uncheck** the **Views** option.

The list is now limited to just sheets set up in your project.

6. Select all the *Sheets* except the presentation sheet 1.0.

7. Click **OK** to close the *View/Sheet Set* dialog.

> *FYI: Once you have selected the sheets to be plotted, you can click Save. This will save the list of selected drawings to a name you choose. Then, the next time you need to print those sheets, you can select the name from the drop-down list at the top (Figure 18-4.2).*
>
> *On very large projects (e.g., with 20 floor plan sheets), you could have a* Plans *list saved, a* Laboratory Interior Elevations *list saved, etc.*

8. IF YOU ACTUALLY WANT TO PRINT A FULL SET OF DRAWINGS, you can do so now by clicking OK. Otherwise, click **Cancel**.

Autodesk can export to PDF. This tool can be found on the File tab, as shown in the image to the right. This is often the preferred format to send clients.

Exporting a 3D DWFx File:

The *3D DWFx* file is an easy way to share your project file with others without the need to give them your editable Revit file (i.e., your intellectual property) or the need for them to actually have Revit installed.

> *NOTE: You can download Revit, from Autodesk's website, and run it in* Viewer *mode (full Revit with no Save functionality) for free.*

The *3D DWFx* file is much smaller in file size than the original Revit file. Also, Autodesk Design Review is a *Viewer* that can be downloaded for free from www.autodesk.com/DWF.

9. Switch to your *Default 3D* view.

 a. You may need to go into *View Properties* and turn off the *Section Box* feature to see the entire model.

You must be in a 3D view for the *3D DWF* feature to work. If you are not in a 3D view, you will get *2D DWF* files:

10. Select **Application Menu → Export → DWF/DWFx**.

DWF/DWFx
Creates DWF or DWFx files.

11. Click **Next...** Specify a file name and location for the DWF file; click **OK**.

That's all you need to do to create the file. Now you can email it to the client or a product representative to get a more accurate cost estimate.

The *3D DWF* file is a meager 5MB, whereas the Revit *Project* file is 50MB. (Your file sizes will vary slightly based on factors like the number of families loaded, etc.) In most cases, an individual's networks and servers are set up so they cannot receive large files via email, so the 3D DWF is very useful.

The *DWF Viewer* is not automatically installed with Revit. Once installed, it can access it from *Start → All Programs → Autodesk → Autodesk Design Review 2013*. You can zoom, orbit and select objects. Notice the information displayed on the left when the front door is selected (see image below). Also notice, under the *Windows* heading, the sizes and quantities are listed. Download: http://www.autodesk.com/products/design-review/overview

> **TIP:** *Autodesk Design Review is a free download from Autodesk. This is their full-featured DWF viewer and markup utility!*

FIGURE 18-4.3 3D DWF in Autodesk Design Review

Exporting the sheets to a DWFx File:

It is also possible to create a DWFx file which represents some, or all, of the sheets in your project. This file is similar to a PDF file but is in Autodesk's proprietary format. However, as just mentioned, the *Viewer* can be downloaded for free and includes markup tools, which allows someone to add notes and revision clouds. The PDF format can have markups but requires Adobe Acrobat Standard (or Professional), rather than just the viewer. The viewer is free, but Adobe Acrobat Standard (or Professional) is not.

The following steps cover how to create a set of drawings in DWFx format.

12. Select **Application Menu → Export →DWF/DWFx**.

13. Do the following on the *Views/Sheets* tab (Figure 18-4.4):

 a. Export: **<In session view/sheet set>**

 b. Show in list: **Sheets in the model**

 c. Click the **Check All button**

FIGURE 18-4.4 3D DWF export options

14. Review the options on the other two tabs (*DWF Properties* and *Project Information*), but do not make any additional changes at this time.

15. Click **Next**… Specify a file name and location for the DWF file; click **OK**.

You now have a DWFx file with all the 2D sheets in it rather than a 3D model. This file can be used to print the project (either in an office/school or sent to a print shop). The file can also be an archive record of exactly what was sent out at each submittal point.

Next you will take a quick look at the markup options in Autodesk Design Review.

16. Browse to the DWFx file just created and double click on it to open it.

 a. Or open Design Review and then use the **Open** command from the *Application* menu.

Notice, in Figure 18-4.5, the *Thumbnail* tab is selected on the left, which shows a small preview for each sheet in the file. Clicking on one of these previews opens that sheet in the larger viewer area.

FIGURE 18-4.5 3D DWF sheets file

17. Make sure the Level 1 Furniture Floor Plan, **sheet A111**, is selected.

18. Click the **Markup and Measure** tab on the *Ribbon*.

19. Use the **Cloud**, **Text** and **Line** tools from the *Draw* panel to add the markups shown in Figure 18-4.6.

20. **Save** the DWFx file (via the *Quick Access Toolbar*).

One unique feature about doing markups in DWFx format, rather than PDF, is that the markups can be referenced back into Revit. This workflow is useful when a senior designer does not use Revit, or a client or contractor is providing feedback on the design and emailing the DWFx file back to you. You will see how this works next.

FIGURE 18-4.6 3D DWF sheet file with markups added

21. Open your *Law Office* file using Revit.

22. From the *Insert* tab, select **DWF Markup**.

23. Browse to your DWF sheet file with the markups.

24. Select the file and click **Open**.

25. Click **OK** (Figure 18-4.7).

26. Open sheet **A111**.

Notice the markups added in Design Review are now in your Revit model (Figure 18-4.8).

27. Select the revision cloud in the sheet view (Figure 18-4.8).

FIGURE 18-4.7 IMPORT markup prompt

Notice the *Properties Palette* provides information about the origin of the markup and its current status (i.e., *None*, *Question*, *For Review* and *Done*).

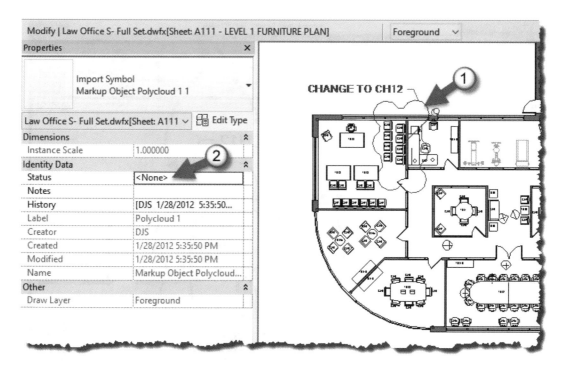

FIGURE 18-4.8 MARKUP overlay – revision cloud selected with properties shown

Follow these steps when you want to remove a DWF Markup link:

- **Manage → Manage Links**
- On the ***DWF Markup*** tab (Figure 18-4.9).
- Select the markup and then click ***Remove***.

FIGURE 18-4.9 MARKUP overlay – revision cloud selected with properties shown

Exercise 18-5:
Managing Construction Phase Revisions

The following is an overview of material as it relates to documenting revisions after the project has gone out for bids. Nothing in this section needs to be done in your *Law Office* model.

Basic Overview of Revisions:

- Specific revisions are managed via the *Revisions* dialog.
 - *Tool location:* View → Revisions
- Revision clouds are added and assigned to a specific revision.
 - *Tool location:* Annotate → Revision Cloud
 - Each cloud can be a separate instance or multiple clouds can be created in a single sketch (per sheet).

Revision Cloud

- A tag is added to a cloud to identify it.
 - *Tool location:* Annotate → Tag by Category (or *QAT*)
- The revision schedule, in the titleblock family, updates automatically.
 - Edit sheet properties to show revision without cloud.
- When a revision will no longer be changed, it is marked as issued.
 - Issued revisions cannot have new clouds added to them.
- Older revisions can be made invisible on the sheets via *Show* options.

In the big picture of BIM, Revit revisions are superficial in that the actual model elements being changed do not know which revision they belong to.

Sheet Issues/Revisions

Sequence	Revision Number	Numbering	Date	Description	Issued	Issued to	Issued by	Show
1	1	Numeric	1/18/2021	PR-001	☑			None
2	2	Numeric	2/12/2021	PR-004	☐			Cloud and Tag
3	3	Numeric	1/28/2021	CO-001	☑			Cloud and Tag
4	4	Numeric	3/2/2021	CO-011	☐			Cloud and Tag

Add

Delete

Numbering
- ● Per Project
- ○ Per Sheet

Row

Move Up

Move Down

Merge Up

Merge Down

Customize Numbering

Numbering...

Arc length

0' 0 3/4"

OK Cancel Apply

FIGURE 18-5.1 SHEET Issues/Revisions dialog

Revisions during Bidding

- Set up the *Revision* dialog for the anticipated change to the model.

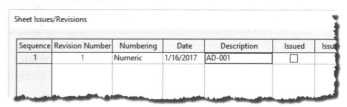

- Make changes to model; e.g., edit *Break Room* (plan and elevation).

FIGURE 18-5.2 EDIT default initial revision

- Duplicate **elevation view** (with *Detailing* or as *Dependent*)
 - With <u>Detailing</u> option:
 - Can move view to another category in *Project Browser*;
 - All view specific elements are now copied; i.e., text, dimensions, detail lines (harder to keep in sync);
 - Crop region can be adjusted if needed.

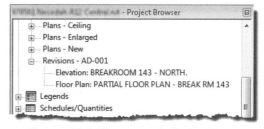

FIGURE 18-5.3 REVISIONS in Project Browser

 - With <u>Dependent</u> option:
 - View stays "close" to the original view;
 - Does not duplicate view specific items (less chance for errors later);
 - Scale cannot be changed from master view.
 - Both scenarios have the following issues:

FIGURE 18-5.4 REVISIONS in Project Browser

 - New view cannot have the same name as the original.
 - Workaround: add a period at the end.
 - A new elevation tag added in the plan views.
 - To hide these (steps required for each view):
 - Select each one and right-click → Hide in view, - *or* -
 - Create a new elevation tag type and filter it out.

- Duplicate **plan view** as *Dependent*
 - View can be cropped.
 - Grids follow crop region.
 - Section heads can be adjusted manually (but do not affect master view).
 - *Annotation Crop* must be turned on.
 - Some annotations extend outside of crop.
 - These can be hidden via select/right-click/*Hide element*.

- o Slight problem for inline posting of plan rack set.

 - o Annotations are not duplicated.
 - o Notice all callouts get reset (and have to be manually adjusted).

- Add revision clouds and tags to each sheet as needed.
 - o Revision clouds/tags can be added directly in most views (rather than placing them on the sheet).
 - o Adjust revision cloud line weight and color (if desired).
 - ▪ Manage → Object Styles → Annotation Objects
 - • Could also change a given view typing VV.

 WARNING: Changing color can affect printing.

- Create a Revision sheet and place duplicated/dependent view on it.
 - o Revit only allows a view to be placed on one sheet.

You should archive your project at end of bidding.

Revisions during Construction:

During construction you will most likely want to clear out all the bidding phase revisions and have the numbering start over for construction phase revisions.

- Archive project.
- Delete all revision clouds and tags from project.
 - o Must un-check the "issued" option first.
- Merge all revisions in *Revision* dialog.
 - o Must un-check the "issued" option first.

 FYI: Revisions can now be deleted starting in Revit 2025.

- Start documenting construction phase revisions (e.g., PR-01, ASI-01, CO-01).
- Consider getting rid of any *Design Options* used for Bid Alternates.
 - o This makes revisions easier during *CA*.
 - o Set selected option to *Primary* and *Accept* it (via *Design Options* dialog).
 - o Delete any unneeded sheets/views related to alternates
 - ▪ Dimensions and tags can be deleted.

 WARNING: This can get messy for large complex alternates.

 - o The deleted options will be available in the archived project.

TIP: There are add-in tools available for Revit which can automatically remove all the revisions in a given project. This can save a lot of time. Some of these tools are free and others cost money. A few examples are:
- CTC BIM Manager Suite
 http://www.cadtechnologycenter.com/store/suites/bim-manager-suite.html
- Ideate Explorer
 http://ideateexplorer.com/

All *Construction Administration (CA)* phase revisions should be made to a single model (i.e., the main model). This way any implications CO-04 might have on PR-57 will be evident to all parties.

In an IPD scenario (or design build) the best approach is to utilize a single model. This can be accomplished utilizing technology such as Remote Desktop (RDP), Autodesk's Revit Server or Collaboration for Revit (C4R).

Example 1: Proposal Request

The biggest challenge related to *Proposal Requests (PR)* is: Should the main model be edited or not? This is due to the fact that the *PR* might not ultimately be accepted due to contract implications (time/cost). If it is highly likely that the *PR* will be accepted (e.g., code or errors/omissions required change), then the main model should be edited.

If the change is not likely (e.g., client is curious what another 50'-0" of building is worth), then a temporary copy of the model could be made (via *Open Detached, Save-As*). However, in the unlikely event the *PR* is accepted, then the main model needs to be updated (some copy/paste might help but will not cover everything).

When in doubt, make changes to the main model and undo them if the change is not accepted. Of course, you can do whatever you want in this case.

Example 2: Schedules

- Make a change to the *Door Schedule*.
 - o Note that revision clouds cannot be added to the view.
- Add a door which causes the clouded items to shift down in the *Schedule*.
 - o Notice clouds do not move with highlighted area.
 - o Designer must manually move cloud when changes such as this occur (if they can remember!)
- Move the revision cloud.
 - o Revision must be unissued before the cloud can be moved.

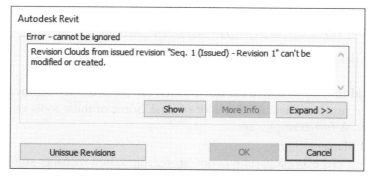

FIGURE 18-5.5 ERROR when trying to change an issued revision

Project Browser Organization

Sorting sheets can be done by creating a *Project Parameter* such as *Sheet Group*. In the image to the right, all sheets listed under **Addendum**, in the *Project Browser*, have their *Sheet Group* parameter set to the word "Addendum" (Figure 18-5.6).

Here is how to create a project parameter associated with sheets:

1. Click **Manage → Project Parameters**.

2. **Add** a parameter (Figure 18-5.7):

 a. *Name*: **Sheet Group**

 b. *Type of Parameter*: **Text**

 c. *Instance*: **checked**

 d. *Categories*: **Sheets**

3. Close the open dialog boxes.

Now all sheets will have a parameter called "Sheet Group."

The following steps show how to sort the sheets by the *Sheet Group* parameter:

1. Right-click on the word **Sheets (All)** in the *Project Browser*.

2. Select *Browser Organization...* from the pop-up menu.

3. On the *Sheets* tab, click **New**.

4. Enter **Sheet Group** for the name. (The name is case sensitive.)

5. Edit the *Grouping and Sorting* tab (Figure 18-5.8):

 a. *Group By*: **Sheet Group**

 b. *Sort By*: **Sheet Number**

FIGURE 18-5.6 PROJECT Browser

FIGURE 18-5.7 CREATING a Project Parameter

FIGURE 18-5.8 SORTING and Grouping the Project Browser

All sheets which do not have a value in the *Sheet Group* parameter will show up under the "???" section; this will be all of the sheets at first. Simply select one or more sheets and edit the *Sheet Group* parameter in the *Properties Palette*. When you click *Apply*, the sheets will automatically move to the specific "folder."

The same process is followed to sort views. It is also possible to use the same parameter to sort both sheets and views. When setting up the sorting/grouping for the views, be careful not to duplicate the "not on sheet" option, as it has views filtered out (i.e., hidden).

Custom Revision Tag

The *Value* in the default revision tag is not editable. However, it is possible to create a custom revision tag in order to control the information displayed. This custom tag has to use one of the built-in parameters (i.e., *Issued to* or *Issued by*) as custom revisions parameters cannot be created. Here is the workflow to create a custom revision tag:

FIGURE 18-5.9 CUSTOM revision tag

1. Open the default revision tag in the family editor.

2. Select the *Label* (i.e. the text).

3. Click ***Edit Label*** on the *Ribbon* (Figure 18-5.10).

 a. Remove "Revision Number" from the right.

 b. Add **Issued by** to the right.

 c. Change the Sample Value to "AD01" (or similar).

 d. Edit the line work if desired.

4. Save as with a new file name.

5. Load the new revision tag in the project.

6. Edit the ***Issued by*** column in the *Revisions* dialog.

FIGURE 18-5.10 Editing the label in the revision tag family

Once the revision tag has been edited and used, the revision schedule in the titleblock can be edited to list and sort by the *Issued by* parameter. The titleblock must be edited in the family editor.

> *FYI: You may want to leave the sorting by "Revision Sequence" so the revisions appear in chronological order.*

| AD01 | 5/16/2011 | Addendum 01 |
| NO | DATE | REVISION |

FIGURE 18-5.11 Revision schedule in titleblock

In newer versions of Revit, it is possible to achieve similar results with the **Customize Numbering** feature, and adding a prefix, such as "AD" in this example; this is in the Sheet Issues/Revisions dialog.

"Magic" Titleblock

Some design firms are using a workflow which eliminates the need to create additional views and sheets above those required for the original construction documents. The "magic" titleblock basically has a masking element within it which hides everything outside the border; we will call this the "*revision titleblock.*"

The process looks like this:
- Make revisions to model (as previously described).
- Cloud and Tag revisions (as previously described).
- Place the *revision titleblock* on a sheet.
 o Main titleblock is not touched.
 o *Revision titleblock* hides everything outside the 8½″x11″ or 11″x17″ border (including the main titleblock).
 o Size and position of crop window can be adjusted within the revision titleblock.
- Delete or hide the *revision titleblock*.
- The PDF is the final record drawing.

FIGURE 18-5.12 Revision titleblock placed on full size sheet

A great 8½″x11″ revision titleblock can be downloaded from AGUI at:
http://forums.augi.com/showthread.php?t=53492

This was created and posted by David Kingham and David Baldacchino!

- * You need to add shared parameter (export from family to new file)
- * You also need to adjust line weights 13 and 14 in project file

Static View Option

Some firms are using a workflow which employs an option similar to the *revision titleblock*, which does not require duplicate/dependent views, but keeps a printable sheet more readily accessible from within Revit. This option basically captures a snapshot of the revised view and places it on a new sheet. The nice thing about this method is it can be quickly printed from Revit at any time, and future revisions will not change the drawing (in relation to how it was issued). The main drawback is this can have a negative impact on the project file size for larger projects.

There are a couple of ways in which this workflow can be accomplished. The first is creating a raster image and the second is using a tool in the Revit *Extensions* add-in (*Freeze Drawings*) available through subscription.

Raster Image Method:

This option involves exporting a raster image file and then linking it back into a new *Drafting View*. This new view can then be placed on a new *Sheet*. The workflow looks like this:

1. Make revisions as previously described.

2. Export the revised drawing to a raster image.
 a. Application Menu → Export → Images and Animations → Images
 b. See notes below on various quality/scale issues.

3. Import the image into a new *Drafting View*.

4. Place the *Drafting View* on a sheet (i.e., a new revision sheet).

The only way to get the image to scale is by using the following settings (Figure 18-5.13):

- *Export Range*: **Current window**
- *Image size*: **Zoom to 100%**
- *Format*: **JPEG** (lossless) and **72** dpi

FIGURE 18-5.13 Exporting a raster image

This works well for details or cropped views, but creates a larger than needed image of an entire floor plan when just a small area is needed. Also, the 72dpi resolution is on the low side. Another option is to select the "*Visible portion of current window*" option and increase the resolution. Either of these changes makes the image not to scale. However, the image can be scaled once placed in the *Drafting View*.

If the revision needs to be changed and reissued, a new image is created and the old one replaced in Revit.

FIGURE 18-5.14 Raster image exported and cropped

Definitions:

- **Addenda** (AD):
 Revision to CDs during bidding.

- **Architectural Supplemental Information** (ASI): Drawing prepared to change the construction contract but does not impact time/cost of the project.

- **Construction Administration** (CA)
 Services performed by designer team during construction phase.

- **Construction Change Directive** (CCD):
 Drawing prepared to change the construction contract prior to acceptance by all parties.

- **Construction Documents** (CD):
 Drawings used by contractor for bidding.

- **Change Order** (CO):
 Drawing prepared to change the construction contract.

- **Proposal Request** (PR):
 Drawing prepared to determine time/cost implications of a proposed change to the construction contract.

- **Project Browser** (PB):
 Organizes all views, sheets, schedules and families in Revit.

Self-Exam:

The following questions can be used as a way to check your knowledge of this lesson. The answers can be found at the bottom of this page.

1. You have to manually fill in the reference bubbles after setting up the sheets. (T/F)

2. *DWF* files are much larger than the Revit file. (T/F)

3. It is possible to see a listing of only the views that have not been placed on a sheet via the *Project Browser*. (T/F)

4. You only have to enter your name on one titleblock, not all. (T/F)

5. Use the _____ tool to create another drawing sheet.

Review Questions:

The following questions may be assigned by your instructor as a way to assess your knowledge of this section. Your instructor has the answers to the review questions.

1. You need to use a special command to edit text in the titleblock. (T/F)

2. The template you started with has only one titleblock to choose from. (T/F)

3. You only have to enter the project name on one sheet, not all. (T/F)

4. The scale of a drawing placed on a sheet is determined by the scale set in that view's properties. (T/F)

5. You can save a list of drawing sheets to be plotted. (T/F)

6. You can be in any view when creating a *3D DWF* file. (T/F)

7. The reference bubbles will not automatically update if a drawing is moved to another sheet. (T/F)

8. On new sheets, the sheet number on the titleblock will increase by one from the previous sheet number. (T/F)

9. Image files cannot be placed on sheets. (T/F)

10. The *Sheet List* schedule never hides data. (T/F)

Notes:

Index

Rendered image of project developed by the reader

Rendered image with Shaw flooring and Steelcase furniture

Rendered image with Shaw flooring, Steelcase furniture, USG ceilings and RPC entourage